The Landscape of Community

The Landscape of Community

A History of Communal Forests in New England

Robert McCullough

University Press of New England
Hanover and London

University Press of New England, Hanover, NH 03755

Printed in the United States of America
5 4 3 2 1
CIP data appear at the end of the book

For Jill and Andrew

Contents

Illustrations

Preface

This has been a thoroughly enjoyable and informative journey from town to town across New England. The burden of many solitary miles was immeasurably lightened by the assistance of countless individuals, regrettably far too numerous to identify here. With a frequency that never failed to surprise, response to my unexpected inquiries transcended polite reply and took on the characteristics of genuine enthusiasm and interest. When that happened, my routine chores were suddenly transformed into rewarding adventures. It is with that spirit in mind that these expressions of gratitude are offered.

Among many such examples, a large number remain vivid in recollection. Paul Collins, formerly chairman of the Parsonage Committee in Danville, New Hampshire, opened his home and spent the better part of an afternoon and early evening helping me pore over committee records. Similarly, Steve Denisburger at Pennichuck Water Works in Nashua, New Hampshire, took time from his busy schedule to drive to his company's reservoir office, where old scrapbooks were stored. Several excellent photographs depicting forestry management on watershed lands were the reward. Foresters J. Wilcox Brown in Dunbarton, New Hampshire; Jack Dirkman in Norridgewock, Maine; William Hall and Eugene Keenan in Essex and Pittsfield, Vermont, respectively; Tom Meiner in Concord, New Hampshire; Professor David Smith in New Haven and Bruce Spaman in Middletown, Connecticut, all devoted valuable assistance.

Without exception, the many foresters and surveyors who supplied useful information patiently responded to questions about their professions, both of which were unfamiliar to me. Their help and suggestions were immensely appreciated. These included Bill Rivers and Darrel Keay in Massachusetts; Roy Boywin, Walter Breckenridge, Ron Duddy, Tim Hill, Ted Natti, Northam Parr, Marshall Patmos, Richard Spafford, and Steve Wood in New Hampshire; Bruce Payton in Rhode Island; and Russ Barrett, David Brynn, Paul Fredericks, Michael Reboin, Steve Slayton, Hallie Whitcomb, and Jim White in Vermont. Mrs. Rachel Keniston in Plymouth, New Hampshire, and Mrs. Edith Chase in Alstead, New Hampshire, graciously permitted access to survey materials compiled over two lifetimes by John Keniston and Heman Chase.

Town clerks, local librarians, and historians were routinely tolerant of my curiosity and provided essential help. The list is long and, with apologies for inevitable omissions, my thanks to the following: In Maine, Priscilla Davis, Brunswick; Anne Gilbert, Sabbathday Lake Shaker Village, New Gloucester; Eric Jorgensen, Brunswick; and Roberta Knowles, Skowhegan; in Massachusetts, Robert Hanson, Dedham; Mary Locke and Karl West, Walpole; Robert Meader, Hancock Shaker Village; Ken Turino, Lynn; and Eleanora

West, Fitchburg; in New Hampshire, Brian Burford, Antrim; Rebecca Courser, Warner; Rene Fox, Canterbury Shaker Village; Sylvia Gettels, Newmarket; Barbara Hill, Newington; Barbara Mann, Stratham; and Carolyn Smith, Enfield Shaker Village; in Rhode Island, Margery Matthews, Foster; in Vermont, Mildred Davis, Westmore; Loretta Laplume, Westfield; and Drucilla Schneiderheinze at the Vermont Conference of the United Church of Christ in Burlington. Flora Nyland at the Terence J. Hoverter College Archives at SUNY College of Environmental Science and Forestry in Syracuse offered enthusiastic assistance. Express thanks to Mrs. Hedwig Bower of Richmond, Vermont, for her translation of Ulrich Meister's *Die Stadtwaldungen von Zurich*, and to John Dolan, of Ipswich, Massachusetts, for information about the history of that state's conservation commission legislation.

Bill Fitzgerald, formerly with the Environmental League of Massachusetts, successor organization to the Massachusetts Forest and Park Association, opened his office's files for inspection; and John Hibbard, director of the Connecticut Forest and Park Association, offered useful comments about forest parks in that state. George Stephens, at the Connecticut Agricultural Experiment Station in New Haven, was equally helpful, as was Hugh Putnam at the New England Forestry Foundation. Clinton Reynolds, the son of Harris Reynolds, kindly permitted inspection of his father's papers. Special thanks to Larry Anderson of Little Compton, Rhode Island, who generously provided a wealth of intriguing information about Benton Mackaye.

Two libraries provided homes away from home. At Vermont's state library in Montpelier, Margie Zunder and Paul Donovan were consistently helpful with all requests. Meg Page cheerfully accepted the challenge, eventually international in scale, of tracking down articles from obscure journals through interlibrary loan systems. At Cornell's Olin Library, Julia Styles accomplished similar feats. So many libraries, historical societies, and archival institutions amiably granted permission to photograph their materials that a complete list is not possible here. Their efforts to accommodate patrons were greatly appreciated. Sincere thanks to Scott Swank at Canterbury Shaker Village in New Hampshire for the opportunity to examine and photograph the McClintock survey and for an enjoyable tour of a fascinating village. Mr. David Stagg, a verderer for the New Forest in Hampshire, and Mr. Chris Webb, warden for Selborne Common, Hampshire, the latter owned by Britain's National Trust for Places of Historic Interest or Natural Beauty, were both immensely helpful, providing key information as well as fascinating tours of their respective domains. Herr Andreas Speich, former director of Zurich's forest service, and ecologist Marcos Christian, were similarly obliging during my trek through the Sihlwald and its archives.

Financial assistance was provided by Cornell's Department of Manuscripts and University Archives and by the Forest History Society in Durham, North

Carolina; appreciation is gratefully acknowledged. Harold K. Steen, the society's director, and Cheryl Oakes, their librarian, gave support and encouragement. Special gratitude, also, to those who waded through drafts of this work and offered suggestions for its improvement. They include David Caffry, Richard Judd, K. C. Parsons, Carl Reidel, John Reps, John Stilgoe, and Michael Tomlan. True appreciation is extended to my parents for their encouragement and confidence in my work and, finally, to Ruth.

Abbreviations

AFA	American Forestry Association
AMC	Appalachian Mountain Club
CCC	Civilian Conservation Corps
CFA	Connecticut Forestry Association, which changed its name to the Connecticut Forest and Park Association in 1928
CFPA	Connecticut Forest and Park Association
MFA	Massachusetts Forestry Association, which changed its name to the Massachusetts Forest and Park Association in 1933. The successor organization is the Environmental League of Massachusetts.
MFPA	Massachusetts Forest and Park Association
RLM	Photograph by the author
RPAA	Regional Planning Association of America
SAF	Society of American Foresters
SPNHF	Society for the Protection of New Hampshire Forests
USFS	U.S. Forest Service. The Division of Forestry was first established in the Department of Agriculture in 1891. In 1901, its name was changed to the Bureau of Forestry, and in 1905 it received its present title, the U.S. Forest Service.
VDFPR-P	Vermont Department of Forests, Parks and Recreation, Agency of Natural Resources, Pittsford, Vermont
VDFPR-W	Vermont Department of Forests, Parks and Recreation, Agency of Natural Resources, Waterbury, Vermont
WPA	Works Progress Administration

The Landscape of Community

Introduction

Forested Communities of New England

The forest that is Hubbard Park in Montpelier, Vermont, blankets a broad hill above the state's capital city. During part of the year, a foliant mix of sprout-hardwoods shelter footpaths and nature trails that link the ways of urbanity below with more densely wooded mountains beyond. Yet even in winter, a plantation of red pine shapes a verdant backdrop to the white-and-gold-domed State House that stands resolutely at the park's base (fig. I.1). The building's quiet gray color understates the enduring mass of its granite walls and Doric portico. Few architectural monuments evoke an awareness of our country's ideals so simply and so elegantly as Vermont's serene temple to democracy. Few settings portray so clearly the intimate bond between forest and community in New England and contributions of the former to the structure and habitability of the latter for a period that fully spans four centuries.

The origins of Montpelier's forest park belong to the early twentieth century, when the land was unused pasture criss-crossed by stone walls. As an urban amenity, it offers mostly aesthetic, recreational, and educational benefits. These uses or functions, together with broader strategies for conservation amid rampant metropolitan sprawl, have grown to characterize town woodlands in New England during the second half of this century. By comparison, a history of the region's community-owned forests proclaims a different story. From their genesis, well before the mid-seventeenth century, New England towns and villages have creatively worked such forests and woodlots to public benefit. Over centuries, utilitarian purpose has achieved impressive variety: building materials and wood products; fuelwood; subsidy for local institutions such as church, school, or poor farm; reclamation of idle land; commercial timber production; and protection of water supplies, to name but a few. Such uses often persist in many locales. When that occurs, forests in all their diversity lie within reach of community inspection, debate, and insight.

Fig. I.1. The Vermont State House below the woodlands of Hubbard Park in Montpelier. Just as classical architecture epitomized the democratic ideals of a young nation during the second quarter of the nineteenth century, forests, too, came to be regarded as America's true antiquity, comparable to Europe's ancient monuments. Classical and romantic themes have found expression throughout the history of New England's cultural environment. RLM, 1993

The ancient continuum of forest and town in New England is a cardinal theme of this history, evident in the progression of five discrete categories of communal forests: common land, public land, town forests, watershed plantations, and forest parks. In many ways, these categories furnish a dynamic context for the appearance of a sixth class of communal woods: areas owned or controlled by local conservation commissions, which became popular after 1960. True, the vast common weald that once enveloped New England's seventeenth-century villages has long since vanished into private ownership. In more than a few towns, however, ancient public woodlands—typically ministry, parsonage, or school lots—span three centuries and remain part of community landscapes, strengthening this sense of continuity. In places such as Keene, Danville, Newington, and Northwood, in New Hampshire and Brunswick, Maine, public lands evolved into town forests as that movement gained ground after 1900. Although town forests, municipal watershed plantations, and local forest parks developed primarily during the twentieth century, their roots are firmly embedded in the closing decades of the nineteenth century. Occasionally, too, forest parks such as Lynn Woods in Lynn, Massachusetts, evince public reclamation of undeveloped land once held as common. There, landscape vistas traverse years as well as distance.

Evolution of a woodland ethic, present throughout New England's history of town-owned forests, is an equally prominent theme. Compelled by material and economic need and in some cases motivated by a desire for long-term community welfare, seventeenth-century colonists very quickly regulated cutting of timber and wood on common lands and levied fines for abusive or wasteful practices. Woodsmanship—an ability to work the forest without destroying trees—was part of a diverse Anglo-Saxon culture carried to New England during the Great Migration by England's rural populations. Glebe or ministry lots, colonial land banks allotted to ease the burden of salaries paid to town ministers, were often managed prudently. Early nineteenth-century timber sale agreements on parsonage lands in Danville, New Hampshire, are superb examples. Poor-farm woodlots, a type of public land, subsidized town welfare and supplied steady rations of fuelwood. With expanding utility of local forests during the twentieth century, collective stewardship has assumed broad dimensions: silviculture to produce commercial harvests, specialized management for protection of water supply and quality, cultivation of wildlife habitats, trail building for recreational use, educational programs for nature study, creation of wilderness patches to monitor ecological health, and simple conservation of undeveloped land to balance irrepressible change in our society.

Collectively, various classes of town-owned woodlands are as much a part of New England's urban and village history as of its forest history; this, too, is a principal theme. As segments of community structure, they represent important but often neglected pieces of the cultural landscape of community. The shifting balance between publicly and privately held land has been a fundamental tension in New England town planning from its inception. Although steadily transferred to individual owners, extensive seventeenth-century common lands—open fields, great woods, and cedar swamps—were an immensely valuable resource, integral to community well-being. Utopian settlements of a different period and ilk (Shaker villages) offer more tactile affirmation of planning methods reliant upon common ownership of land; several built thriving timber economies. Public land allotments required by town charters are tightly bound to community origins in New England, much as ancient common lands in England embody village antiquity in that country. Where a modicum of wilderness exists on these public lots, a similarly fitting antiquity for New England towns can be observed. Greater heed should be paid to historic significance of these lands in the future. The course of twentieth-century forests has influenced community structure dramatically. Town forests sparked a momentous return of private land to the public sector and began a trend that has scarcely slackened. Watershed reserves and wildland parks are part of physical and spiritual community makeup. Imaginative planners such as Benton Mackaye offered visionary ways to blend forest and community. As state and federal efforts to acquire private lands for public use have lost momentum, towns, many acting through their conservation commissions, now shoulder the campaign.

The spirit of this history resides among New England's less familiar towns, out-of-the-way communities that have quietly achieved an unpretentious harmony with their sylvan landscapes. Many a forested hilltop or secluded patch of woods bear uniquely descriptive place-names, such as Brunswick Commons in Brunswick, Maine; Kilburn Crags in Littleton, New Hampshire; Meeting Waters Town Forest in Springfield, Vermont; the Children's Wood in Jaffrey, New Hampshire; and Barrack-Matiff Town Forest in Salisbury, Connecticut. Contributions of these New England places to the region's vernacular culture is a salient theme as well. The many sketches of local woodlands offered here help to define the period of origin and character for each class of community-owned forest. Examples thus become measuring sticks against which the history of countless town woods yet to be explored can be gauged. Distinctions between public lands, town forests, forest parks, and conservation spaces can sometimes blur, and the value of these numerous illustrations—to the region as well as to individual states—should not be discounted.

Secondary topics warrant introduction as well. The influence of European practices—in village structure, emergence of professional forestry, and birth of the town forest movement—is conspicuous. Credit, too, is given to numerous foresters, surveyors, and conservationists who have trekked through the history of local woods. Many have generated lucid thought on matters of conservation and have translated it into visible practice, leaving meaningful legacies for their communities.

One additional introductory point. Throughout most of their history, the character of New England's town-owned forests has been guided by utilitarian purpose. Such use is not inconsistent with the recreational, educational, and ecological goals that have eclipsed cultivation of timber and wood during the past half-century. Recent trends threaten to extinguish the cultural diversity once readily accessible in town forests and village woodlots. It is essential to maintain a vestige of utilitarian character on these wooded lands to better understand the natural and cultural history of New England forests. More precisely, it is vital to keep that information in the fore of our collective outlook. Recent interest in preserving historic landscapes suggests an opportunity for an alliance between groups advocating conservation of natural and cultural resources. Such a coalition, in the past elusive, can find common ground in the quest for habitable living environments.

A few organizational details require brief explanation. Separation of community-owned forests into categories has forced some overlap in the development of substantive context for the origins of each. This is particularly true for town forests, watershed plantations, and forest parks. Within the latter, evolution of recreational and educational forest activities is sufficiently diverse to cause separate chronological overlap as well as partial restatement of subject matter appearing in other chapters. On an entirely different matter, the fragmentary quality of seventeenth-century town records and the tremendous diversity of culture and land-use practices evident in these communities allow only very broad conclusions about stewardship of common woodlands. The presence of identifiable types of common forests or woodlots and their routine appearance from town to town is clear. However, further investigation is necessary to determine just how consistently woodsmanship was employed on these lands and whether certain forested areas—swamps, for instance—retained their communal character longer than other types of common land. Finally, connecting local events to broader trends that may, or in specific cases may not, have been influential presents logistical difficulty. On the one hand, an understanding of context is essential if one is to grasp broad developments in communities and their use of forests. The danger, however, is to become lost in a tangle of analytical undergrowth and fail to observe the simple individuality of each place. Hopefully, a fruitful balance has been found between the two.

Chapter 1

From the Common Wood

New England's history of communal forests begins not on American soil but with a pastoral English landscape of labyrinthine detail. In that country, aged patterns of field division, methods of cultivation, rituals for gathering wood and timber, and practices of animal husbandry on broad meadow or under forest cover all helped to sculpt the land. At the core of many English communities lies an ancient system of open-field farming that relied on common arable, pasture, woodland, bog, and waste. Clusters of dwelling houses provided the nucleus of a village form that evolved over the centuries and eventually surfaced in New England.

The path from the English common field and wood twists and turns along its way to the uplands, cedar swamps, and great woods that became New England's first communal forests. For one thing, England's dispersed regions have vastly different traditions of land ownership and use, and cultural belongings ferried to America were anything but uniform. For another, communal rights to English lands rested, at times precariously, on the sanction of nobility. Woodland domains were enveloped by a pervasive body of law and administration aimed at sustaining the king's deer. The word *forest* acquired a very specific, much despised meaning and more than occasionally encompassed both weald and open country. Moreover, by the time of the Great Migration, a period from 1630 to 1642, communal rights to field and forest had been extensively limited by a legal process known as enclosure. Yet embattled by adversity and clutched tenaciously by a rural society, common rights endured and continued to yield meager repayment to those who worked the land. In these time-worn customs, cast into stone by human need, lies the seed for collective environmental stewardship. That, too, was transplanted into New England soil.

Fig. 1.1. Brunswick Commons, legacy of the Pejepscot Proprietors to the town of Brunswick, Maine. Although much diminished from its original thousand acres, the commons has survived as public land for a period that spans three centuries. RLM, 1992

By Hook or by Crook: Common Fields and Woodlands in England

It is reasonable to suggest that some form of communal or shared interest in forest resources and tillable land has existed in England since the time of early tribal gatherings. Despite a shroud of mystery obscuring concepts of property during England's prehistory, there is evidence that a dispersed agrarian society of Celtic tribes established small, scattered farmsteads and began to claim the land. Social nucleation advanced to a stage where settlements included populations too large to be composed of mere family units. Distinctive field systems began to emerge as arable segments delineated by rudimentary boundaries were farmed year after year. In these villages, a collective agrarian effort became essential amid a harsh landscape marked by dense woodlands, heather-blanketed moor, and coastal marsh. Yet if communal effort gave birth to community, the extent to which this labor was formalized in harvesting of forests and crops is not known.[1]

Although Roman occupation of Britain, begun in A.D. 43, introduced a fairly regimented system of land tenure, formal town planning, and advances in agricultural practice, the impact on remote villages and any related common land was negligible. With disintegration of the Roman Empire and its dispersal from British soil by the close of the fourth century, rural settlements continued to evolve. This process was accelerated by invasion of Anglo-Saxon tribes from northern Europe during the fifth century. The village form that emerged consisted of a cluster of dwellings surrounded by open land, forests and waste. The nucleated village—a term eventually applicable to New England's first communities—became characterized by a system of open-field farming, where plowing, sowing, and harvesting were undertaken in concert by community members. Whether first examples of open-field farming were introduced by Anglo-Saxons and are Germanic in origin or the process was an evolutionary one bearing traces of Celtic practice has been a subject of considerable discourse and difference of opinion.[2]

Despite uncertainty caused by the lack of any recorded history, it is possible to sketch a partially complete picture of these self-contained agrarian communities. Tilling large areas of open land by manual labor and with primitive implements became a communal endeavor, and some form of administration was necessary to achieve consensus. Cultivation depended on the plow, which cut open fields into strips of land that offered a ready means of division. Each strip represented a single day's work by a laborer and a team of oxen. Lands of one person adjoined those of another as strip after strip was plowed. The size of open land, its contour, and soil texture determined the shape and length of strips. Community boundaries were measured loosely by the area of land cultivated and adjoining woodlands or wastes.[3]

As crop succeeded crop year after year, soil fertility became depleted, and open lands were divided into well-defined fields. Initially, a two-field system was sufficient, with one field alternately left fallow after harvest to be manured by grazing livestock. This evolved into a three-field system, one-third sown in autumn with crops needing a long growing season (wheat or rye, for instance), one-third sown in spring with crops needing only several months to mature (such as beans, peas, or barley), and one left fallow. Cattle and other livestock were grazed in common on fallow, unplowed lands called sikes, turned into scattered meadows, or allowed to roam woodlands and wastes. The latter, despite an unpromising title, were far from sterile and consisted of gently sloping hills wearing mantles of fern, heather, and gorse and often sustaining scrub forest. These features—open fields, woods, and clustered dwellings, together with the ways and cart paths that connected them—defined community structure.[4]

Nearby woodlands were shared in common and provided indispensable resources: wood for light construction, fuel, fencing, farm tools, and household implements, as well as timber for more substantial buildings; wild honey was gathered and used in mead, a fermented drink. Woodsmanship became an integral part of village culture. Trees were "coppiced," cut close to ground level so that the stump, or "stool," could generate new growth, called underwood. Valuable tracts of underwood measured by the "rindge," roughly equivalent to an eighth of an acre, could be harvested at regular intervals. Poles cut from hazelwood contributed to construction of "hurdles," woven frames used for gates, fishing weirs, and thatching wood. Fibrous lime trees provided rough cordage. Hornbeam was cut for "billets," short lengths used for fuel and charcoal. Ash and maple furnished wood for tools, simple furniture, and implements such as bowls. Timber, primarily oak, elm and ash, was used for planks and beams in buildings. Distinctions between wood and timber were important and remain so in our language today, for we speak of timber frames but wood fires.[5]

Wooded pastures also were essential to village economy. Swineherds fattened pigs on harvests of acorn and beechnut mast for several weeks in autumn prior to slaughter. Cattle and sheep consumed whatever woodland foliage was within reach. To protect stands of timber and wood from grazing, trees were "pollarded," "lopped," or "cobbed" at a sufficient height above ground to place new growth beyond reach of cattle, sheep, and deer. Shredding, the removal of only selected branches to allow growth of new ones, achieved a similar purpose. Twigs and leaves became fodder, and branches too small to burn separately were bundled together to form "faggots," used to produce quick, hot fires.[6]

While collective use of woodland and open fields sustained a primitive agrarian economy, other forces also shaped these villages. Emerging feudal tenure forced servitude but also shaped land tenure and defined communal rights. Although woodlands were usually owned by kings or lords and timber was coveted, commoners acquired rights to underwood through grant or prescription. As tenure became more clearly defined, so too did rights to take wood for fuel (fire-bote) and cut poles for fencing and hedging (hedge-bote) or for repair of farm equipment (cart-bote). Timber also could be cut for the repair of buildings (hous-bote), but these rights were frequently exercised on hedgerow timber and did not necessarily indicate common rights to forests. Its derivation long lost, the privilege of gathering what boughs and limbs could be reached "by hook or by crook" eventually became part of village dialect.[7]

Important, too, is a certainty that agricultural practice and methods of woodsmanship were not uniform throughout England during this period. Contrasting landscapes promoted pastoral husbandry in one region and arable in another, and village structure naturally adapted to the needs of each. Two- and three-field systems of open-field farming were confined to a large midland area. In western territories, Celtic influence tended to confine communities to small hamlets, and plots individually owned were not unusual. Land was often subdivided and conveyed to heirs upon an owner's death, and holdings thus became scattered. Other domains, Kent in particular, never developed the extensive common fields found in midlands areas but instead generated noncontiguous rectangular plots that were eventually enclosed and held in severalty. In such areas, evolution of property rights appeared to be more advanced. Practices also tended to blur where regions converged.[8]

In forested areas, settlements, or "dens," evolved from common rights as herdsmen established squatters' claims. In regions where species of trees needed for primitive enterprise were plentiful, such as hazelwood for hurdle weaving or oak for barking, villages sometimes developed around these industries. Absence of forest cover in other districts forced villages to develop a trade for wood products, and woodmongers negotiated the countryside. Such diverse traditions are significant for New England communities because a large share of immigrants who journeyed to America during the Great Migration came from rural areas of England. They carried with them a cultural potpourri that quickly shaped New England's towns and landscape. Nucleated settlements that appeared in New England during the seventeenth century owe much of their constitution to Anglo-Saxon villages.[9]

During the ninth and tenth centuries, England succumbed to invasions of Danes and Norsemen. Ensuing political upheaval, however, was confined to the highest levels of society, and village life changed very little. Noteworthy, though, was the introduction of a method of measuring land based on cultivation rather than tribute. Anglo-Saxon land division was based on "hides," a unit of land varying in size from 60 to 120 acres but calculated to represent actual or potential resources for which taxes could be levied. Scandinavians instead measured land according to areas capable of cultivation by an ox-drawn plow, and grants of land were expressed in terms of plowland, or "carucatas." This innovation established a system, measurable on the ground, by which common rights and services owed to a lord could be gauged accurately.[10]

Norman conquest during the eleventh century precipitated refinement of feudal society, and land tenure became more rigidly defined under a strong central monarchy. William the Conqueror's Domesday Book, a survey of every estate and manor granted to his knights, served as a complete record of taxable lands in 1085 and checked the power of feudal baronage. An extensive body of common law gradually incorporated ancient rights to open fields and woodlands. Norman kings also created royal forests, primarily for preserving game and only incidentally the flora that offered sanctuary to hunted prey. The term *forest* became a legal one, encompassing grasslands, moors, villages, and towns and seldom heeding any natural woodland boundaries. These forests were created by edict following the appointment of commissioners who viewed land by "perambulation" and bounded it with "meers," or fences. Hunting without permission from the king was forbidden, and transgressions were met with severe penalties. Scores of king's foresters became part of an elaborate administrative and judicial system.[11]

Nobility eagerly sought social status associated with forests, and substitute forms of game management were awarded to the privileged. Favored subjects were occasionally granted a "chase," an unenclosed area of woodland unprotected by forest laws and often subject to common rights. "Parks," sanctuaries encircled by deer-proof fences, were given to private owners, who thus acquired a ready source of meat. Parks, too, were free from forest laws, but owners exercised a greater degree of control than did chase owners, and commoners were less apt to exercise privileges. Within lesser tracts called warrens, acquired by grant or prescription, keepers assumed exclusive right to kill fowl or game. Unlike parks, warrens remained unenclosed.[12]

At their thirteenth-century pinnacle, royal forests were an economic boon, and income flowed from nearly every direction: "herbage" charged for pasturage of domestic animals, "cheminage" exacted as tolls on carts and wagons plying forest roads, and "farms" paid for use of parks or other special areas. Plots of cultivated land called assarts were wrested from forest edges upon annual payment of rent. Iron and charcoal industries contributed their duty. As the economic importance of royal forests increased, greater emphasis was placed on balancing profit against loss of cover for deer. Licenses placed limits on harvests of underwood and timber and required construction of low hedges encircling cut-over lands to prevent cattle and pigs from foraging on new growth.[13]

As forest boundaries arbitrarily stretched across England's landscape, conflict between crown and nobility smoldered. Struggles culminated with the

Forest Charter, signed in 1217, two years after the Great Charter. This document returned extensive lands to baronage, released private domain from forest laws, checked the power of foresters over individual property, curbed abusive tolls, and abrogated capital punishment for killing deer. Costs of administering extensive royal forests contributed to their decline in economic importance during the fourteenth century, when they were eclipsed by taxation as a primary source of revenue.[14]

Common rights persevered turmoil of the age and stubbornly buttressed community structure. With lords otherwise occupied in their quest to subdue the power of the king's authority, opportunity for enterprising villagers was at hand. Although fines were imposed for illegal felling of timber, tribute amounted to little more than payment for downed trees. Grazing was permitted, and formal distinctions between land uses trickled into common law, the result of efforts to protect valuable open fields, or "champion country," and forest growth. "Common appendant" permitted liberal grazing of animals essential to husbandry: horses, oxen, cattle, and sheep. "Common appurtenant" was more limited, acquired only through long use or direct grant and usually applicable to smaller stock—goats or asses, for example. "Pannage," grazing of swine on acorns and beechnuts and technically payment by the pigs' owner for the privilege, was used to measure the size of woodlands. "Estover," the right to take wood for fuel, was well accepted. Common wastes continued to provide essential materials: clay for mill dams, bracken for litter, and sedges for thatch.[15]

From the Domesday Book to the Great Migration, England's agrarian society remained largely medieval, and lords managed their estates by collection of rents, fines, and sale of lands. This period, however, was also one of significant change, to which common lands slowly succumbed. Virgin forest and waste yielded to ax and plow to satisfy the demands of growing populations during the twelfth and thirteenth centuries. As long as rents were paid, assarted lands lay outside common grasp and offered villagers a chance to select crops and methods of cultivation. Collective farming persisted in some regions. In others, plots were cultivated according to individual choice, although villages without common grazing were rare. With agricultural success came an increase in the number, scale, and efficiency of farming operations. "Engrossment," amalgamation of separate plots, became commonplace. Scattered parcels proved cumbersome, and new farm complexes took shape in more convenient locations, changing patterns of settlement within villages. Freeholders and yeomen amassed large tracts, which they in turn could lease or sell. Lands once cultivated in common were replaced by large private holdings; small villages, no longer functional, vanished.[16]

Much in demand, forest resources too became more carefully managed and conserved. Encroachments of homesteads within woodland boundaries (purpresture) were often fined. Wood pastures were frequently "compartmented," enclosed by fences after cutting to sustain new growth. Woodlands became capital reserves, generating income with minimum expenditure. By the fifteenth century, primitive regulations had been introduced by legislation, prompted by increasing shortages of wood and timber, which eventually became severe in many regions. "Grubbing out," or clear-cutting, was prohibited; woodlots were required to contain a minimum number of trees per acre; cutting was limited to trees of a certain dimension; and compartments were mandated.[17]

Enclosure did not occur suddenly. Where waste and forest were plentiful, loss of plots here and there raised little concern. Only in regions where cultivation occurred on a grand scale did removal of cherished pasture rankle. Still, hardships tended to multiply over time. With increased populations came larger numbers of livestock and a need for greater areas of pasture. Those who tilled the soil relied on their animals, which, in turn, depended on sufficient forage. Plagues that swept England during the fourteenth century caused labor shortages and forced wage increases. Lords were compelled to secure new sources of income and did so through sale of their lands. The timely arrival of the wool trade offered a convenient remedy, and common fields were peddled, enclosed, and devoted to pasture for sheep.[18]

Tenants rebelled against enclosure as harmful effects began to multiply. In response, the Statute of Merton was enacted in 1236 by England's Parliament and was followed nearly fifty years later with a second law, the Statute of Westminster II. Together, these legislative acts sanctioned enclosure of waste land only if sufficient property remained to satisfy tenants' communal rights acquired over time. Still, legislation was easily circumvented, and enclosure continued to gain momentum. Lords sought to bolster their quest for enclosure with arguments that common-land husbandry was archaic and inefficient, stymied individual progress, encouraged the spread of disease among poorly bred livestock, and attracted an undesirable human element from villages. Thomas More's *Utopia*, which appeared in 1516, publicized the plight of the peasantry and the growing loss of common land, but the enclosure movement continued to reduce common lands up to and after the Great Migration.[19]

The diversity of England's rural society extended far beyond agrarian custom. Populations also toiled under the weight of an ecclesiastical yoke, and parish districts, well defined by boundary, served as administrative units for enforcement of tithes. These levies encumbered whatever scant property villagers managed to accumulate, attaching "praedial tithes" even against common rights to pasture, waste, or cut wood. Parishes, too, overlapped town organization in New England and were often influential in the formation of splinter communities and allotment of lands to support a minister and church, (glebe lots). Nor was evolution of common rights in England consistently vernacular. Experiments with town planning began much closer to home than North America and can be traced to thirteenth-century bastides established in England, Wales, and France by Edward I and to the colonization of Ulster province during the seventeenth century. Granting of common rights was recognized as essential to community stability, and ancient custom became formalized in town planning; yet both parish administration and the planting of new towns belong to a context more closely aligned with the origins of public lands in New England.

Ancestral habits collected over centuries are not easily flung aside. Little wonder that New England's first settlements functioned amid common rights to open field and wood, that patterns of cultivation and land division recalled ancient methods, or that practices varied from town to town. Memories of adversity and repression are sometimes equally difficult to erase. Colonists' attitudes toward New England's unbounded resources are better understood within a context of odious laws governing royal forests, the imprint of feudal subjugation and parish collar, and competing quests for land. At the same time, America's wilderness and unfamiliar ecology created new hardships, bending aged protocol and forging new patterns of land use, some of them uniquely American. New England's seventeenth-century villages were an uneasy blend of rural custom overlapping a calculated experiment by London investment companies. In common woods one can observe a dawning tension between the quest for individual property and the birth of collective stewardship as relentless clearing quickly dwindled supplies. Custom and practicality were soon beckoned by necessity; resources were marshaled, and laws restricting the cutting of town wood and timber were enacted.

New England's Nucleated Villages

Excluding native American settlements, New England's first villages were a unique blend of economic enterprise and utopian idealism, in which extensive forests and open lands became communal resources, shared in both ownership and use. Title to lands in New England was acquired by the English Crown according to rights of discovery and possession, with only superficial recognition of prior aboriginal claims. Through a series of complex and sometimes overlapping royal grants, land was conveyed to quasi-public investment associations, such as the Massachusetts Bay Company, that were formed to create plantations. These organizations were given authority to govern and legislate and were vested with exclusive power to acquire and dispose of land and to create towns.[20]

Acting in a legislative capacity through their general courts, colonial governments made grants of land to individuals and, more often, to groups of individuals who established communities and became town proprietors. These grants, or town charters, encouraged physical occupation of unsettled land and contained certain inducements, as well as covenants requiring occupation and improvement of the soil. Failure to accomplish these objectives within a given period resulted in charter forfeiture. General courts exercised strict control over distribution of land, and charters contained covenants restricting acquisition of property by newcomers without court consent. In addition, they usually required newly formed communities to provide for a minister and to construct a meetinghouse. Towns became legal entities defined by the boundaries contained in their charters.[21]

Proprietors were left with the responsibility for determining village form and shaping surrounding lands. Their tasks included surveying boundaries, assigning individual house lots, locating and constructing roads and highways, and designating land and water rights to entice millers. They also allotted specific parcels to serve community needs, including meetinghouse, school, parsonage, cemetery, and a central common used for militia training and as a temporary gathering point for livestock. Remaining lands were held in common, technically a form of undivided ownership, with each owner sharing a proportional right to use the whole subject to limitations against abuse or waste. Clustered against the unknown, villages consisted of a tight nucleus of individual house lots encircling, often haphazardly, a central open area or large meetinghouse lot. Surrounding uplands, meadows, swamps, marshes, and wastes were owned in common and used for cultivation, grazing, and felling of timber and wood needed for buildings, fencing, or fuel.[22]

The earliest settlements were confined to shorelines or located inland on principal rivers and were, for many years, sustained primarily by agriculture. Among a progression of fifty or more towns founded before 1642 were Braintree, Dedham, Dorchester, Duxbury, Hingham, Ipswich, Lynn, Manchester, Muddy River (Brookline), Newtown (Cambridge), Plymouth, Reading, Rowley, Roxbury, Salem, Taunton, Watertown, Wenham, and Weymouth in Massachusetts Bay and Plymouth colonies; Guilford, Hartford, New Haven, Saybrook, and Wethersfield in Connecticut; and Portsmouth, Providence, and Warwick in Rhode Island. Many others organized soon after that date, among them Andover, Marshfield, Medfield, Topsfield, and Groton in Massachusetts Bay, and Farmington and Norwalk in Connecticut. Although these villages demonstrated certain general characteristics, it would be far from accurate to describe them as a homogeneous collection of settlements. Methods and practices of common land ownership varied from town to town, in some cases displaying a melding of more than one system and in others reflecting many patterns of their English counterparts.[23]

At first, proprietors comprised the greatest part of towns' populations, coexisting in mutual agreement and united by economic incentive, religion, or ancestry. During this period, distinctions between proprietors and town were not especially important. An idealistic sharing of communal resources in spiritual confrontation with the wilderness made New England's early villages unique. Common land was an integral part of these utopian settlements, and it assumed a quality, also unique, that persisted for as long as its use furthered community welfare. Despite the pervading influence of religion in many aspects of village society, it did not bind villagers to communal ownership of land. In fact, proprietors quickly dismissed any theological doubts about the ability of individuals to own property. Instead, commonly held resources furnished a practical means for ensuring survival in an alien environment. As circumstances changed, so too did village structure, and common land steadily vanished into private ownership.

Common lands were often assigned specific uses, depending on a variety of factors such as topography, extent of forest cover, accessibility, function, drainage, or proximity to bodies of water. New England's woodlands were an ecological patchwork with great diversity both in range and in location of tree species. Frequently, specific tracts were assigned as woodlots or identified by a dominant tree, reflecting this variety; but in other cases, timber and wood were conveniently harvested from nearby uplands. Meadows and cow commons were often partially wooded, and ubiquitous swamps provided a dependable source of cedar, pine, and swamp oak. Various and clearly defined categories of common land, specifically communal woodland, can be documented consistently from town to town. These mark a beginning of New England's unbroken tradition of communal forests. Beneath these categories, however, lay variations in land ethics that undoubtedly affected the appearance of the landscape and contributed in part to the duration of its communal nature.[24]

Common and Undivided Lands

Following assignment of house lots and public lots for meetinghouse, school, parsonage, and cemetery, town proprietors reserved outlying regions or uplands for the common benefit of town inhabitants. Town records refer to these lands as "common and undivided lands" or simply "the common" and indicate that they were opened to grazing, cultivation, and felling of timber and wood for countless purpose: framing, lumber, clapboards, shingles, staves, fences, bark for tanning, pitch, or fuel. Initially, permission to cut timber was freely granted, with little reason to monitor individual consumption in towns where woodlands were plentiful. However, restrictions against sale of communal timber resources beyond town borders quickly appeared within a few decades. Conveniently accessible wood resources became scarcer as New England began to build an economy based on timber, and communities quickly learned to protect their common wood.[25]

In Massachusetts, Braintree voters agreed in 1646 "that every man that is an inhabitant of the Towne shall have Liberty to take any timber off the Common for any use in the Towne [provided] so they make not sale of it out of the Towne."[26] In Dedham, town members specifically voted in 1656 that "the woodlands aboute the Towne shall lye as thay doe in comon till the Towne see case to alter them."[27] Cambridge selectmen, possibly after a winter storm in December 1658, granted permission to nine town members to cut timber from the common to repair houses, fences, and outbuildings.[28] At a town meeting in 1667, Reading voters agreed that privileges of land, timber and common should belong only to those houses already erected.[29]

Costs, if any, of removing wood from common land varied considerably. In 1672, a citizen of Dorchester was allowed to "take trees for sels and plates for a barne out of the Comons beyond the blew hils and alsoe liberty to take some Cedar trees for Cloboards out of some of the Swamps in [or] land beyond the blew hills." He was also specifically authorized to haul trees felled by residents of other towns, but payment for these privileges was not mentioned.[30] Six years later, in contrast, selectmen in Muddy River (later Brookline) "granted to Thomas Pecke, Senr. Ten loade of crooked timber in Muddy River Comon land, such as Deacon Eliot shall approve of and upon such termes as they can agree for, he beinge willing to give for it as he buyes the like of Deacon Eliot or any other man."[31] Common woodlands occasionally yielded hidden reward. In Connecticut, Farmington citizens agreed that "if any man belonging to the towne shall find a Bee tree in any of our commons and shall sett the two first letters of his name on it faire in vew it shall be accounted his pries."[32]

Communal Woodlots

While many common woodlands were informally bounded and identified simply as uplands or common, proprietors also set aside specific parcels as communal woodlots for themselves and eventually for town inhabitants who owned no proprietary interest. Occasionally, ownership of these lands became clouded, and disputes arose between proprietors and nonproprietors as populations increased and newcomers acquired greater influence over communal resources. Accessibility was undoubtedly a factor in choosing the location of woodlots, and selection was probably achieved with a modicum of care.

In Massachusetts, Salem proprietors set aside a tract of land in 1636 "along the shore on Darby's fort side . . . to run along toward Marble Head 1120 pole" to supply wood and timber for town commoners.[33] Five hundred common acres in Dorchester were reserved for exclusive use of noncommoners.[34] Plymouth, in 1674, designated a large area of uplands and swamps bounded by a series of brooks, rivers, and ponds to be common for general use of wood and timber by town inhabitants but prohibited the felling of trees for bark or removal of wood from the town.[35] Communal woodlots occasionally fit neatly within natural boundaries, as in Braintree, where the town's wood on Rock Island in Quincy Bay was eventually sold in 1757.[36] In Rhode Island, Providence sequestered a large area of land "in the Neck between Pautuckett River, and Moshosick River" in common for timber and fuel wood.[37]

Great Cedar and Pine Swamps

Settlement patterns in New England's nucleated villages initially followed a path of least resistance dictated by an uncharitable landscape. Protected coves and inlets or river shorelines offered the best locations for village centers and often displayed expansive open areas suitable for cultivation or pasturage. William Wood's *New England's Prospect*, first published in London in 1634, described the practices of Indians, who regularly fired woods to clear undergrowth, which both impeded travel and provided sanctuary to hunted animals. Lands freed from this thick vegetation were more easily cleared and cultivated and were thus more valuable. However, great areas of swamplands—some be ten, some twenty, some thirty miles long being preserved by the wetness of the soil wherein they grow"—were spared from Indian torches and displayed a primeval tangle of forest and small growth.[38]

Swamps often presented formidable obstacles to clearing and were widely tapped as communal woodlands, described by names such as Cedar Swamp, Pine Swamp, Ash Swamp, Rail Swamp, or simply Great or Common Swamp. In 1682, inhabitants of Lynn, Massachusetts, felled timber for a meetinghouse from a remote and rugged bog, a spot that became known as "Meetinghouse Swamp" in casual honor of that event.[39] In varying degrees and depending on factors such as elevation and soil quality, white pine, balsam fir, red spruce, alders, and willows could be found in these wet soils, as could cedar or white oak, sometimes dubbed "swamp oak," which offered excellent wood for fence posts and rails. Smaller growth provided a ready source of fuelwood during winter months when deep snow prevented gathering of old wood in upland regions.[40] Possibly because swamps were easily identifiable by topography and forest cover, town records offer consistent evidence of their communal use.

In Massachusetts, Ipswich selectmen in 1652 granted liberty to inhabitants to fell firewood from the swamp "that lies inbetween Timber Hill and Bush Hill provided that no man take above 2 rods in breadth and to fell all and clear as they go across the swamp."[41] Medfield voters granted a villager rights to cut two hundred cedar planks in a common swamp as long as he removed them within six weeks,[42] and a Weymouth townsman was given two acres of swamp wood to make charcoal.[43] In 1667, Plymouth set apart certain areas of uplands and swamps to remain forever common for fencing and firewood,[44] and Salem designated a town swamp as a common woodlot for firewood, provided all wood was removed the same day.[45] In 1676, Andover granted rights to remove fallen cedar from the town's Great Swamp and to sell it out of town, provided a fee was paid. However, the town prudently guarded its communal resource. Standing trees were excluded, and choppers were cautioned against damaging a nearby rail swamp. Years later, a committee was appointed to inspect all shingles, boards, and timber taken from that lowland area.[46]

Common swamps were not unusual in other colonies. At a town meeting in 1665, residents of Providence voted "that the pine Swampe lieing by Swassacutt pond south west wardly: and also the Cedar Swampe neere Wyunkeake" should not be divided but instead remain common.[47] For a time, proprietors in Saybrook, Connecticut, allowed removal of cedar wood from their swamp but required that notification be given to a designated committee when it occurred.[48]

The pervasive use of swamps as communal forests naturally raises questions about the duration of their collective nature. Increasing populations and pressures for land eventually changed settlement patterns in these early villages. Common and undivided lands were soon apportioned among proprietors and passed into individual ownership, referred to as severalty. Lands differed in value according to location, elevation, soil quality, and access to meadows, and fairness dictated that parcels distributed be of equal worth. Consequently, individually owned lots appeared in almost random fashion and fractured large areas of remaining common land.[49]

Although depletion of forest reserves in these uplands was rapid and methodical, swamps remained a valuable and protected source of timber and fuel in many towns while common lands steadily disappeared. Restrictions against cutting swamp wood were customary, and in some towns swamplands were among the last to be divided, surviving as common well into the eighteenth century. Duration doubtless depended on a great many factors affecting value, such as proximity to village centers or land already divided, availability of waterpower, extent of depleted forests, and existence of particularly valuable stands of timber. With pasture and meadow in great demand, those swamps capable of easy drainage and with limited wood were probably divided fairly quickly.

As early as 1657, two men were appointed to view the cedar swamps at Muddy River to determine their value, and a year later the lands were leased for a period of fifty years for the sum of forty shillings in wheat and pease.[50] Selectmen for Medfield authorized the sale of certain swamp timber in 1663 to accommodate grazing for cattle.[51] In 1666, members of Reading agreed that all great swamps fit for conversion to meadows be laid out. To encourage clearing, wood growing on individual lots remained free for the cutting, provided butts and tops were removed.[52] A parcel of common swamp in Cambridge was awarded to one of its citizens in 1663 as reparation for woodlots given to an adjoining farm.[53] A Dedham committee appointed in 1665 painstakingly apportioned a common swamp and compensated owners whose shares contained no islands.[54]

These scattered examples of divisions are balanced by recurrent mention of restrictions against cutting swamp timber and by divisions that occurred much later. By 1658, Cambridge residents were controlling destruction of timber in their Great Swamp by permitting felling only in areas appointed by selectmen, prohibiting cutting of second growth, requiring clearing of cut timber and slash, and setting fines for violation of these regulations.[55] Cambridge selectmen closely scrutinized the removal of wood from common swamps and carefully recorded the labor of one man who hauled "a jagg of wood with two cattle" from a common swamp bordering the Menotomie River.[56] However, earlier restrictions against cutting wood on common land had specifically excluded trees felled from the Great Swamp, implying that forest growth was dense and warranted clearing.[57] In 1677, selectmen in the town of Wenham prohibited cutting of timber from town swamps by any town inhabitants other than householders and limited consumption by them to no "more timber in one yere than will make two thousand in inch Boards." Trees already cut were forfeited to the town.[58]

Divisions, when they occurred, were sometimes achieved in stages. Reading's cedar swamp was partially divided prior to 1670, and cedar trees lying outside divided areas were offered as a means of equalizing shares. However, conditions were placed on the size of cedar trees cut, and land and other types of timber were reserved. Wood on Reading's great ash swamp was not completely sold until after 1733.[59] Eighteenth-century divisions of swamp lands were not unusual in many towns founded before 1642. Taunton undertook division of its Great Cedar Swamp (fig. 1.2) in 1711,[60] and Braintree continued to own a town swamp until 1740, when it was finally leased for clearing.[61] Saybrook, Connecticut, eventually voted to lay out a cedar swamp in 1769.[62]

A plan of Walpole, Massachusetts (fig. 1.3), illustrates ragged patterns of land apportionment as well as more regimented partitions. Originally part of Dorchester's grant, the region was surveyed in 1696 and 1697 by John Butcher who identified twelve separate divisions by Dorchester proprietors. His plan was revised in 1726 by James Blake, shortly after Walpole was chartered, to show vacancies between divisions (referred to as waste lands) and to mark lots sold both before and after the twelve portions, in some cases referred to as "dividends." It was recopied and annotated in 1891 by Frederic Endicott to show land ownership based on Blake's survey. Endicott's recapitulation reveals several common swamps at a time when almost all land had already been divided (see figs. 1.4 and 1.5). A 106-acre common swamp borders the Punkapoag Plantation, and the Mashapaog Cedar Swamp, partially common,

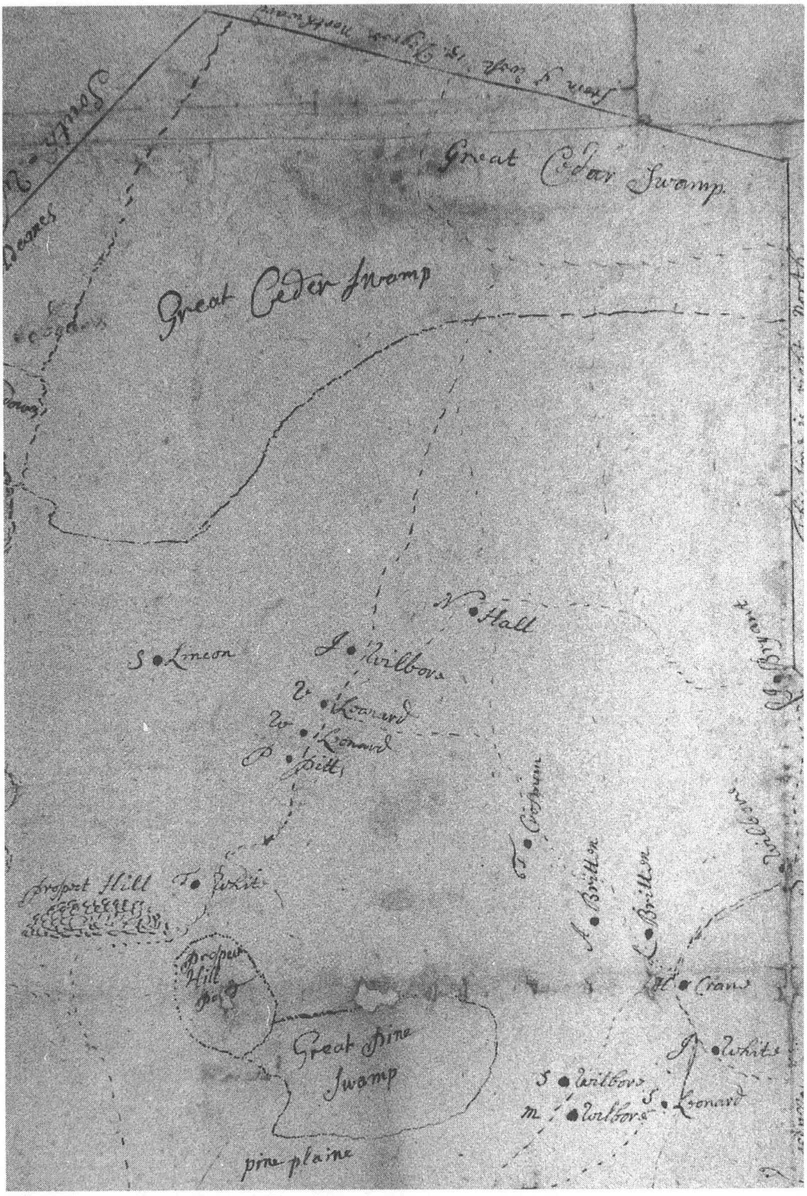

Fig. 1.2. Great Cedar Swamp and Great Pine Swamp in Taunton, Massachusetts, from a plan of the town drawn by Morgan Cobb in 1728. Common swamplands were noted frequently in early town records and supported a variety of tree species utilized by colonial villagers, including cedar, pine, ash, and white oak, sometimes dubbed "swamp oak." Massachusetts State Archives

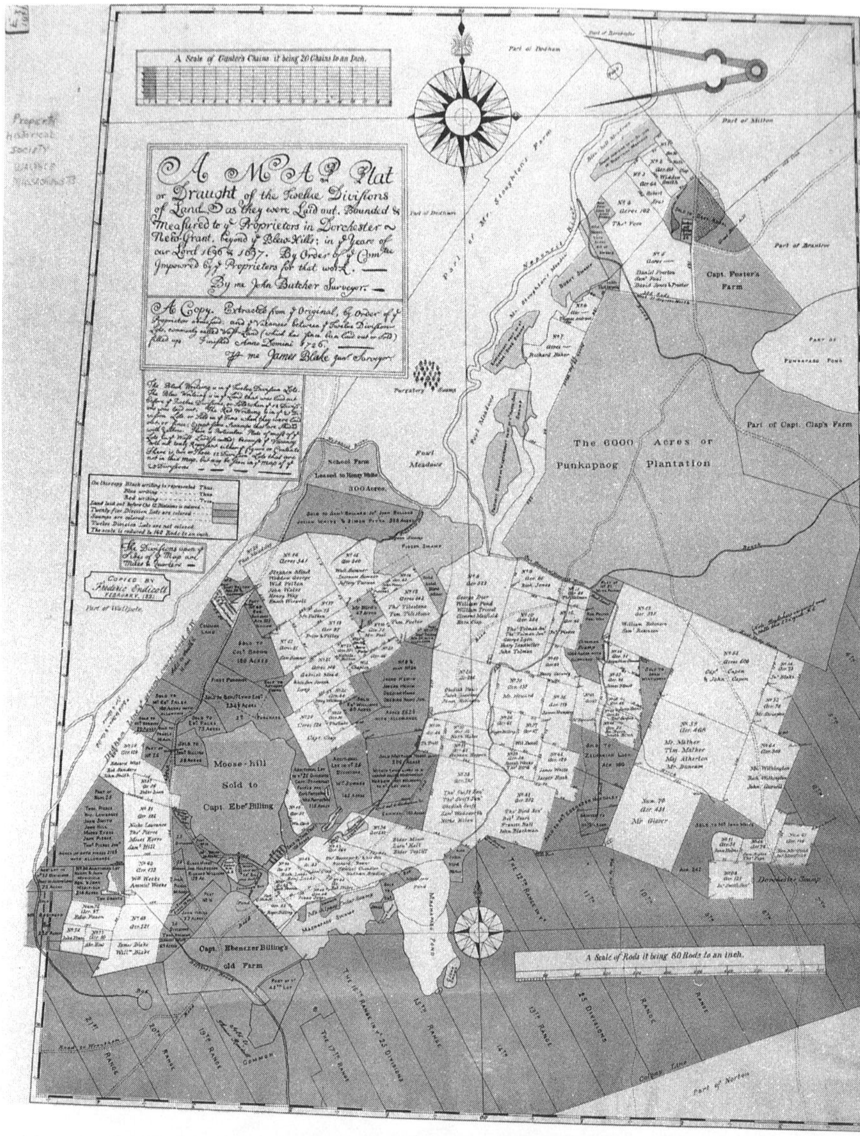

Fig. 1.3. *A Map Plat or Draught of the Twelve Divisions of Land as they were Laid out, Bounded and Measured to the Proprietors in Dorchester—New Grant, beyond the Blue Hills.* A plan of Walpole, Massachusetts, by John Butcher in 1696–97. Butcher's work was copied by James Blake in 1726 and recopied with annotations by Frederic Endicott in 1891. The plan illustrates ragged patterns of land division and the relative size of individual grants. At a time when most of the town had been divided among individual owners, several swamplands remained common. Walpole Historical Society

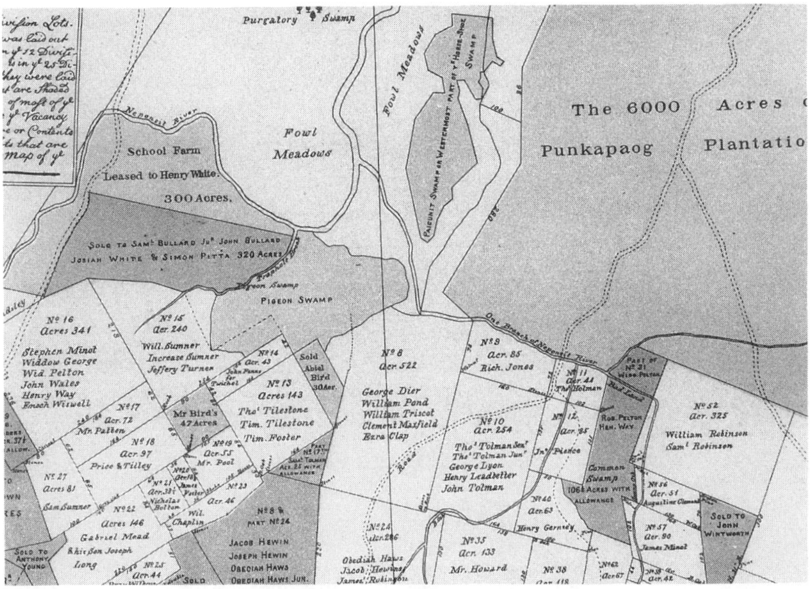

Fig. 1.4. Detail of John Butcher's plan of Walpole. A 106-acre common swamp lies just south of Punkapaog Plantation near a branch of the Neponsit River. The ownership of several other swamps surrounding Fowl Meadows—Purgatory, Pigeon, and Pacunit or Horse Shoe—is not clear. Walpole Historical Society

lies to the west of Mashapaog Pond in Mashapoag Swamp. Whether others—Burnt, Horse-Shoe, Pigeon, Purgatory, and Blew-Hills Meadows—lie within divisions of land or remain unclaimed is not clear. A few, Dorchester Swamp and two small swamps northeast of Burnt Swamp, appear to be divided.

Cow Commons, Ox Pastures, and Great Meadows

An important category of common lands developed among proprietors who were allotted portions of arable land variously identified as cow common, cow pasture, common meadow, ox pasture, or common field. Title to these plots of meadow or pasture was often vested in individual proprietors, who owned the land in severalty but in practice allowed common cultivation, pasturage, or removal of timber and stones at some point during the year. Where rights of pasturage were shared among town members, on occasion both commoners and noncommoners, the practice was called stinting. Such stinted lands were considered a valuable property right and were frequently employed as inducements to encourage new settlers.[63]

Proprietors were entitled to cultivate their individual plots, bounded by

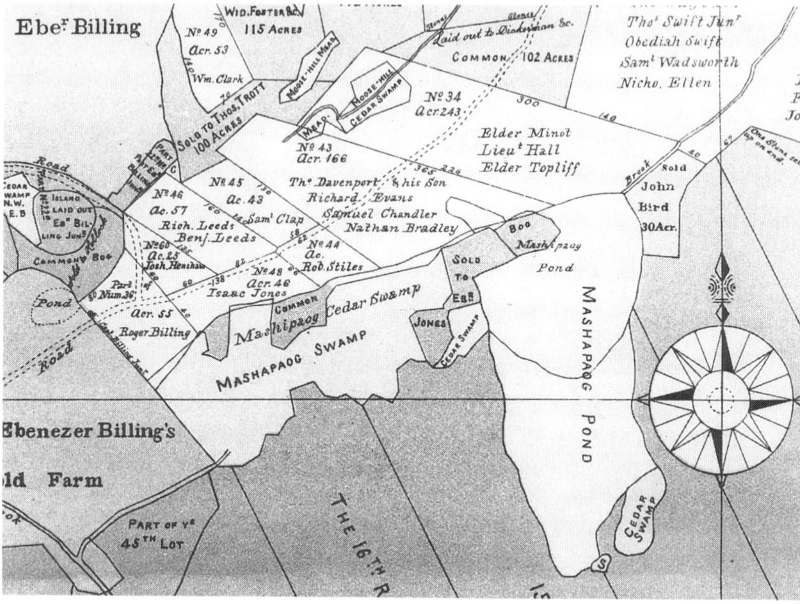

Fig. 1.5. Detail of John Butcher's plan of Walpole. The Mashapaog Cedar Swamp is shown as common, and a bog to the west appears to have remained common while surrounding lands were divided. The status of Mashapaog Swamp and nearby cedar swamps is uncertain, but Moose Hill Cedar Swamp had been sold. Walpole Historical Society

either merestones or fences, and animals were fenced out rather than in. Strict regulations required maintenance of fences, fence viewers were appointed as minor town officers to observe upkeep, and fines were imposed for trespass resulting from poorly maintained fences.[64] Frequently, these common fields were opened after growing season for pasturage and herbage, "commoning time" or "winnowing time." Limits based on proprietary shares were placed on the number of animals permitted to graze. Common meadows and pastures were sometimes located in lowlands bordering rivers and varied considerably in size. Land capable of cultivation without extensive clearing was rigorously sought and usually was devoted to crops or grazing. The scattered locations of these meadows, many of them formerly cultivated by Indians, may have been a considerable factor in determining the locations of early settlements.[65]

Often these open lands contained substantial forest cover, and the culling of timber and wood was commonplace. In 1651, for example, proprietors of New Haven, Connecticut, detailed instructions for partition of the town's ox pasture into a planting field. Town members were told to clear and plant their proportional share for a period of five years and then "to leave it to the oxe pasture againe, not in Indian hills, but plaine, that it may be the fitter for pasture."[66] Those who failed to clear lands forfeited their proprietary interest. Remaining portions were left for grazing, and orderly cutting of wood was permitted on both ox pasture and a nearby cow pasture. Woodlots were laid out by acre or half-acre, depending on need, and a number of restrictions were imposed. Tops and bodies of felled trees were to be removed within a year, subject to rights to leave two trees on an acre of land, probably for shade. All brush wood had to be cleared, but cutting was prohibited within six rods of ox pasture fences. Hoarding or cutting for sale to another was prohibited, and wood on the town's side of the planting field was specifically reserved for elders.[67]

Practices in other towns followed similar, if less specific, patterns. By 1656, Ipswich residents needed to improve grazing land for their day cattle and selected an area identified as "the thick woods lying beyond the cow common." A committee was appointed to locate fences, with costs to be apportioned among cattle owners.[68] Ipswich also established a sheep common and prohibited anyone from topping or felling any logs or trees on the common or "sheep walks."[69] In 1667, Cambridge divided its "common woods and timber with in the Cow comon beyond winottime" and appointed surveyors to measure a line between wood lots and swamp lots in the ox pasture. One town member was ordered to pay a fine for removing fuelwood from a town lot in the ox pasture without permission.[70] Salem records make reference to land divisions in its "pine meadow" in 1643, and in 1670 town voters ordered that any trees felled and left unclaimed on the cow common for more than a month were to be forfeited and used for a meetinghouse.[71]

Timber Rights

In addition to land specifically assigned as common, certain communal rights for expressly limited purposes also existed in New England's nucleated villages. On occasion, such rights were similar to those privileges that ripened over time into an explicit and well-recognized property interest in England. Stinted lands are probably the clearest example. However, communal rights to stinted lands usually failed to mature into a dominant interest but instead were eventually absorbed by underlying ownership and were thus extinguished. Sales, reservations, or outright grants of timber rights to commoners and noncommoners alike are another example.

Individual grants of timber rights on common land occurred routinely. In 1664, Topsfield, Massachusetts, voters divided standing timber on five hundred acres of common land designated to remain common in perpetuity and gave a share in both land and timber to each town member who paid a portion of the minister's rate.[72] At Hingham in 1679, the town agreed to sell wood and timber growing on its common lands near Rattlesnake River for a sum of thirty pounds, with seven years to cut and remove trees.[73] Marshfield voters sold rights to remove ore and cut pine wood for charcoal from common lands to support the manufacture of iron, but their grant was made subject to a condition that only town inhabitants be employed by the industry to cut wood.[74] In Connecticut, New Haven gave a tanner rights to cut trees on common land within two miles of the town to obtain the bark needed for his trade.[75]

When proprietors sold land but reserved timber for communal advantage or placed restrictions on its use that somehow benefited their community, situations were reversed. In 1658, for instance, Cambridge selectmen carved a one-hundred-acre parcel from undivided common land for one of its town members but reserved wood and timber "to ly in Comon to the use of the Towne."[76] Reading citizens sold a thirty-acre parcel of land in 1671 but reserved one half of the timber.[77] Occasionally, timber rights were reserved for only a small group, as in Weymouth, where a three-acre parcel was sold subject to privileges of neighbors to gather wood.[78] To encourage manufacture of much-needed metals, Dedham offered land, wood, timber, and stone from common areas to any town members willing to embark on mining ventures, but they reserved oak trees designated by selectmen and exacted payment into a treasury amounting to one tenth of any profits.[79]

Management of Forest Resources

Exploitation of New England's forests began very quickly after settlements first appeared. Primeval woodlands were a novelty to British immigrants whose islands had been gradually stripped of their forests; in fact, a critical shortage of wood supplies had developed by the late sixteenth century. Colonists who arrived in America during the Great Migration rapidly absorbed existing open lands. In many towns, clearing land became necessary to build tiny settlements, provide space for planting, and establish routes for transportation between villages. Mystery and superstition associated with forests, together with more tangible perils such as wolves, snakes, and hordes of insects, also helped to make woodland more foe than friend. Initially, survival in the wilderness—shelter and food—were paramount concerns. The first dwellings were ramshackle assemblages of bark and saplings chinked with mud and straw. These were quickly replaced by more substantial abodes, which copied methods of English construction, half-timbering filled with mud or wattle and daub. However, such buildings proved unable to withstand harsh New England winters. Eventually, dwellings evolved into an all-wooden style aptly described as the "Forest Vernacular."[80] Shingles for roofing and siding; timber beams for framing; slabs of roughly dressed wood used as floorboards, called puncheons; clapboards for wainscoting; and wooden bolts to join parts together were all acquired from nearby woodlands. As populations continued to grow, housing became a thriving industry. Meetinghouses were an essential part of each community and contributed to the building boom.

An immediate need for food afforded villagers little opportunity to painstakingly remove all trace of forest from cultivated lands. Trees were girdled and left standing, their bare limbs affording sufficient light to sustain a crop of corn. As soil grew unfertile from overuse, the process was repeated, and worn-out lots were converted to pasture. Tree stumps were left protruding from the ground. Clearing land by these means consumed much less time and expense, and both concerns were critical. Burning underwood and ground vegetation was practiced to produce arable land, much as the Indians had done, but fire also was employed to eradicate whole forests. Ashes became a valuable by-product: When left on the ground as fertilizer, nutrients were returned to the soil; when processed into potash, a product used in the manufacture of soap and gunpowder could be sold.[81] In 1653, townsmen in Norwalk, Connecticut, were directed to burn woods "in the fitting season" and to give convenient notice so that inhabitants could secure their fences.[82] In New Haven, similar notice was provided by the beating of a drum.[83] In Massachusetts, Weymouth citizens paid a man by day to burn town woods in 1651.[84] Beginning in 1654 and continuing for almost thirty years, Medfield residents routinely appointed a burning committee each February, along with other officials such as fence viewers and hay reeves.[85]

Once New England towns had become sufficiently well established to provide for basic wants of their members, an economy based on land and timber resources began to develop. Barns were needed for threshing, storage of hay, and protection of cattle. Farm implements and equipment were made of wood, and fences were essential to successful cultivation of crops. With the expansion of commerce, warehouses became necessary for storage of goods. Bridges, wharves, highways, wheeled vehicles, and sleds for winter travel all consumed quantities of wood, and sawmills appeared in nearly every village. Bark was utilized for tanning; pitch wood served to make candles, tar, and coal; a cooperage industry depended on white oak for barrel staves. Brick and lime kilns required vast quantities of wood for fuel, as did the iron works that soon developed.[86] In Rhode Island, Providence authorized a resident to operate a lime kiln on common land and to remove stones and wood to build and fuel his business. The lime rocks were soon declared to be perpetually common.[87]

Demand for finished goods required a commodity that could be exported in exchange, and timber resources were readily available. Timber markets were established first in wine-producing countries, particularly Spain, Portugal, and Madeira. These regions rapidly became dependent on New England's white oak for staves used in making casks. In Massachusetts, Ipswich restricted the felling of white oak without license from a committee.[88] Cotton from the West Indies, sugar cane from Barbados, and fish from Newfoundland were all purchased with New England timber. White pine was coveted by England's Royal Navy for masts, and colonial governments consistently reserved the largest of these trees. Except for England, which imported much of its wood from northern Europe, trade with these regions established reciprocal expanding economies, causing even greater reliance on forest products.[89] Ships were needed to carry merchants' trade, and a thriving industry developed in towns such as Salem. Shipbuilding required immense quantities of timber, and lumbering became New England's dominant industry during the second half of the seventeenth century. Settlement expanded northward into New Hampshire and Maine in an effort to tap virgin forests in those territories.

The scarcity of wood in England must have served as a reminder to many that supplies were finite. Colonists quickly observed depleted woodlands and adopted measures to limit waste, distribute resources fairly, and prevent other towns from harvesting local timber. During this period, woodland management was dictated by economic incentives—stewardship made fiscal sense—and the need to apportion resources equitably among owners. Although evidence of cutting methods is scant, trees selected for wood products from common land were probably culled singly or in small groups. Size of trees and availability of equipment and oxen to cut and move heavy timber made logging difficult and time-consuming. Moreover, practice forbade waste or abuse of resources owned in common.[90] Cutting of second growth was often prohibited, and persons removing wood from common lands were required to clear slash to reduce the risk of fire. In Connecticut, New Haven allowed town members to cut firewood on the town's ox pasture but limited the quantities taken. They also provided incentives for avoiding waste by allowing anyone to carry out trees and wood lying on the ground. Employing woodsmanship transplanted from England, townsmen prudently cautioned: "therfore if they medle with great trees, it was better they lopped them."[91] Unchecked cutting also left impenetrable tangles of wood and brush, and some regulations focused on this problem. In 1643, Guilford required choppers to cut up felled trees within six months or lose them to anyone who cut and carried them away.[92]

Attempts to avoid waste and to make common woodlands accessible were customary throughout New England colonies. In Warwick, Rhode Island, townsmen required residents who felled trees on common lands, whether for timber or fuel, to "cleave or square the timber or fetch it home uncleaft within the space of three monthes." Violators forfeited their timber to any who would take it, but spruce was allowed to remain for a year before it became common property."[93] In Massachusetts, Weymouth voters mandated that trees on town commons within six miles of the meetinghouse could not be cut more than a foot above their bases.[94] Eventually, regulations occasionally extended beyond purely economic concerns and recognized the importance of trees as town amenities. In 1693, members of Reading acknowledged a need for shade trees in their village and ordered that "no person shall fall, girdle, hack or loap" any marked trees.[95]

Restrictions generally fell into one or more of several categories and usually affected common land. Typically, town selectmen prohibited the cutting and transporting of timber without permission or demanded that it first be offered to town inhabitants. They also placed claims on all cut wood left uncleared and levied fines for violations. In Massachusetts, as early as 1635, Salem prohibited cutters from leaving tops and bodies from felled trees in cart paths or traveled ways.[96] Less than a month later, town officials recorded a statement of concern for disappearing woodlands: "we have found by experience that the transporting of boards and clapboards from our plantacon hath not only bared our woods verie much of the best tymber trees of all sorts, but bereaved also our inhabitants of such boards and clapboards whereof they stand in need."[97] Watertown recorded waste of its timber on common lands in 1635 and fined abusers who felled trees without permission.[98] In 1639, Ipswich prohibited felling of timber for commercial sale from its common lands and required permission to cut fuelwood. Those cutting timber for their own use were required to remove it within a year.[99] Cambridge halted unchecked felling of young trees for fuel by requiring the consent of townsmen for cutting of all timber, excluding old and decayed trees.[100]

Occasionally, limitations were especially detailed or extended beyond ordinary practice. In 1650, Providence decreed that no person take any common fencing, bolts, pipe staves, clapboards, shingles, pitch lights, or building timber out of the plantation without authority. Fines were set in amounts of ten shillings for every ton of fencing or building timber and for every hundred clapboards, two shillings sixpence for every hundred shingles, three shillings for every hundred-weight of pitch wood, and ten shillings for every hundred pipe staves.[101] A later order established individual rations of ten gallons of coal or tar made from pitchwood, mandated personal use, and forbade sale of both goods.[102] By 1667, shortages of wood had become so severe in Cambridge that proprietors were restrained from selling shares in woodlots separately from house lots without consent from selectmen.[103]

As town administration became more sophisticated, control of forest resources began to serve a more public function, not just a means to protect individual wealth. Sale or taxation of wood products taken from common land helped to meet communal expenses. In Massachusetts, Manchester placed a series of taxes on ship masts: five shillings eightpence on bark masts and one shilling for shallop masts. A tax of one shilling was also levied on each tree used for clapboards, pike staffs, hogsheads, and barrels. An officer was appointed to examine special items such as bowsprits.[104] Plymouth, too, introduced a system for taxing wood products cut from town commons, including pine and spruce planks, cedar clapboards and shingles, cedar bolts, fencing, and tar made from knots.[105] In Rhode Island, Providence sent its town lumber to a local sawmill and then transported it to a water site, where it was held pending shipment.[106]

Enforcement of regulations and taxes required the appointment of officials who were charged with various responsibilities. Surveyors of wood, or woodreeves, identified trees eligible for felling on town and common lands in Cambridge. They were paid fourpence an hour or twopence for each tree felled, together with half the amount of fines levied for any violations.[107] Corders of the wood measured firewood sold in Salem.[108] As the timber industry expanded, official positions were created to accommodate trade, including cullers of staves and hoops and sealers of shingles and wood.[109]

The relationship between nucleated villages and the depletion of forest resources is an ironic one. These clustered settlements contributed to exploitation of woodlands by encouraging clearing in a single large area. As populations in that area increased, splinter communities formed, new villages were cleared, and surrounding forests were consumed. Depleted timberlands then hastened change from communal ownership because cleared land was more valuable and subjected proprietors to greater pressures to sell. Irony exists because America's history of urban settlement has relatively few examples of villages begun with extensive communal lands. Modern planning considers such public land to be one of the most effective means for controlling land use.[110]

Communities of Change

For as long as sharing of communal resources proved an equitable and practical means for wilderness survival, common land in its assorted forms continued to characterize New England's nucleated villages. However, a multitude of causes, among them proprietary conflict, economic pressures, speculation, political and religious unrest, and evolving agricultural practices, all helped to shatter the fragile harmony that existed in these communities. Before the close of the seventeenth century, such pressures had forced changes in patterns of land use and in village form itself. These constraints were also responsible for a subtle but relentless transformation of common land from a communal resource to a commodity belonging to a select group of town members.

Town proprietors in New England's early plantations were initially eager to increase village populations, and they offered inducements, usually valuable rights to common pasture and woodland, to newcomers. However, the steady arrival of immigrants and their demands for a share of undivided lands eventually led to conflict. As some proprietors accumulated wealth, undivided lands became a source of friction between a closed circle of proprietors, precursors of an emerging landed class, and the remaining inhabitants. Cottagers, the poorest of town populations, were usually permitted to rent small plots of land within common areas, along with rights of commonage for pasture and wood. These residents eventually sought to perfect their claims into full ownership and added to the conflict. Faced with attempts by noncommoners to wrest control over common lands, proprietors became organized as a formal body, strengthening their title to undivided lands through civil complaint and legislation.[111]

Questions involving land use and distribution eventually became separated from towns' political affairs, and noncommoners were forced to contend with only the latter. Formal organization by proprietors failed, however, to quell discontent as noncommoners began to greatly outnumber proprietors. Envy of political privilege and power, squatters' claims against common lands, increased absenteeism among proprietors or their heirs, and the emergence of towns as distinct political entities capable of a concerted voice against a privileged class, all challenged a proprietary system of government. Residents of Duxbury, Massachusetts, boldly nullified division of certain common lands because "common lands were never granted to the owners of Court grants, but to the inhabitants of the town" in perpetual succession, rather than to particular persons.[112]

Complex economic issues also contributed to change in these communities. A burgeoning economy based on trade and largely dependent on timber resources lacked sufficient currency to operate efficiently. Land continued to provide the surest potential for accumulation of wealth. Fortunes, in turn, encouraged investment, and land was readily available to secure credit. Effective control of these investment holdings was linked to division of lands into towns. As speculation intensified, absentee proprietors multiplied in number. Lands acquired were held for profit, with sale to be accomplished as quickly as possible. Proprietors conducted business from distant points and frequently reneged on the financial and contractual obligations associated with settlement. Expensive tasks such as roads, public buildings, and land divisions were often left uncompleted because absentee proprietors had failed to pay taxes levied against their unsold lands.[113]

At the same time, colonial governments attempted to produce currency by auctioning unappropriated lands and by securing disputed boundaries through settlement. Both added more energy to the push for land. Government and speculator were joined to a common task, and large numbers of town charters were granted. However, strict controls once exercised by general courts over methods of town settlement began to erode. House lots were larger, farmsteads became dispersed and isolated, and outlying lands were quickly assigned to individual proprietors rather than held in common. Restraints on alienation without the consent of proprietors, so important to early communities, were noticeably absent. As a result, newly formed towns often lacked the physical cohesiveness that characterized older villages. Concern for subordination of individual interests to those of the community as a whole eventually gave way to policies that responded to complex political and economic developments.[114]

The period's unsettled political and religious climate was also a factor. As farmsteads became dispersed, new communities began to take shape. Arduous travel between farmstead and distant meetinghouse or school inevitably led to construction of a new, more accessible meetinghouse and appointment of a minister to serve its parish members. Ecclesiastical independence fostered political independence, and disputes over taxes frequently ensued. The expense of supporting a minister created considerable burdens for town members, and splinter communities quickly sought to avoid the taxes levied by their parent villages. Ultimately, requests for township charters were made to the general court, which possessed sole authority to create towns and empower them to govern their affairs.[115]

In the midst of these internal and external forces, agricultural practices were gradually improving and forcing marked changes in land division. During a period when subsistence farming and defense against attack by Indians were crucial to survival, compact settlements served a necessary function. However, scattered distribution of separate lots assigned to different uses proved cumbersome and inefficient, requiring daily treks to and from these outlying areas. In response, land holdings were steadily consolidated, and farmsteads became more dispersed. House lots and farm lots were combined to avoid lengthy journeys between the two. Consolidation of valuable land holdings improved economic standing and permitted more efficient husbandry, which in turn contributed to greater profit. Accumulation of wealth by a few helped to undermine the communal spirit once nurtured by a common struggle against wilderness and guided by dogmatic religious and educational beliefs. Common areas, tactile monuments to a communal temperament that fired these nucleated villages, succumbed to the push for consolidation of farmsteads. Both residual undivided tracts and common pastures, the latter often valuable because of their status as cleared lands, methodically passed to individual ownership. As common lands dwindled, reflecting a fading communal spirit, proprietors as a group ceased to be a viable force in town government. Economic necessity, which had earlier dictated collective management of communal resources, soon faded in favor of a more certain path to prosperity through individually owned land.[116]

Common Woodlands after 1700

The characteristic that distinguished common land, wooded or open, belonging to early nucleated villages from common land associated with later settlements was proprietary intention. During a period when groups of individuals organized to form their own self-governed societies, land they owned as tenants in common was truly a communal resource. However, as populations and demand for land increased, land divisions in existing villages occurred steadily, while new towns proliferated and pushed back the wilderness. In both cases, common land assumed a different identity, that of a commodity able to produce individual wealth. This distinction is important because it helps one focus on the communal spirit that set apart the earliest villages from later towns and made the former unique, both in patterns of living and in shaping a landscape.

The close of the seventeenth century provides a handy, albeit somewhat arbitrary, marker for the appearance of new town forms that became standard throughout the next 150 years, until most lands in New England had been allotted into towns. Nevertheless, it is probably unfair to assign this distinction concerning common land to a specific period. Variations in cultural traditions among colonists and adaptation of those traditions to America's challenging environment caused significant variety in town and landscape form. As early villages matured, some clung to common lands during the eighteenth century and, in a few cases, well into the nineteenth century. Whether this can be attributed to a tenacious communal spirit or to other, more tangible influences requires close scrutiny and comparison of individual town histories. In contrast, remote frontier settlements in areas of Maine, New Hampshire, and Vermont during the eighteenth and early nineteenth centuries, often initiated by purely speculative motives, were located by survey grids that facilitated rapid division and sale of lands. These towns rarely displayed any sizable tracts of common land except in cases where survey errors left gaps or gores and allotment to individual proprietors was overlooked.[117]

Late common woodlands appear in a variety of contexts, many similar to those in nucleated villages. General common lands sometimes remained open to residents for timber and wood, even though division of those lands had begun. Division frequently occurred in stages, with allotment of proportional shares in specific tracts assigned to individuals who continued to share ownership for short periods. Separation of these tracts into individually owned parcels followed when surveys had been completed and boundaries established. Frequently, division of common lands formalized and solidified proprietary rights, thus foreclosing the ability to gather forest resources to all but a select few. As clearing for agriculture, building materials, fuel, and export trade gradually depleted local woodlands and as distances for transporting wood to needed locations continued to increase, woodlands became increasingly valuable. When permission to cut wood on common lands was withheld, timber trespass frequently ensued. Constant occurrence of complaints by proprietors against other town members suggests that trespass was an acknowledged practice and represented a calculated defiance of proprietary rights. In such instances, divided common lands did not lose their communal status entirely.[118]

Specific tracts of common land continued to be set aside as woodlots in towns such as Manchester, Massachusetts, where proprietors voted in 1706 to "lay out 50 or 60 Acors of land at the west end of our common for a wood lot" and ordered that "no wood nor timber shall be cut nor carryed to any water side or landing place and exposed to sale out of our town of manchester upon the penaltie of paying the sum of ten shillings a cord for every cord of wood and ten shillings for every tun of timber." [119] Common swamps existed in some towns chartered during the eighteenth century, such as Lunenberg, Massachusetts.[120] Division of common land occasionally resulted in the sale of timber and wood separately from ownership of the land. In 1753, Braintree selectmen appointed a committee to determine the best method for dividing the town's common land, and four years later they recorded the sale of the town's wood.[121]

In newly established towns where lands were quickly assigned to individual proprietors and held for sale, common lands occasionally materialized. In such instances, valuable stands of timber were sometimes protected. In Warner, New Hampshire, several proprietors' lots were designated as common land, and the town appointed a committee of three to take care of wood and timber on this land until it could be sold.[122] However, there is little to suggest that this common land was open to communal use. Similarly, proprietors in Kittery, Maine, surveyed large blocks of land well prior to sale. As a result, early maps reveal extensive lots identified as common, but public rights in these areas were probably limited. Large, unsettled townships in northern Maine are still owned in common, often by timber companies, but towns have not been incorporated, and there is no communal interest.

From Common Land to Public Land

One must recognize that, in very early settlements, ownership and control of both land and community were vested in town proprietors. Apart from certain restrictions, proprietors' dominion over their land was generally unfettered by any governmental oversight other than their own. As populations increased, however, towns as distinct entities capable of representing broad community interests through the voices of appointed or elected officials began to take shape. Amid a press for individually owned land, plots were set aside to sustain these growing political bodies and their institutions. Yet the ability of towns actually to acquire or convey title to land appears much later. Thus, muniments of early grants were usually confined to simple votes recorded in town records. In rare cases in which deeds were executed, they sometimes ran to selectmen in trust. Consequently, these plots are more properly classified as public lands, simply to make clear the general absence of any underlying proprietary interest. Public lands were assigned for sundry purposes, most often to encourage settlement by clergy. Millers, too, were in great demand, and mill lots to subsidize saw or gristmills became an ordinary part of town structure. Other types—a potash lot to stoke an ashery in Warner, New Hampshire, for example—appeared with less frequency. Many such parcels were awarded subject to a condition that grantees remain for an agreed period, and towns typically retained a measure of control over these lands until conditions were satisfied.[123]

During this period of murky town development, common lands were occasionally assigned by proprietors for general community benefit. Parcels set aside as "forever common" or "common in perpetuity" imply a concern greater than immediate economic or material need. Although usually described as common, such lands hint at transition to public lands, and their stewardship assumes a slightly different character as well. Although desire for economic return from land and its resources remains part of that stewardship, the context suggests a focus on horizons more distant than just present yield. Sincere interest in long-term communal welfare probably factored in land allotments for a great many towns. In Massachusetts, as early as 1659, Weymouth voted that part of its commons from the "head of lower Fresh Pond to Carpenter's Swamp remain in common and not be granted out by townsmen." [124] Medfield set aside a number of parcels in 1670, to lie common forever, and Topsfield assigned land to remain common in perpetuity.[125] In Connecticut, Lyme townsmen designated land to remain perpetually common for the town's use.[126] Despite good intentions, the altruistic motives of proprietors were not always inherited by successions of town officials, and such lands often vanished into private ownership.

Brunswick Commons

Maine's Brunswick Commons is one of the best examples of an early-eighteenth-century proprietors' grant for unspecified public benefit in New England. Unlike so many early public lands, Brunswick's property still retains its communal character, although its area is much diminished. Originally a thousand acres, the parcel was unusually large for a voluntary endowment by absentee proprietors. Equally remarkable is the fact that charter company owners were responsible for preserving Brunswick's legacy as a community resource, despite repeated attempts by town members to sell it.

Brunswick was founded by the Pejepscot Company, a Boston investment firm that acquired a large coastal area between Casco and Merrymeeting Bay on both sides of the Androscoggin River in 1714. Company owners recognized the good sense, fiscal as well as common, in planning a stable, well-ordered community, and they developed their lands slowly. Several towns were carved from the Pejepscot Purchase, all part of Massachusetts Bay Colony. Brunswick received its charter in 1717, and two years later proprietors voted to set aside a

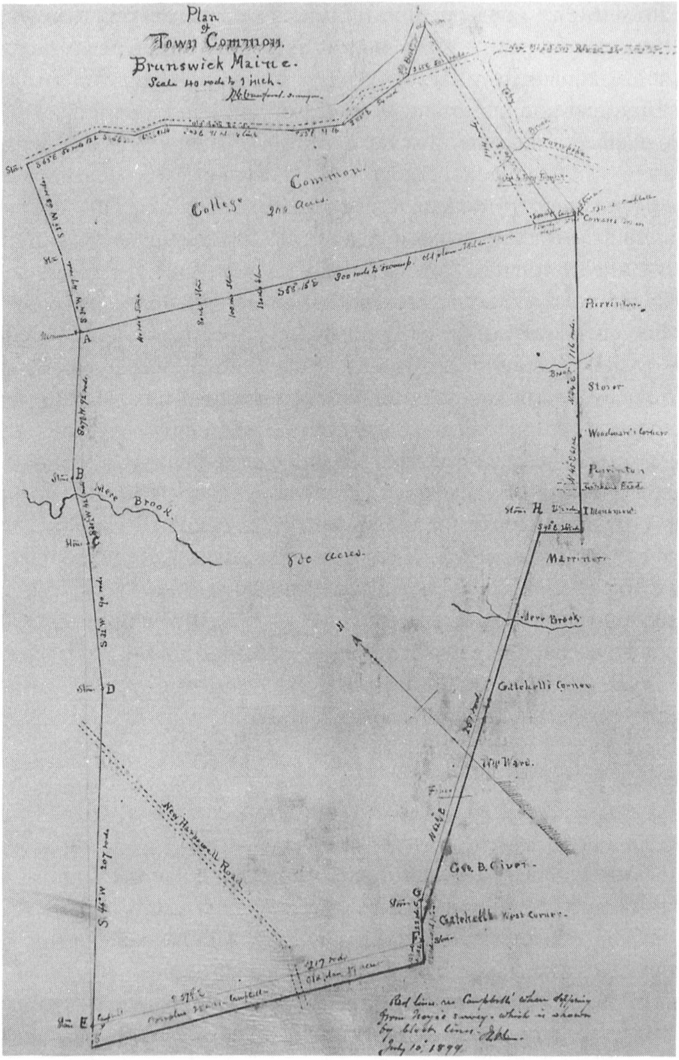

Fig. 1.6. Plan of Town Commons, Brunswick, Maine, drawn by J. B. Crawford in 1899. Crawford's survey was limned during a period when concern for stewardship of the commons was gaining momentum. With assistance from foresters Austin Cary and James W. Toumey, the Brunswick Village Improvement Society began a plantation of white pine in 1900. Office of the Brunswick Town Clerk

thousand acres of land to be held in general and perpetual commonage for the town. That vote was confirmed in 1742 and again in 1771, when the company appointed an agent to execute a deed. In 1773, town voters agreed to accept the land, but controversy over its boundaries caused delays until 1779, when a deed was finally presented. Granted to town selectmen in trust, the property was conveyed for the use and improvement of town inhabitants forever. That same year, selectmen took steps to secure lumber growth for the town's benefit.[127]

The commons occupies a sandy plain stretching between village and sea and was probably heavily forested when the town's charter was granted. Between 1719 and 1779, the land maintained an uneasy existence, its resources gleaned by any who made the effort. Indecision among town officials before 1779 became indifference afterward, and neglect governed during most of the nineteenth century. Early surveys were haphazard and perimeters not well marked. Encroachments steadily chipped away at boundary edges, and a two-hundred-acre parcel was set aside for Bowdoin College in 1811. Fire took its toll, and trees managing to escape flames were cut for fuel by the town's poor. Attempts were alternately made to sell the tract, convert it to agricultural production, or use it as a poor farm. These were consistently rebuffed by company owners, who exercised exceptional foresight. Frustrated by unsuccessful attempts to secure a public benefit, a committee appointed in 1857 to oversee the property complained that the expense of maintenance far outweighed any community value.[128]

Toward the close of the nineteenth century, however, interest in the parcel was rejuvenated. A new survey, by J. B. Crawford, was prepared in 1899 (fig. 1.6), supplementing an 1858 survey by Charles J. Noyes. In 1900, Brunswick's Village Improvement Society solicited advice from the U.S. Division of Forestry and began a plantation of white pine that year. With assistance from foresters Austin Cary and James W. Toumey, Brunswick Commons became one of the earliest planted town forests in New England. A municipal airport constructed in 1934, later a naval air station, reduced Brunswick Commons to is current size of fewer than one hundred acres. However, ninety-year-old stands of white pine remain, and the parcel continues to pay tribute to the Pejepscot Proprietors' extraordinary commitment to town planning.[129]

Shaker Timber Communities

Common land associated with village settlement in New England reappeared in an entirely different context during the late eighteenth and early nineteenth centuries as utopian societies embarked on unique ventures in town planning. These idealists sought purity in life through strict obedience to an assortment of quixotic teachings. Although tenets varied, many groups adopted communal ownership of property as part of their social or religious doctrine. Where organizations assumed characteristics of small villages, common land became a purely communal resource, much as it had been in New England's nucleated villages.[130]

New England states captured a fair share of these communities. A few, such as Brook Farm, founded in 1841 in West Roxbury, Massachusetts, by transcendentalist George Ripley, were thoughtfully conceived. None, however, proved as enduring or successful as the Shakers, who established nearly a dozen communal villages in Massachusetts, Connecticut, New Hampshire, and Maine and several more in a number of other states. Many became stable, prosperous, and highly self-sufficient. The Shakers, whose formal name is the United Society of Believers in Christ's First and Second Appearing, originated in England and emigrated to America in 1744. Their doctrine espoused common ownership of property, celibacy, nonresistance, spiritual power over physical disease, and the bisexuality of God, which had the effect of promoting equality between the sexes. Their distinctive name is derived from the outbursts of shaking and trembling that accompanied religious worship. During the nineteenth century, two dozen communities existed in eastern and midwestern states; more than half endured into the twentieth century, and at their peak membership approached six thousand.[131]

Consideration of Shaker villages in the context of New England's communal forests is important for several reasons. Communal ownership and management of property endured as an essential part of these successful villages, contrasting sharply with the emphasis on privately owned land that quickly engulfed New England towns. As a result, they suggest striking points for comparison to early nucleated villages, particularly their agricultural and timber economies, common land systems, utopian idealism, and sense of community. Moreover, Shaker communities that survive today offer visible evidence that communal lands can be used as a planning tool to shape or control village form. In so doing, they precede early-twentieth-century planning theory by more than a hundred years. The fact that many of these communities developed thriving timber economies also underscores the viability of collective forest-resource management. Stewardship, too, extended well beyond immediate return, and several communities employed silvicultural practices fairly advanced for the period. Finally, accounts by some Shakers offer a rare glimpse into forest landscapes and logging practices at the time.

Land in Shaker villages was owned and administered by a group of trustees. Members' needs were supplied through agriculture, animal husbandry, and small industries powered by an extensive variety of mills. Essentials that could not be grown or manufactured were purchased with funds generated by the sale of Shaker products such as seeds and wooden articles, which became

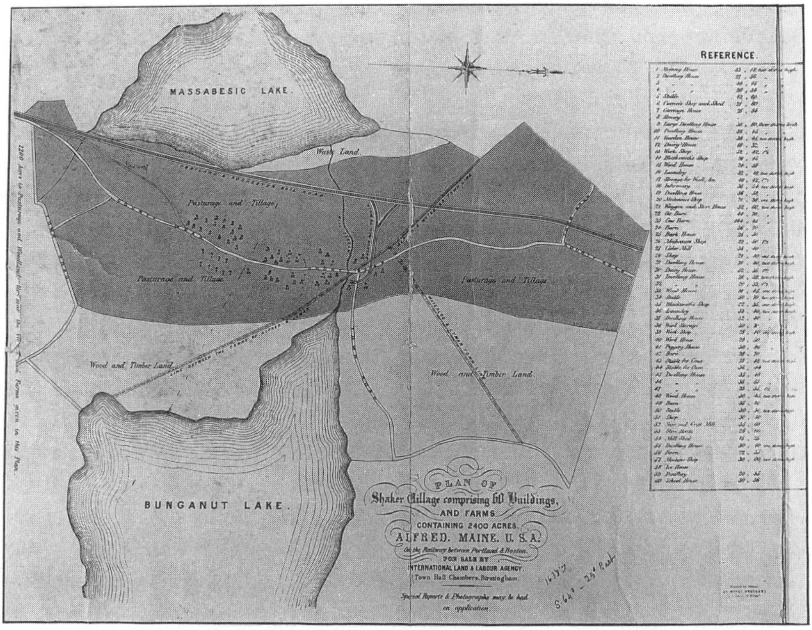

Fig. 1.7. *Plan of Shaker Village Comprising 60 Buildings and Farms Containing 2400 Acres, Alfred, Maine.* An undated, unsigned plan of the Alfred community as it appeared at the time of its sale in 1932. The village nucleus is surrounded by common pasture, tillage, wood, and timberlands. United Society of Shakers, Sabbathday Lake, Inc.

well known for their simple design and careful craftsmanship. Villages were planned with equal simplicity and care and were separated into divisions called families. Each family developed a cluster of dwellings, workshops, and mills that gave these units a degree of physical autonomy similar to splinter communities in early New England towns. Surrounding fields, pastures, orchards, and woodlands were meticulously tended in quiet isolation from the outside world.[132]

A plan of Maine's Alfred Community, depicted shortly before its sale during the Depression, showed distinctive village clusters surrounded by communal pasture and woodlands (fig. 1.7). To a degree, settlement was linear and extended along a road that bisected the community. Such a feature was typical of many Shaker villages, yet groups of buildings were usually sufficiently dense to balance any linear influence. Among Alfred's sixty buildings were a sawmill and a gristmill (52); five wood houses, presumably for keeping fuelwood dry (15, 33, 38, 40, and 48); a bark house (25); and a wood storage shed (38). A rail spur penetrated village perimeters to serve mill and machine shop. Wood and timber lands formed an eastern backdrop, partly in the neighboring towns of Lyman and Waterloo, and encircled the westerly end of Bunganut Lake, where its waters spilled into a millstream. An extra twelve hundred acres of pasturage and woodland lay south of areas shown by the plan.

Agrarian activities were consistently present in all communities and dominant in most. Depending on the period, logging in surrounding forests was certainly present in most and dominant in some. At the very least, members cut and split cordwood from nearby woodlots and hewed or milled timber into lumber for a multitude of village buildings. Timber also supplied raw materials for a countless variety of wooden products. Barrels, spools, sieves, buckets, tubs, churns, cheese hoops, noggins, firkins, seed boxes, brooms, rakes, scythe snaths, and wheelbarrows were just a few of the many items produced and marketed by Shaker mills.

Successions of water-powered sawmills were an integral part of village morphology, as were related but secondary buildings such as kilns and woodsheds. Many villages had several sawmills reserved for different purposes: timber and lumber, long lumber, turning, and wood or bark. With more sophisticated mill systems, the demand for raw materials increased, and a few communities engaged in the purchase of timber rights from ever greater distances. Loggers were sometimes hired to transport heavy timber to mill sites. Milling, too, became a commercial venture and served neighboring landowners, as well as an occasional company in need of specialty lumber.

The extent to which these general characteristics were more or less present in New England's villages depended on a number of factors. Location and its many complex economic variables were important. Communities, too, were far from static and encountered periods of decline and prosperity in ideology, economics, population, and membership composition. Shakers recognized the importance of economic productivity as a means of bonding members, yet profit posed certain ethical dilemmas. When prosperous conditions, rather than ideology, attracted new members, community social structure was weakened. Quandaries of conscience also fostered conflict between otherwise devoted members. At Sabbathday Lake in New Gloucester, Maine, an elder accused trustees of denuding forests to carry on a business, and a sister who remarked that the community's money was "in the woods" was ridiculed for her materialism.[133]

Among the dense pine woods of New Hampshire and Maine, logging and milling were especially lucrative. At the Shaker village in Canterbury, located in central New Hampshire and established in 1792, a reservoir and canal were engineered in 1800 to supply saw and grain mills erected at about the same time (fig. 1.8). At the North Family's village, a second sawmill was built in 1814 to saw firewood and was improved after 1900 by erection of lum-

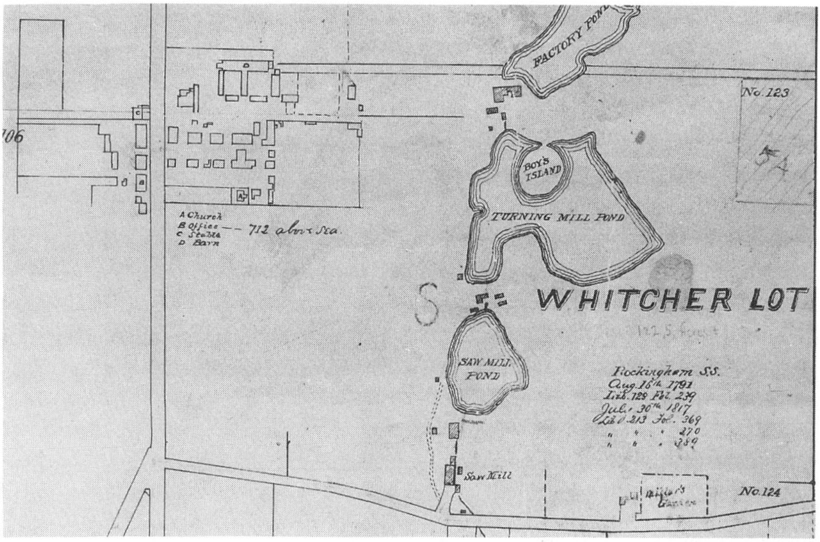

Fig. 1.8. The Whitcher Lot, from a plan of Canterbury Shaker Village drawn by John A. McClintock in 1879. Shakers in Canterbury engineered a series of millponds and waterways to power a variety of mills located near the Church Family village, shown here. Several other mills were erected near the North Family village. Canterbury Shaker Village, Inc., Canterbury, New Hampshire

ber rolls and a wood pulley. A third mill turned woodwork for furniture and tool handles. Other woodland resources were not overlooked. From sugarbush flowed saccharine return that spilled over into by-products in the shape of sap buckets, assembled by Shakers in countless numbers.[134]

Canterbury Shakers worked a number of distant woodlots, among them a tract in neighboring Gilmanton identified as "pine lot" (fig. 1.9) and another in Loudon and Canterbury with the title "bug lot," which leaves little doubt about its most abundant resource. Records of timber harvests between 1830 and 1835 are fairly complete and range from a low of sixty logs and three thousand feet of timber from the pine lot in 1831 to a high of three hundred logs in 1832. A sampling of other production statistics reveals that four hundred cords of wood were cut in 1850, and 379 bushel barrels of sap were collected from the sugar camp (fig. 1.10) to make 2,214 pounds of sugar. In 1853, maple sugar production fell to 1,850 pounds, but 1,400 dozen brooms and 152 pails were assembled.[135]

There is little to suggest that any significant forest cultivation occurred at Canterbury. If one can ascribe significance to the absence of any records showing time-consuming forest planting, then cut-over lots were generally left to

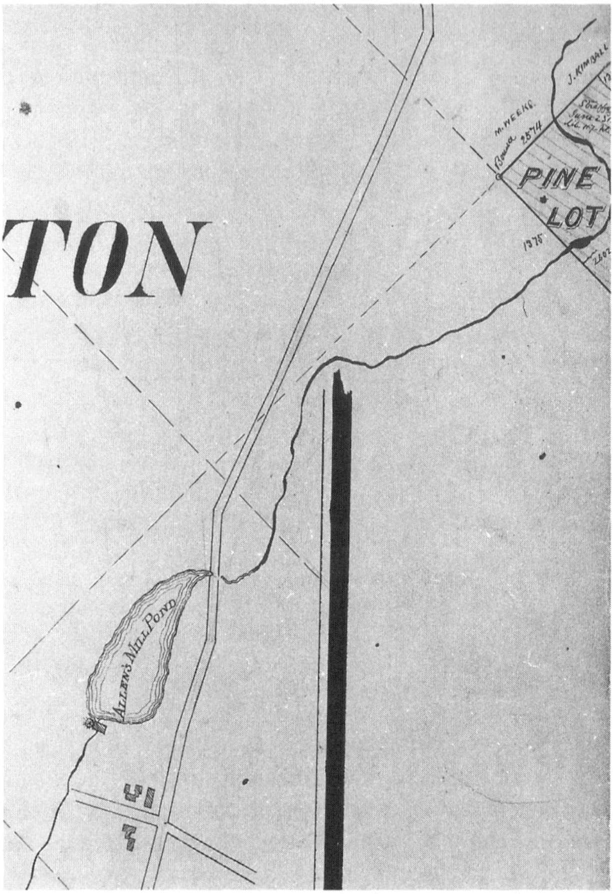

Fig. 1.9. The Pine Lot in Gilmanton, New Hampshire, from John McClintock's 1879 plan. Canterbury Shakers procured remote woodlots or timber rights in several neighboring towns. The Pine Lot, shown here outside the ink border of McClintock's immense and superbly detailed plan, was acquired in 1823 and yielded steady supplies of sawtimber. Canterbury Shaker Village, Inc., Canterbury, New Hampshire

natural reseeding. In 1864, however, fifty young maple trees were set out at their sugar camp, and forty coniferous and shade trees were transplanted around village buildings three years later. Elder Henry C. Blinn, a prominent community member for many years, also began a small arboretum. When control of white pine blister rust became necessary in 1923, Canterbury's trustees entered into an agreement with New Hampshire's forestry commission for

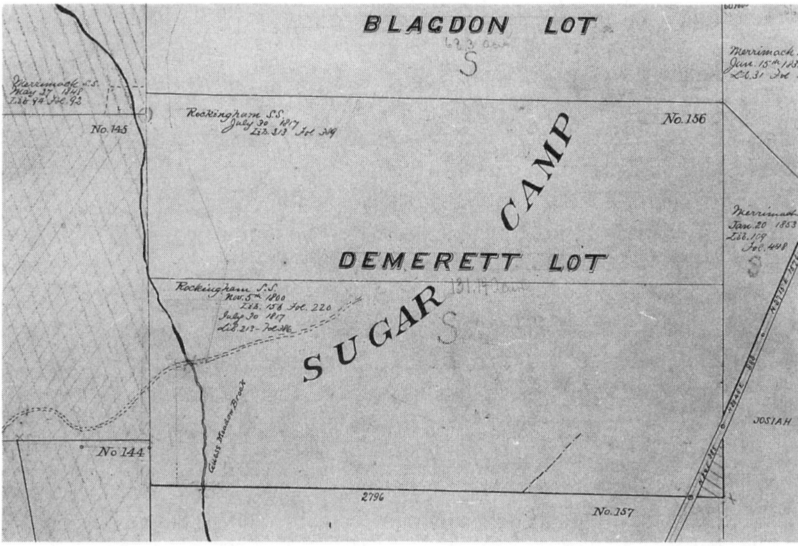

Fig. 1.10. The Sugar Camp at Canterbury Shaker Village, from John McClintock's 1879 plan. Maple sugar production contributed substantially to Canterbury's economy, as did by-products such as sap buckets. Although Shakers generally left cut-over timberlands to natural reseeding, fifty young maple trees were transplanted at the sugar camp in 1864 to help sustain future yields. Canterbury Shaker Village, Inc., Canterbury, New Hampshire

eradication of currant and gooseberry bushes fostering the disease. Work on the village's three thousand acres became the commission's largest single cooperative undertaking, and the state government paid 20 percent of the cost. Shaker trustees appropriated $1,000, a clear commitment to their timber holdings.[136]

Whether from decline in members capable of working with heavy timber, preoccupation with other matters, or simple financial need, the community began to sell timber rights on its land holdings after 1891 and continued to do so well into the twentieth century. Written agreements were reasonably well prepared and usually identified species and dimension of trees to be cut. They also established a time within which cutting and hauling had to occur, required removal of slash, and generally permitted assembly of portable steam mills. Nevertheless, their own mill continued to operate.[137]

Enfield's settlement, situated along Lake Mascoma in western-central New Hampshire near the Connecticut River, was organized at about the same time as Canterbury and maintained similarly vigorous wood and lumber industries. As in Canterbury, a number of mills were constructed; some were specialized, including one capable of sawing forty-foot timbers. Sugaring and production of barrels, wooden tools, and sap buckets were also an important part of village economy. Although neighboring landowners had cut off most old-growth white pine by 1850, Shaker woodlots held a few surviving stands. Accounts of these ancient forests provide a revealing glimpse into lumbering practices of the day. Occasionally, remaining trees grew in clumps of three or four, but more often they stood alone, with substantial distances between. Butt diameters seldom measured less than two feet and more often ranged from four to five. Limbs were confined to treetops, eighty to one hundred feet above the ground. Trees were prone to windfall, and sapwood rotted quickly; however, heartwood remained sound for many years.[138]

Henry Cumings, long a member of Enfield's community, left to become editor of a local newspaper, the *Enfield (New Hampshire) Advocate*. Writing in 1908 with what must have been a clear memory, he recalled a cold February day in 1857, when he joined a logging crew as they headed into pine woods. Trees, two to three centuries old, 125 to 160 feet tall, were felled by four men, "two on each side, standing facing each other, one chopping right handed, the other left." With sections sawed into sixteen-foot lengths, each tree yielded seven to nine logs. On that day, choppers discovered a tree blown down years before. Cumings's sketch of the scene is descriptive: "The sap was all rotted off the outside, and a yellow birch tree about one foot in diameter and perhaps forty feet high was growing on the upturned earth; yet when tried with the saw it proved to be perfectly sound, it was about two and one-half feet in diameter, and we got five or six nice logs and the timber was beautifully white and clear."[139]

In Maine, Sabbathday Lake's wood industries were the community's most profitable for periods during the second half of the nineteenth century. Barrel shooks, shingles, lumber and a great number of assembled wooden implements often netted better returns than did farm produce. Although an unfortunate episode of speculation and embezzlement dampened a brief period of economic prosperity, lumbering activity remained fairly constant. Construction of a "Great Mill" in 1853 introduced this era of rising industrialization, and a great mill it was—three stories tall and sixty feet long (fig. 1.11). The mill's basement contained saws for dotting and splitting as well as a machine for edging and cutting barrel staves. Separate planers for timber planks and boards operated on the second floor, and turning lathes, circular saws, and planers for thin materials such as sieve rims were all assigned to the upper story. A shingle machine was added to the mill in 1886. Crews kept detailed accounts of milling activity. For example, by May 15, 1887, sawing had ended for the season, and blades had spit out 100,000 board feet of lumber: 75,930 feet for the village and the balance for neighbors.[140]

Dimensional lumbering eventually overtook milling of barrel staves at Sabbathday Lake, and loggers were hired to pull timber to the village mill pond by ox and horse teams. Diaries are filled with entries showing full days spent

Fig. 1.11. The Great Mill at Sabbathday Lake Shaker Village in New Gloucester, Maine. Erected in 1853, the building's three stories housed a variety of milling equipment, including splitting and circular saws, planers, turning lathes, a shingle machine, and a special invention for edging and cutting barrel staves. United Society of Shakers, Sabbathday Lake, Inc.

sawing wood. Concern about overproduction and its ethical consequences prompted one elder to estimate the size of New Gloucester's timber resources and to set an annual quota. However, this rudimentary policy of sustained yield management was abandoned by the mid-1890s. In 1918, milling activity had increased to 123,000 board feet for the village and 60,000 for neighbors. A substantial network of accounts with lumber companies in Portland, Lewiston, Freeport, and Auburn, was established, and invoices from 1903 record shipments of pulpwood to a Yarmouth paper mill.[141]

Although very little is known about Shaker woodsmanship, it is fair to surmise that at least some form of elementary forestry management existed. Were circumstances otherwise, one would question why a careful order that surrounded the lives of community members was not carried into the woods. At Canterbury, cultivation of sugar maples, treatment of blister rust, and protective clauses in timber-sale agreements reflected a genuine concern about forest resources. At Sabbathday Lake, timber production and ideology collided, resulting briefly in a primitive form of sustained yield management. In Enfield, New Hampshire, old-growth pine was conserved long after nearby lands had been cleared of this indigenous tree. Blow-downs, too, were salvaged when heartwood remained sound. At Enfield, Connecticut, village member Omar Pease planted 150 acres of worn-out sand plains with white pine seed he had patiently gathered from cones. Sprouts were protected by rye sown at the same time. Pease died in 1883, unaware of the need for thinning young plantations. Nevertheless, his experiment was inspected by a trained forester in 1895, and silviculture was begun soon after.[142]

. . .

From town to town, consistent reference to common woodlands in early records makes clear the character and fundamental value of these lands to their communities. Mention is frequently specific enough to justify arrangement into discrete classifications. Broad conclusions about the purposes underlying use of common wood and timber are also possible. The presence of collective stewardship, essential to the economic and material wants of town citizens, is certain. Apparent, too, is extension of that stewardship beyond immediate, mundane demands to foster long-term community welfare. More cautious assumptions can be drawn about the presence of simple woodsmanship—protection of underwood or lopping of large trees. Unfortunately, town records are nearly always silent as to all but the sparest of detail. The precise nature of the diverse practices that existed from one town to the next—traditions of woodsmanship in particular—is thus concealed. Sound conclusions may be reached only after the enormous picture puzzle of town landscape has been pieced together from very detailed study of specific communities. The influence of rural English customs will undoubtedly be prominent. Relationships between common forests and community structure in early New England towns also are inextricably linked to English antecedents. The significance of that relationship becomes fully apparent during the early twentieth century, when New England towns turned to Germany where communal forests had evolved from comparable village origins into carefully managed resources.

Chapter 2

Sylvan Public Lands

In New England's nucleated villages, common lands comprised the greatest share of forests utilized for communal benefit. Simultaneously, a second class of communal property, public lands, began to develop as specific parcels were allotted to support community institutions, primarily church and school. Lacking currency to pay salaries for ministers and teachers or to purchase books, colonists resorted to abundant land and its resources as a means for subsidizing these expensive obligations. Parcels sequestered for the benefit of church, minister, school, or the public at large were generally free from any remaining proprietary rights. Ownership sometimes accrued to schools and churches or their ministers, sometimes to towns through their selectmen; disputes on the point were routine. While common land sustained communities as a whole, public lands supported essential pieces of community structure and were a key element in town planning (fig. 2.1).

Distinctions between common and public land are subtle but important nonetheless. They hinge on the extent to which communal benefit dominates any underlying individual ownership. At times, transformation from common to public was casual, almost imperceptible—as where public forces, represented by a concerted voice or commitment, eventually outweighed any outstanding claims of proprietors. When that happened, common ownership ripened into public title, usually controlled by town officials, and private rights were ultimately extinguished. Town commons, central to many New England villages, are sometimes very good examples. In such cases, the terms *common* and *public* are synonymous, the nature of communal benefit complete in both. Certain assignments of common land for public use (woodlots, for example) may also have signified a complete conveyance. Yet in such cases, too, underlying title may have remained with proprietors, who only wanted to clear their land. In that event, communal and proprietary rights overlapped, and land remained common.

The collective stewardship evident on common lands also was employed on public lots. Although the character of use was usually compelled by economic

Fig. 2.1. The village library and church in Newington, New Hampshire. Parsonage and church lots owned by the town have surrounded the village center from the early eighteenth century and have helped to preserve its physical integrity, all the while yielding periodic supplies of wood and timber. Foresters working for the U.S. Forest Service became intrigued with Newington and publicized it through their community forest program in 1938, the approximate date of this photograph. New Hampshire Division of Forests and Lands, Concord

or material need, supervision often fell to town selectmen or to individuals who were directly linked to benefiting institutions. Stewardship thus assumed a different dimension—a short step removed from individual interest and sometimes more consistent than that offered by proprietors of common land. In contrast to common lands, too, many public lots have withstood centuries of change and remain community property, testimony to successive generations of local stewards. A surprising number survive as twentieth-century town forests and impart a dynamic sense of historical continuity. Tracing ancient origins of these public lots, including, briefly, their English antecedents, helps one grasp this continuum. England has long since recognized the significance of its ancient common lands, and New England's public lots are a comparable link to its own town origins.

To the Town Belonging: Land Allotments in English Urban Planning

Town planning in New England seldom displayed the very formal characteristics employed by English settlements in other regions of America—tidewater or southern colonies, for example. Geometric grids and strategic axial spines appeared infrequently, roads usually advanced according to irregular patterns, and buildings were placed subjectively, albeit functionally, their locations often dependent upon relationships to traveled ways, bodies of water, central commons, fields, or prominent structures such as meetinghouses. Village plans were simple and yet developed an innate and harmonious variety as their citizens adapted rural habits and building traditions to a new environment. Beneath rustic custom, however, lay a calculated experiment in the planting of new towns. Civil prosperity, stable communities, and a strong economic base offered the surest means to assimilation of new territories. This, in turn, ensured profitable dividends to both the English crown and the privately chartered investment companies underwriting these ventures. The prospect of wealth was employed to entice individuals and groups into the unknown. Institutions such as churches and, eventually, schools were essential to community makeup and were fully sanctioned by investors. Land and its resources served the aims of all concerned. As did common land, public land in New England communities stemmed from English antecedents.

New Town Plantations

English exploits in colonization began a distant continent from North America and preceded the Great Migration by almost four centuries. A remarkable number of new towns were established in England, Wales, and southwest France during the reign of Edward I (1272–1307). These plantations, identified as *bastides* in France, became part of a grand scheme by England's king to secure control of recently conquered territories. Military strategy influenced site selection and design for these towns to some extent, but other factors were equally important. Towns with stable economies generated revenue for their founders in the form of rents and taxes; offered a market for produce, which also could be taxed; improved the regional economy; and increased the value of surrounding land.

Grants of communal rights to woodlots and open fields were included in some of these towns, a formal recognition of provincial routine. Whether these rights were an integral part of town planning or an afterthought, a simple recognition of practical necessity, is not entirely clear. Many towns were erected on manor lands, often occupying wastes or fringe areas where common grazing brought paltry rents. In such cases, land outside towns remained under the control of manor lords. In other instances, particularly in northern Wales, planted towns possessed extensive fields of their own, together with common rights over pasture and nearby woodland.[1]

Where Welsh forests were abundant, timber rights were frequently granted to burgesses who acquired lots in new towns. The charters of Okehampton and Rhuddlan made provision for common use of forest and wastes, and at

Flint, Edward graciously accommodated town members: "that they may have all their necessaries both for the founding lead ore and for other business, without cost, from the woods and underwoods of eight Welsh townships adjoining, as well as common pasture therein."[2] In England, the amount of land incorporated into town plantations depended on whether a new town clashed with an older village. Where conflict was absent, large areas of surrounding waste and pasture were subject to common rights to graze livestock and gather fuel and building materials. At Newborough, cattle were permitted to graze in Needwood Forest; at Newport and Maldon, common rights were granted in moors and marshes.[3]

Ulster Colony

Whether Edward's experiences in town planning were incorporated into collective knowledge and later recalled is not known. However, another experiment, colonization of Ireland's Ulster province early in the seventeenth century, helped to prepare England for its more distant ventures soon to follow. Settlement of Ulster was executed by a complicated process of town planning in territory newly acquired and inhabited by an increasingly hostile population. The relevance of these events to the origins and development of New England's villages is considerable when one considers the many physical and human barriers linked to organized formation of towns.

England's incursions in Ulster culminated with the "Flight of the Earls" in 1607, and the lands of these Irish lords were confiscated by the English Crown. An ambitious regional plan encompassing six counties was proposed in 1609 and modified a year later. In its revised form, forfeited lands were divided into categories identified as great, middle, and small proportions. Property was distributed by lottery to freeholders, who were required to establish manors and erect castles or fortifications, and a number of towns were planned for each county. Inducements such as duty-free imports and exports were offered for a period of seven years, and free use of the Crown's timber was authorized for construction of buildings.[4]

An especially large area of land incorporating part of several counties was redefined as the County of Londonderry, and two important towns, Derry (later renamed Londonderry) and Coleraine were proposed. The immense financial burdens tied to these bold plans forced the Crown to engage London to assist with development. Extensive lands for both new towns were granted to that city in perpetuity. Barren mountains as well as bogs, the latter often containing significant stands of timber and wood, were not calculated as part of allotted acreage and were granted as a boon. Remaining county lands and woods were held by the Crown, and various fishing rights, admiralty rights, and commercial licenses were granted. The felling of timber was permitted for construction of plantations, but private sale was restricted. Eventually, lands encircling both towns were conveyed to freeholders.[5]

London quickly formed an association, the Honourable Irish Society, to manage its plantation and proceed with tasks of raising capital. Londonderry's charter, executed in 1613 by the Crown to ratify title and establish a governmental framework for the town, specifically granted four thousand acres to the new settlement: "and also all those lands next adjacent to the aforesaid city or town of Derry, lying and being on or towards the west part of the river of Lough Foyle, containing by estimation, four thousand acres, besides bog and barren mountain, which said bog and barren mountain may be had and used as waste to the same city belonging.[6]

Eventually, remaining county lands were assigned to the Irish Society, and London's livery companies—with appellations such as Drapers, Fishmongers, or Salters—were enlisted to form associations and undertake development. Each association received proportions in county lands and established manor estates. Freehold rights were conveyed from these manors and in turn leased to British subjects or to Irish natives. Rights to timber for building houses, repairing dwellings, and fashioning tools or farm implements were reserved by company agents and were employed to promote settlement. The Irish Society reserved similar timber rights.[7]

Circumstances surrounding Ulster's plantations were complex and, in many ways, peculiar to that region. Although they yield few definitive conclusions about structure of investment in North American colonies, certain aspects invite comparison. The importance of land division in town planning—more precisely, the manner of apportioning its wealth between inhabitant and investor and the relationship of that partition to the success of new communities—is fundamental. Lands surrounding both Londonderry and Coleraine were ultimately conveyed in freehold estates rather than assigned in common for future use of town members. Nor did creation of manors by livery companies in outlying regions promote development of extensive common rights. Traditions of individually held property, strong in Celtic heritage, proved unbending. In short, control of Ulster's lands typically resided with absentee owners, comparable to proprietors of later New England towns who were motivated primarily by speculative intent.

Colonization of Ulster was a genuine experiment, in truth a financially difficult one, but assumptions that public concerns were neglected would be inaccurate. The town of Londonderry was attentively planned on an axis, with a large central square for a town hall. Views to that public space from cardinal directions were framed by carefully placed houses. Limited common privileges were granted, timber rights among them, and bogs or waste may have retained a communal character for extended periods. Planners also acknowledged ecclesiastical concerns as a vital part of these communities and set aside lands to support church and clergy. Glebe lots were apportioned in Londonderry and other counties, in the latter amounting to sixty acres for each thousand-acre estate. Allotment of glebe, or ministry, lands also became a standard condition for town charters in New England.

Glebe Lands and Rural Parish

Religion in English society, its influence on community landscape, and its contribution to the planning of towns in New England are sufficiently important subjects to justify a cursory glance at parish organization in rural England. The origins of these ecclesiastical precincts are obscure but probably date from the seventh or eighth centuries, the missionary period of Anglo-Saxon rule, when minster churches became centers for religious activity. Minster fortifications also provided a haven from strife, and village settlements often materialized nearby. With the spread of Christianity to the countryside, practical concerns made clear a need for places of worship within increasing numbers of villages. Whether medieval parishes were preceded by a similar division, primarily secular and incidental to methods of taxation by civil authority, is unclear. Despite uncertainty about their precise derivation, parishes indisputably became provincial units within which entire parochial systems were established during the tenth and eleventh centuries. Ironically, Danish invasions during this period helped to strengthen the roots of local churches by weakening central cathedrals and minsters, which had accumulated wealth and invited attack.[8]

During the late Anglo-Saxon period, parish churches also became symbols of status and capital worth capable of elevating wealthy landowners to higher ranks of nobility. Lords who established local churches exercised control over the appointment of clergy and were entitled to a share of parish revenue. A body of theological law emerged during the tenth century, tithes became compulsory, and parish boundaries were more clearly defined in order to smooth means of collection. By time of the Norman Conquest, parishes had become complete units where reciprocal duties of clergy and parishioner were well established. Following unification under William the Conqueror, a surge in parochial church building began, and consolidation of the central church prompted reform measures that reached parish levels. Claims to church income by nobility became more tenuous, and ownership of church lands accrued to priests, who also received a percentage of tithes, together with certain fees such as church-scots or plough-scots, a levy on plow teams.[9]

As England's economy expanded during the twelfth and thirteenth centuries, parish churches began to exert a stronger influence on rural landscapes. Tithes increased, encumbering everything from minerals, mills, crops, and cut wood (praedial tithes) to common rights in pasturage or waste lands (agistments). Tithe barns needed to store hay, crops, and wood became part of rectorial complexes. Church endowments in land and personal property began to increase as well. During the fourteenth century, however, harvest failures, plagues, and oppressive taxation to underwrite war with France all contributed to a reversal of fortunes. Parish priests, already under scrutiny for their burden of tithing, were forced to compensate for lost income by enforcing statutory rights to tithes. Plagues were ecumenical, and churches compensated for a shortage of clergy by assigning two or more congregations to one individual. Nonresident clergy were added to the growing list of complaints.[10]

These events hastened a stronger role in church affairs by parishioners, and the office of churchwarden emerged as a means for voicing community concern. Churchwardens became local representatives for all church business and guardians of the parish domain, eventually holding glebe lands in trust. As conflict between the office's religious and secular demands mounted, the position increasingly drifted toward the latter. By the close of the sixteenth century, clerical influence over churchwardens was nominal. By the time of the Great Migration, parishioners had already established a strong degree of control over church affairs, and the practice of holding glebe lands in trust was well accepted. In 1647, following abolition of England's ecclesiastical courts, churchwardens were placed under the jurisdiction of local justices of the peace.[11]

Connections between England's rural parishes and New England's early settlements are significant in several ways. Parishes quickly became discrete precincts in many New England towns, eventually receiving separate town charters. Disputes over taxes imposed to support clergy in parent communities eventually led to enactment of provincial laws granting full autonomy to parish districts within towns. Parish inhabitants were entitled to appoint officials to conduct everyday affairs and levy taxes against members' property. Land was routinely set aside to support church or minister. Valuable open fields were leased for pasture or cultivation; woodlands supplied revenue from sale of timber, fuelwood to supplement wages, and materials for construction of church or meetinghouse. Moreover, conflict between churchwarden and parish in England portended similar battles in the colonies as clergy and parishioner vied for control over these lands. Assertion of public interest through a secular voice frequently prevailed.[12]

Woodlots for Minister and Church

In New England's nucleated villages, parcels assigned to ecclesiastical purpose were identified by a variety of names, usually either ministry, glebe, parsonage, or church lots. These terms were used interchangeably to describe at least three separate categories of church-related lands: parcels intended to help towns pay for church and minister, building lots on which church and parsonage were actually constructed, and grants aimed at enticing ministers to accept a community's calling. In the latter case, ministers sometimes acquired title; but disputes were not unusual, and towns often retained ownership, especially where tenure of a particular minister was brief or controversial. Now and then, a single large parcel accommodated all three objectives, but more often at least two separate lots were designated. Woodlots could be found in each category, but many were also isolated on separate tracts, primarily to furnish fuelwood required as part of salaries. On other parcels, harvested stands of timber yielded immediate revenue and released land to agricultural use, which produced higher rents.

Town members in early Massachusetts villages set about the task of locating and designating ministry lands in the same matter-of-fact way in which other town affairs were conducted. Charters did little more than require general provision for minister and church, leaving means and dimension to selectmen's discretion. Yet the communal spirit that bound proprietors to town affairs ensured that allotments were adequate for a community's needs, were selected thoughtfully, and were located in accessible regions. Committees were sometimes appointed to choose lands, and parcels were often taken from common land. Occasionally, officials were simply charged with finding an appropriate lot within a designated perimeter around their village. Sizes of woodlots varied considerably: some were small (40 acres and less), others large (sometimes as much as 500 acres). From its uplands and meadows, for instance, Topsfield surveyed four parcels totaling fifty acres and assigned them to ministerial use.[13] On occasion, purchase or trade with individual owners was necessary in order to acquire land in a certain location. Such was the case in Marshfield, where the town purchased a house lot, meadows, and uplands for its minister in 1657 and debated ownership for the next two decades. In 1677, town members finally confirmed an original intent to acquire those lands for use by successive ministers and to remain so forever, thus foreclosing any individual claim by the town's minister.[14]

Payment of ministers' salaries was typically measured in currency but delivered in its equivalent in products such as wheat, peas, rye, barley, Indian corn, cattle, butter, or English goods and usually a year's supply of fuelwood. By 1711, Ipswich had converted a share of its ministers' salary from wood to currency, but a portion was still paid in grain.[15] Lands acquired to pay for ministers were rented for grazing, cultivation, hay, or timber, and lease agreements sometimes required similar forms of payment. The town of Rowley also purchased lands for use by its ministers, and in 1662 selectmen rented half, together with upland meadow, gates, and commonage, for the sum of four pounds ten shillings, to be paid in corn and cattle at the time the ministry rates were due.[16]

Tenants who rented ministry lands received the benefit of whatever forest resources may have existed on their leasehold estate, both for their own use and presumably for any quotas of fuelwood to be delivered to their minister. In Reading, every man was required to carry a fair share of winter wood to the minister's door by the last day of October (later commuted to the end of November), with a penalty of double the minister's rate if tardy.[17] Towns often assigned more than one minister's woodlot to guarantee a dependable yearly supply. In 1673, Cambridge selectmen appointed a committee to locate a woodlot for their minister's house from common woodland within five miles of the town center but left its size to committee judgment. Worried about conserving the town's ever dwindling forests, selectment then prohibited the sale of any wood, timber, or woodlots within that five-mile perimeter. In 1682, five hundred acres of land were set out for ministry use, and nine years later an additional forty-acre woodlot was created.[18] Weymouth proprietors voted in 1697 to preserve all wood and underwood on extensive common lands for ministry purposes during a term of twelve years and prohibited all cutting on those lands.[19]

While common lands were being systematically divided in early villages during the seventeenth and eighteenth centuries, public lands for church and minister continued to be set apart and managed with care. Plymouth, in 1701, sequestered a forty-acre parcel from its common land for that purpose and ordered that no wood be cut for a period of ten years. In 1709, ministry lands located near the town pound were enlarged, and in 1728 a committee was appointed to review the bounds of the ministers' woodlot.[20] Topsfield inhabitants took the precaution, in 1722, to order that no trees growing or standing on parsonage lands be cut, and they imposed a penalty of trespass for any violation.[21] Of course, yearly cutting of timber and wood on ministerial lands hastened conversion of wooded land to open field, but lands capable of cultivation could be either rented or sold. In Manchester, two ministry lots were leased in 1744, and a woodlot in Millets Swamp was sold in 1766, but rights to cut wood for their ministers was reserved for a period of twenty years.[22] Church lands were sometimes divided as splinter parishes formed. Weymouth, for instance, was split into north and south parishes; in 1727, town voters apportioned ministry woodlands between clergy for both parishes.[23]

Occasionally, designation of ministry lands resulted in surveys. William Godsoe, surveyor for Kittery, Maine, laid out two parcels for use by the town's ministry. Boundaries for the first, located on Pine Point, were delineated in 1701, and the parcel contained three separate tracts totaling 139 acres (fig. 2.2). Two of the pieces were carved from common land. As he did on many of his surveys, Godsoe sketched the location of nearby houses and identified a thickly wooded section by depicting trees, presumably pine. The second parcel was located in a central part of Kittery and totaled 109 acres. Godsoe's 1702 survey appropriately bears the title: "Land Ordained for Ye Ministry."

As forests disappeared from coastal villages, searches for church woodlots

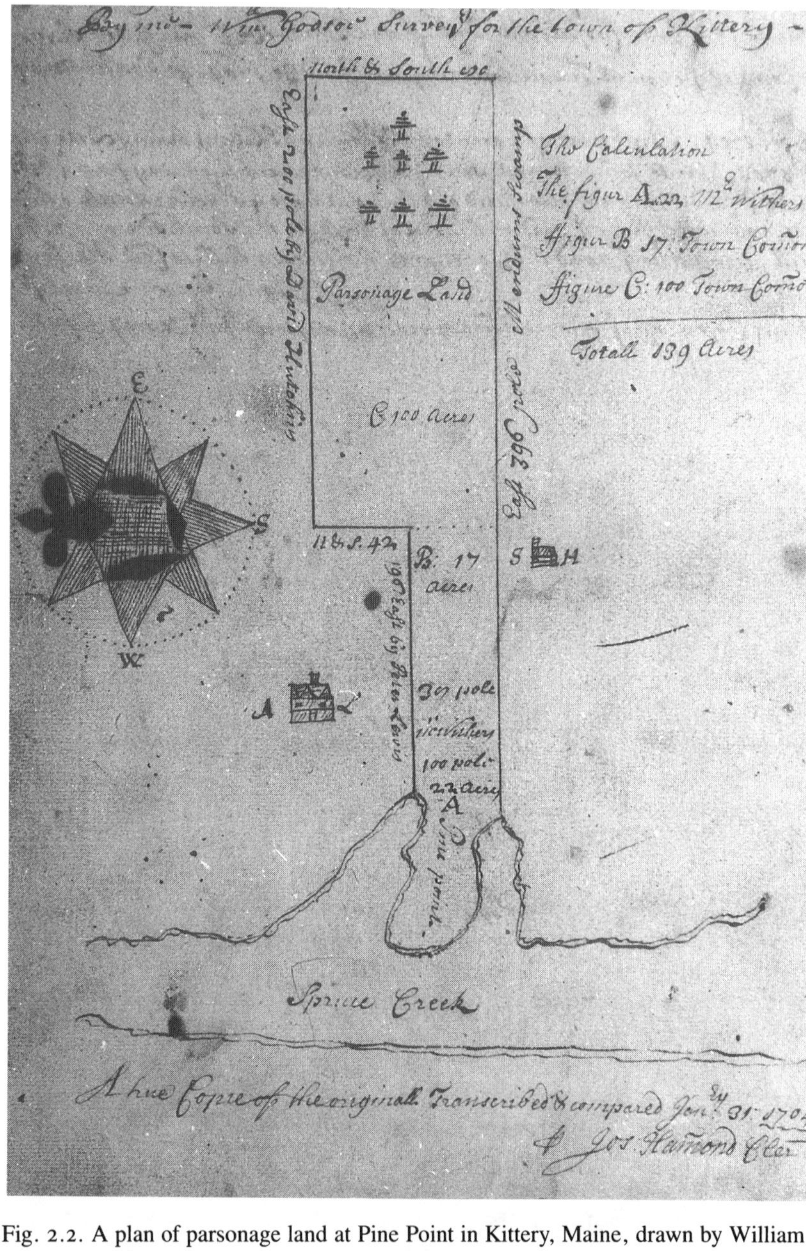

Fig. 2.2. A plan of parsonage land at Pine Point in Kittery, Maine, drawn by William Godsoe, town surveyor, in 1701 and copied by Joseph Hamond, town clerk, in 1704. As he did on so many of his surveys, Godsoe depicted a wealth of detail, including building and ground-cover descriptions. Although the largest portion of Kittery's parsonage lot was assigned from common land, twenty-two acres fronting Spruce Creek were purchased from a private owner. Office of the Kittery Town Clerk

sometimes extended beyond town boundaries. In 1737, Brookline selectmen were asked to find a woodlot for town benefit but were apparently unsuccessful until 1759, when Samuel White, a Brookline resident, sold a twenty-acre parcel of woodland located in the nearby town of Needham. White was paid forty pounds for his trouble, and his deed recites the transaction's purpose: "for a Woodlot to Supply The Minister or Ministers that may be Settled in Said Town from time to time." Selectmen voted to accept the deed in 1761 and two years later viewed bounds and removed blow-down trees. The Needham woodlot served its purpose, and when town members accepted the ministry of Rev. John Pierce in 1796, he was offered an annual salary calculated at four hundred dollars and sixteen cords of wood delivered to his door. In 1805, the town sold standing wood on the lot but reserved sixteen cords to Mr. Pierce for the coming winter. A committee was appointed to supervise the sale, and proceeds were paid into the town treasury and applied toward purchase of their minister's fuelwood during future years. By 1836, the woodlot had generated another crop, which was divided into 150 lots and sold at public auction; proceeds were divided equally between the towns' two parishes. Ten years later, however, a third parish had formed, and selectmen were authorized to sell the land at auction and to allocate proceeds equally among all three. Undoubtedly, a similar fate befell many such lots.[24]

Minister's Woodlot, Keene, New Hampshire

Generally, the size of public lots, their location, accessibility, methods of designation, length of tenure in the public domain, and management practices varied considerably among communities and continued to change as town planning became more regimented. With some exceptions, towns were free to sell or exchange public tracts, and many did. Later, towns were often quickly surveyed, mapped, and divided among absentee proprietors to promote rapid sale of lands. Resulting plans show tidy geometric divisions of land from which individual farmsteads took root. Not surprisingly, topography and survey did not always coincide on many of these grids, and boundaries of granted lands often differed from those appearing on plans. Public lots were usually delineated when maps were prepared or shortly after. In some cases, lands within a town were divided in stages, with groups of proprietors holding small portions in common until partitions could be surveyed and mapped. Frequently, these later divisions also reserved public lots.

To encourage settlement in its western regions, Massachusetts authorized the opening of four towns in 1732, one of which, Upper Ashuelot, later became Keene, New Hampshire, located in the state's southwestern corner. A plat, intended to be six miles square, was submitted in February 1734, and it designated sixty elongated house lots of equal size for settlers (fig. 2.3).

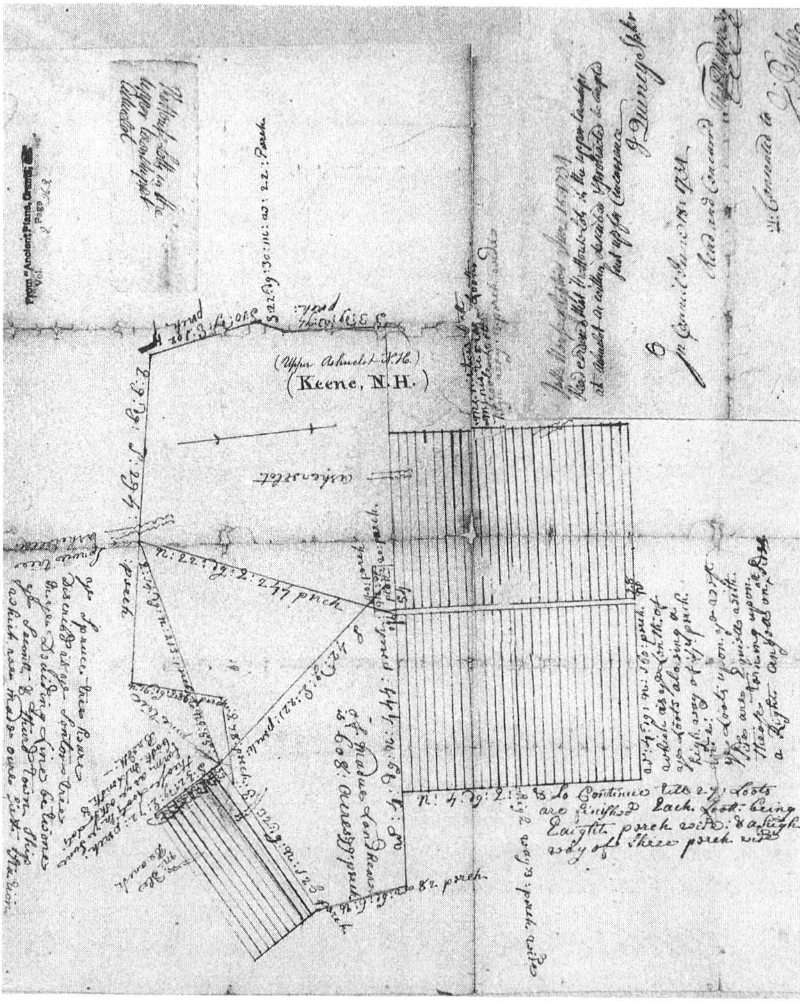

Fig. 2.3. An unsigned 1734 plan of Upper Ashuelot, New Hampshire, chartered by the Massachusetts General Court two years earlier. The town was later renamed Keene and received a second charter from Governor Benning Wentworth of New Hampshire. Three elongated parcels, a ministerial lot, a lot for the first settled minister, and a school lot were placed at village center in the plan's southwest quadrant, adjoining the axis formed by intersecting highways. Simple grid plans were frequently employed by absentee proprietors in New England after 1700. Massachusetts State Archives

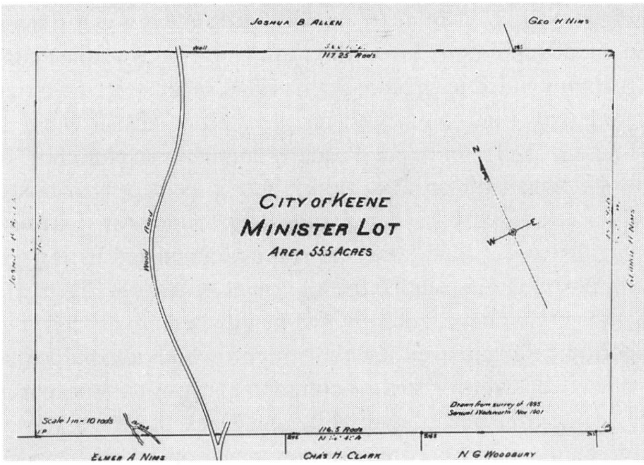

Fig. 2.4. A survey of the Minister Lot in Keene, New Hampshire, drawn by Samuel Wadsworth, city engineer, in 1895. Acquired in 1787, the land became the site of logging bees where parishioners assembled to cut and haul their minister's winter wood. The land is still owned by the city and is currently under forestry management. Wadsworth Collection, Cheshire County Registry of Deeds, Keene

A ministerial lot, a tract for the first settled minister, and a school lot also were assigned. A majority of parcels filled quadrants neatly divided by two highways that intersected at the town center. A smaller tract of nine lots was segregated by extensive meadowlands. Following establishment of a boundary line between Massachusetts and New Hampshire in 1741, the latter became a separate province, and a new charter for Keene was issued by Governor Benning Wentworth in 1753. This second charter, also six miles square, reserved five hundred acres for Wentworth and three separate shares (each almost 400 acres): a glebe lot for the Church of England, a lot for the first settled minister, and one for the Society for the Propagation of the Gospel, England's missionary society. Public lots assigned by the Massachusetts charter, along with the glebe and first minister's lot under the Wentworth charter, eventually passed into private ownership.

Finally, in 1787 proprietors laid out a fifty-acre tract of land in the town's northeastern hills, to be kept as a woodlot for the gospel minister and to be managed by town selectmen. An ample growth of pine covered the parcel, and by 1792 the annual chore of cutting and hauling a yearly supply of fuelwood had become a festive "logging bee." Parishioners assembled at the woodlot with tools and sleds and hauled loads of wood to the minister's house, where food and cider were proffered. Keene's woodlot was shared with the Baptist church, which organized in 1816, but by midcentury town officers became concerned about supplies of wood and took precautions against cutting young growth. Custom yielded to practicality in 1863, when selectmen voted to create a separate trust fund into which proceeds from sale of wood and timber could be funneled and later divided equally among town churches. During this period, blow-downs from an 1854 storm, shook for barrels or boxes, and oak lumber produced respectable sums for the community. An 1850 survey added five acres to the parcel, and a second survey, conducted in 1895 by Samuel Wadsworth, city engineer, settled the lot size at 55.5 acres (fig. 2.4).[25]

Keene's selectmen have faithfully retained ownership of their minister's lot and have periodically garnered saw timber, cordwood, and pulpwood. Almost a century after Wadsworth's survey, common hardwoods—primarily red oak, red maple, white ash, beech, and white and black birch—have replaced old growth, and hemlock has become the dominant coniferous species. Management plans were drafted by private forestry consultants in 1979 and again in 1985, both recommending selective cutting of mature growth to improve future stands. The scar of an electrical power line now slashes across nearly nine acres of the parcel. Still, neither culturally assisted and natural evolution of tree species nor intrusion of a modern utility easement has seriously altered the character of Keene's two-hundred-year-old public woodlot. Most of the land remains forested, as it did in 1792 when parishioners gathered to haul their minister's wood.[26]

Parsonage Lands in Newington and Danville, New Hampshire

Two extraordinary examples of villages whose forested public lands have supported church and minister exist at Newington and Danville, New Hampshire, both towns located in the state's southeastern corner. Newington's parsonage and church lots have historically surrounded a tiny and secluded village center, composed of town hall, school, library, church, and parsonage (fig. 2.5). Apart from producing steady revenue from sale of wood or rental of pasture for almost three centuries, Newington's modest public lots peerlessly illustrate how community-owned land can be used to encircle village centers and contain sprawl. Such a theory of town planning did not originate in America but developed in England toward the close of the nineteenth century. Newington's public lots, historically at least partially forested, are thus unique in American planning history. Danville's parsonage lands, only slightly less intriguing than Newington's from a perspective of village planning, nonetheless offer a singular demonstration of attentive forestry management by town officials at a very early date. Such meticulous care of forest resources underscores the strength of collective stewardship in some New England villages and confirms that these traditions have been carried forward on community initiative.

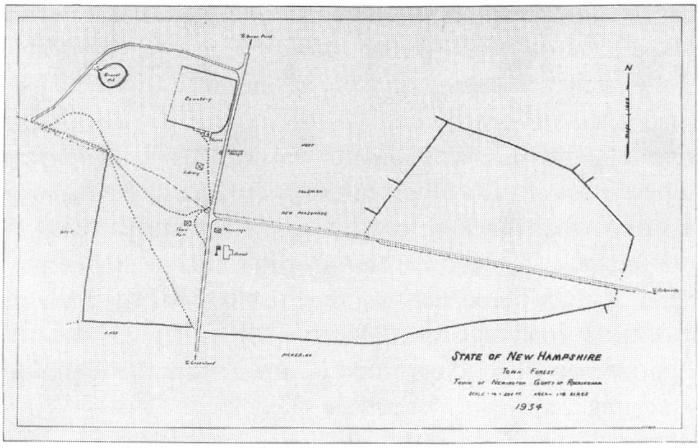

Fig. 2.5. A plan of the Newington Town Forest, drawn by B. Gate for New Hampshire's state forestry department in 1934. Newington's public lands became a town forest as that movement became popular during the early twentieth century. New Hampshire's foresters energetically promoted local forestry and compared Newington's example to Europe's ancient community forests. New Hampshire Division of Forests and Lands—Concord

Newington village occupies a thick neck of land reaching out into New Hampshire's Great and Little bays just west of Portsmouth. At one time part of Dover, Newington emerged as a separate parish whose settlers occupied a region known as Bloody Point. In 1710, members of Dover voted to set apart fifty or sixty acres of common land for ministry use in that ecclesiastical community and appointed two men to establish boundaries. The church lot, as it became known, is a forty-two-acre parcel that encompasses the town cemetery. It is the oldest of three contiguous parcels that now comprise town-owned land and is occupied by the Congregational church (dating from 1712), a town hall built in 1872, and a library constructed in 1892. Culling of timber and wood from this parcel prior to the Civil War appears to have been limited, for the land was kept primarily open as pasture and cropland.[27]

A second, twenty-acre parcel located in "pitch pine plains" and separated from the church lot by the road to Greenland, was purchased in 1765, together with house and barn, from Newington innkeeper Nicholas Knight. Acquired specifically for use and support of the town ministry, the parsonage lot provided annual income from lease agreements, which indulged an assortment of uses. A new barn was raised in 1786 with timber from this lot, and improvements such as stone walls and a town pound also were erected. Disputes about ownership and use of the land began with the ordination in 1788 of Rev. Joseph Langdon, whose ministry proved to be anything but peaceable. Langdon's claims to the parsonage house as well as to adjacent woodlands were particularly irksome to some parishioners, who practiced a defiant and calculated trespass and regularly gleaned fuelwood and fencing material. Town members ultimately prevailed, Langdon was dismissed, and the parsonage lot was returned to town oversight.[28]

Sorely tempting divinity after their ordeal with Mr. Langdon, town selectmen gave permission in 1819 for a tavern to be kept in the parsonage house, hardly an orthodox manner of gaining spiritual remedy. After 1825, lease agreements for the parsonage lot customarily reserved timber and fuelwood rights, and in 1842 the house and barn were put to service as a town poor farm. By midcentury, warrants for town meetings typically queried whether voters would authorize selectmen to cut wood for town use at the school, poor farm, and town meetings, as well as for private sale.

A third parcel, the Downing lot, was apparently added at about the same time the parsonage lot was purchased, but exact date of acquisition and its size are unclear. A type map prepared in 1934 by Warren F. Hale, then an assistant state forester with New Hampshire's State Forestry Commission, shows two tracts totaling seventy acres, but boundaries between are not evident (fig. 2.6). Also puzzling is the fact that town records consistently document management of the parsonage lot during the nineteenth century but make no specific references to the Downing parcel until 1874, when a small area was cut. Given similar dates of acquisition, contiguous boundaries, comparable forest cover, and administration by town selectmen for like purposes, it seems probable that both parcels were considered together, and references to the parsonage land at times must have included some or all of the Downing lot as well.[29]

Although a segment of parsonage land had been cleared for meadow in 1787, abundant stands of wood and timber grew to maturity during the second half of the nineteenth century. For a brief period, Portsmouth became a regular customer for cordwood. However, dissension between town and church over ownership of the parsonage land, a dispute ultimately decided in favor of the town, caused a temporary lull in logging activity on some of the property. Nevertheless, a substantial cut produced more than $1,000 in 1894 and helped to defray costs of building a library. Sale of timber in 1912 netted almost $1,800, considerably more than needed to pay for the installation of running water in town buildings; wood and timber marketed in 1920 contributed about $1400 toward the expense of a new stone schoolhouse. Steady sales of cordwood after 1890 also added to town coffers.[30]

Management of Newington's forested parsonage lands prior to 1900 appears to have been thoughtfully supervised by town officers, with selective cutting in small lots as stands matured. New growth took root through natural seeding until 1915, when reforestation of cut-over lands began with slash-burning and transplanting of eight thousand pine seedlings.[31] Instances of waste or mismanagement are not evident, and although a fire destroyed a young growth of pine

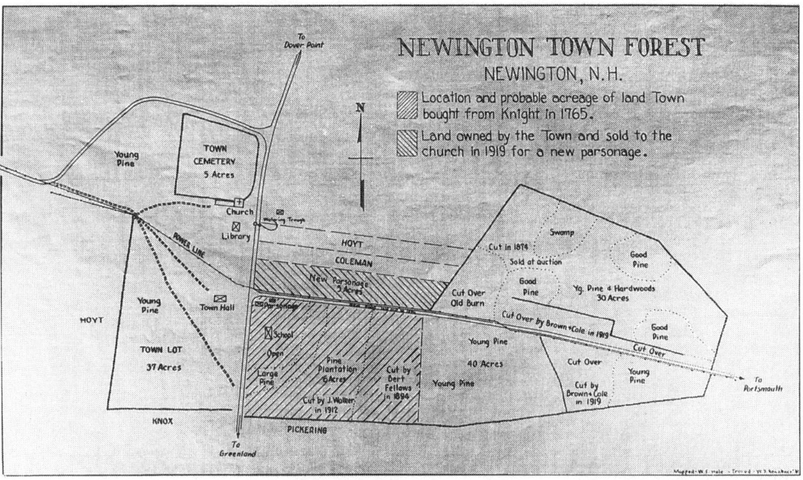

Fig. 2.6. Forest type map of the Newington Town Forest, drawn by Warren F. Hale, a state forester, and traced by Walter P. Knoettner in 1934. Although nearly one hundred acres of Newington's land was acquired—and its trees felled—when construction of Pease Air Force Base began in the early 1950s, recent closure of that airfield has given the town an opportunity to reclaim its forest. New Hampshire Division of Forests and Lands—Concord

in 1889, such hazards are common in forestry management. Timber and wood consistently produced substantial community benefits, and town ownership helped to preserve the integrity of Newington's village center.

With the development of professional forestry and the emergence of a movement for commercial town forests during the early twentieth century, Newington's public lands assumed new significance as a model town forest. Much of the writing about municipal forests during this period cited Newington as America's shining example to rival Europe's ancient community forests. During Franklin Roosevelt's administration, the U.S. Forest Service also took special interest in Newington's example and recounted its history in a number of their publications. Sadly, nearly a hundred acres were acquired by the federal government in 1953, when Pease Air Force Base was constructed. Nevertheless, recent closure of that base has given town members an opportunity to reestablish their important forest.

In contrast to Newington's public lots, Danville's clearly bounded parsonage lands are not contiguous and do not encircle a village center (fig. 2.7). Still, certain similarities do exist. Hawke parish, which became the town of Danville in 1836, was originally a western settlement in Kingston. Travel to Kingston's distant church became onerous for inhabitants, and a separate charter was

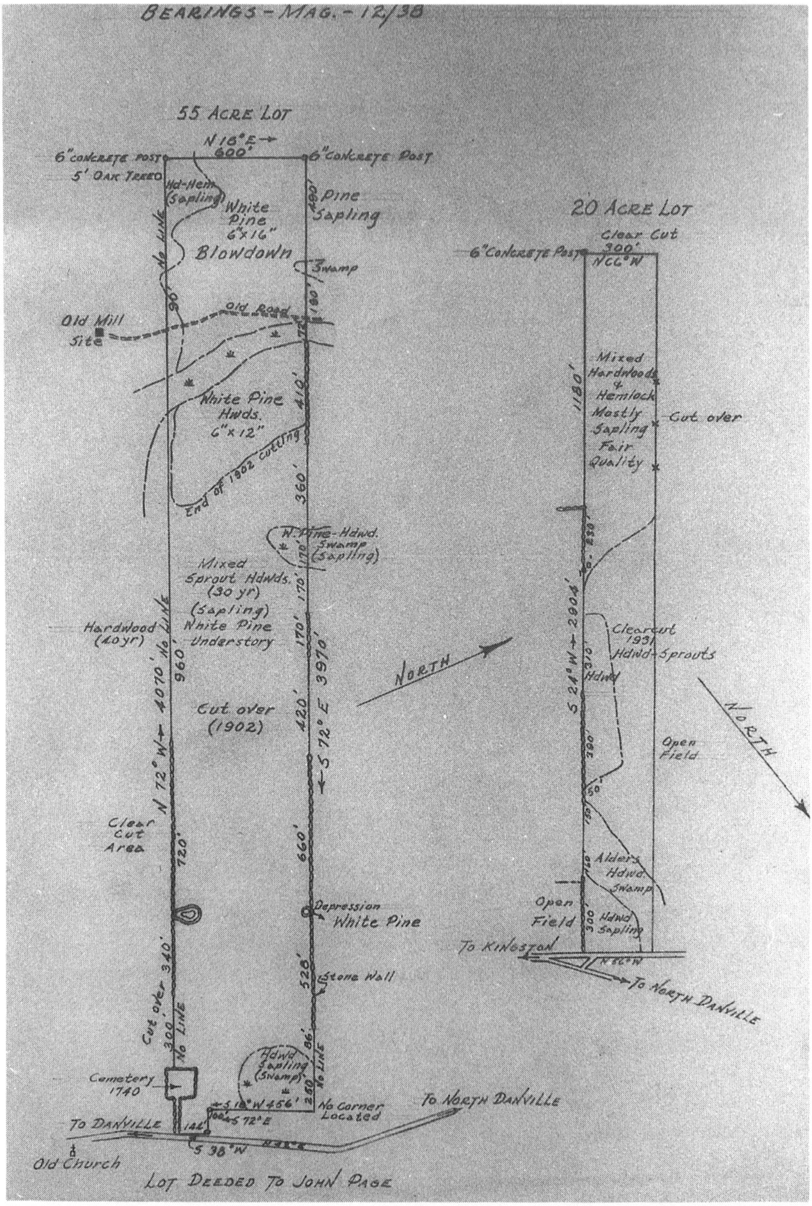

Fig. 2.7. Forest type map of parsonage lands in Danville, New Hampshire, drawn by the state's forestry department in December 1938, several months after a devastating hurricane blew down a stand of white pine. Acquired in 1761, Danville's public lots have been meticulously managed by the town's parsonage committee for timber production and grazing. The larger parcel is currently part of the Danville Town Forest. New Hampshire Division of Forests and Lands—Hillsborough

Fig. 2.8. Meetinghouse in Danville, New Hampshire, erected in 1760 and framed with timber cut from the town's parsonage lands. Originally part of the town of Kingston, the community was chartered as the parish of Hawke that same year and later given the name of Danville. RLM, 1993

granted to the parish in 1760; predictably, it required the new community to support ministry, school, and their poor. Two separate parcels, fifty-five acres and twenty acres, were acquired in 1761 for ministry support in a complicated exchange of land set aside in Kingston for benefit of Hawke parish. In that year, construction of a meetinghouse was begun, with timbers cut from parsonage lands and hauled to the site by parishioners' oxen (fig. 2.8). In 1763, parish members ordained Rev. John Page, but unlike the stormy tenure of Newington's Rev. Joseph Langdon, John Page served the parish steadfastly until his death in 1782 following a smallpox epidemic that decimated the community. In addition to a generous endowment, his annual salary was placed at forty-five pounds, use of parsonage lands, and twenty five cords of wood cut and delivered to his door.

While Newington's church and parsonage lands are significant for their enveloping bond to the village center, Danville's parsonage lots are noteworthy for an entirely different reason. Meticulous stewardship of Danville's lots has

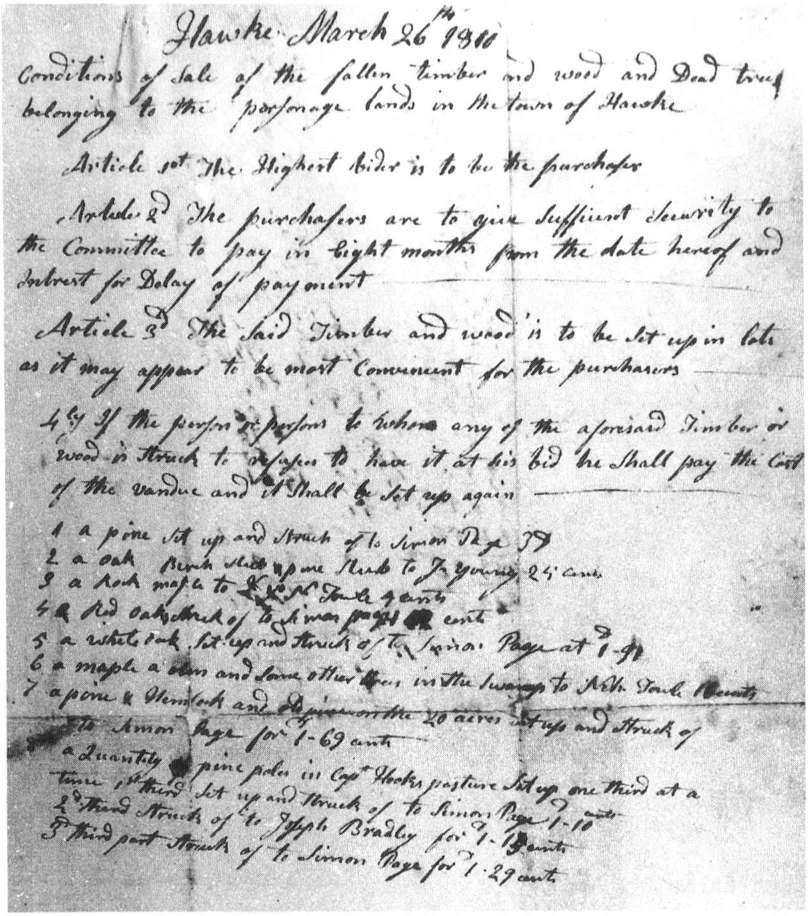
Hawke March 26th 1810

Conditions of Sale of the fallen timber and wood and Dead trees belonging to the parsonage lands in the town of Hawke

Article 1st The Highest bider is to be the purchaser

Article 2d The purchasers are to give Sufficient security to the Committee to pay in Eight months from the date hereof and Interest for Delay of payment

Article 3d The said Timber and wood is to be set up in lots as it may appear to be most Convenient for the purchasers

4th If the person or persons to whom any of the aforesaid Timber or wood is struck to refuses to have it at his bid he Shall pay the Cost of the vandue and it Shall be Set up again

1 a pine set up and struck of to Simon Page 39

2 a oak [illegible] to J. Young 25 cents

3 a Rock maple to [illegible] Towle [illegible] cents

4 a Red oak struck of to Simon Page [illegible] cents

5 a white oak set up and struck of to Simon Page at 1-91

6 a maple a elm and some other trees in the swamp to Nath. Towle [illegible] cents

7 a pine & Hemlock and [illegible] on the 20 acres set up and struck of to Simon Page for 1-69 cents

a Quantity [illegible] pine poles in Capt Hook's pasture set up one third at a time, 1st third set up and struck of to Simon Page 1-10 cents

2d third struck of to Joseph Bradley for 1-1[illegible] cents

3d third part struck of to Simon Page for 1-29 cents

Fig. 2.9. Timber sale agreement dated March 26, 1810, covering fallen timber and dead trees on Danville's parsonage lands. Trees were sold individually or by small lots to the highest bidders, who were required to demonstrate their ability to pay the town. A number of similar agreements have been carefully preserved by the parsonage committee and, together with receipts and other memorabilia, are stored in an ornately carved wooden trunk acquired in 1830. Parsonage Committee Records, Danville, New Hampshire offered a unique glimpse at forestry management by a New England town during the nineteenth century. Selectmen were initially charged with care of these lands until 1790, when a separate committee of three, later known as the Parsonage Committee, was appointed. This committee, only recently dissolved, has faithfully entered proceeds from timber sales and rental agreements in town ledgers stored, along with receipts, in an ornately carved wooden trunk obtained in 1830.

Danville's timber sale agreements depict auctions at which trees were sold individually or in small groups, with conditions attached to each sale. While practical concerns for moving heavy timber may have been a factor, this method also provided an easy way to ensure that only mature growth was cut. For example, an auction conducted in March 1810 (fig. 2.9) organized trees into lots convenient for purchasers and required successful bidders to provide security for payment within eight months, with interest for delayed payments. Individual pine, red and white oak, and red maple trees were set up and struck off to successful bidders. More extensive offerings included a swamp stand of unidentified trees, along with one maple and one elm, hemlock and pine from a twenty-acre parcel, and three divisions of pine poles in Captain Hook's pasture. Prices offer a glimpse at the relative value of tree species during that period: 4¢ for a red maple, 39¢ for a pine tree, and $1.91 for a white oak. An 1828 auction of old wood and timber prohibited successful bidders from carrying away any fallen wood caused by their cutting. An 1834 auction demanded cash for bids less than a dollar and required buyers to collect their wood within a year or pay a fine of 25 percent of their bid.

Parsonage lands were consistently rented, undoubtedly for grazing because tenants were prohibited from plowing, planting, or sowing the property. An 1821 agreement granted permission to cut fencing material and to erect fences, but restrictions against cutting anything fit for boards or timber were clear. An 1838 agreement required tenants to burn brush where wood had been cut and to sow grass seed supplied by the parsonage committee (fig. 2.10). Significance of auctioning individual trees and forcing bidders to clear their wood becomes apparent in a context in which land was also managed for grazing. In Danville, use of public woodlands for pasture, fence wood, and timber appears to have been informally balanced in a manner similar to the way communal rights to forests for grazing and wood were separated from timber ownership in feudal England. Although management of these lots may have exhibited a land ethic that recalled ancient European customs, community members nevertheless received full advantage of undivided ownership. In this way, Danville demonstrates a unique adaptation of ancient custom to a New England village environment. It seems fair to suggest that simple patterns of multiple land-use management must have appeared in other towns, such as Newington,

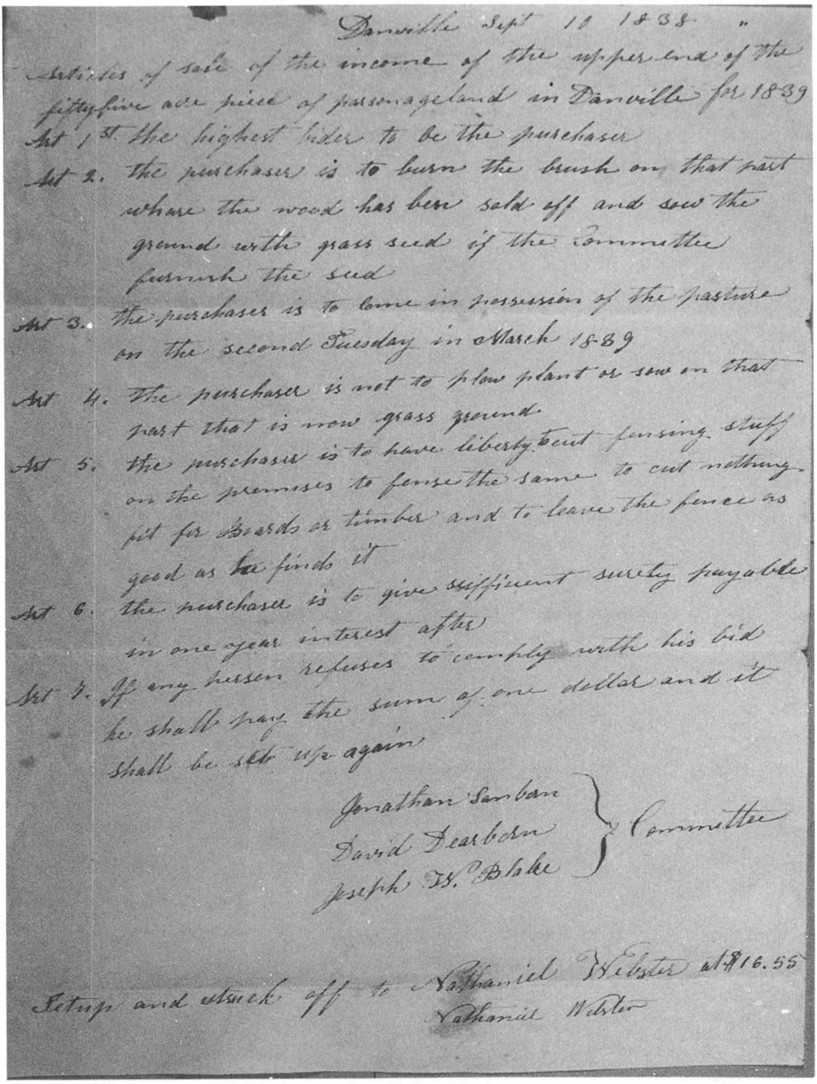

Danville Sept 10 1838

Articles of sale of the income of the upper end of the fiftyfive acre piece of parsonageland in Danville for 1839

Art 1st the highest bider to be the purchaser

Art 2. the purchaser is to burn the brush on that part where the wood has been sold off and sow the ground with grass seed if the Committee furnish the seed

Art 3. the purchaser is to come in possession of the pasture on the second Tuesday in March 1839

Art 4. The purchaser is not to plow plant or sow on that part that is now grass ground

Art 5. the purchaser is to have liberty to cut fensing stuff on the premises to fense the same to cut nothing fit for Boards or timber and to leave the fence as good as he finds it

Art 6. the purchaser is to give sufficient surety payable in one year interest after

Art 7. If any person refuses to comply with his bid he shall pay the sum of one dollar and it shall be set up again

Jonathan Sanborn
David Dearborn
Joseph W. Blake } Committee

Setup and struck off to Nathaniel Webster at $16.55
Nathaniel Webster

Fig. 2.10. Rental agreement dated September 10, 1838, leasing part of the fifty-five acre parsonage lot in Danville for one year. The land was opened to grazing, and tenants were required to burn brush and sow grass seed in areas where timber had been sold. Plowing and planting were prohibited, and existing fences were to be maintained. Toward that end, permission to cut poles for fencing material was granted, but felling timber fit for boards was forbidden. Parsonage Committee Records, Danville, New Hampshire

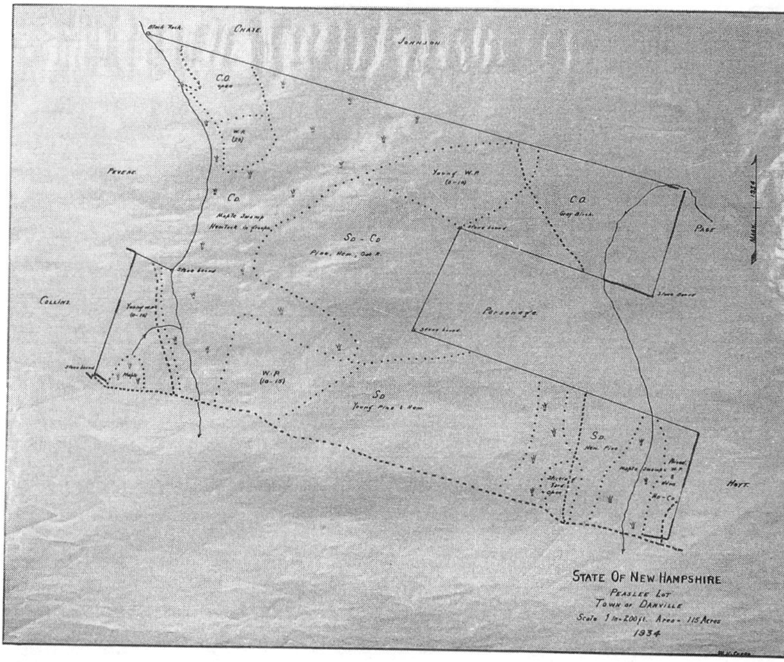

Fig. 2.11. Forest type map of Danville's Peaslee Lot, drawn by W. Harry Croke for the state's forestry department in 1934. The 115-acre parcel adjoins the 55-acre parsonage lot and, with two other parcels, is part of a 300-acre town forest. New Hampshire Division of Forests and Lands—Concord

where public lots were important to village economies and where lands were invariably rented for agriculture as well as harvested for timber.

Benefits of Danville's multiple land-use practices are extremely well documented, and ledger books show yearly entries after 1830 for either rental income or sale of wood and timber, or occasionally for both. Rents were steady, with amounts generally ranging between $10.00 and $25.00 before 1850 and slightly more after that date. Sporadic sales of rye produced $17.35 in 1836, $60.15 in 1849, and $34.25 in 1865. Timber sales also yielded returns: sums such as $179.95 in 1830 and $146.74 in 1834 seem typical. An 1837 sale netted $319.48, and profits were used to pay for repairs to the meetinghouse and to build a fence around the town's cemetery. Significant sales of timber and wood occurred in 1847 ($1,218.43), 1863 ($1,107.40), 1895 ($1,975.00), and 1902 ($4,500.00). Parsonage committee members prudently invested a hefty portion of these earnings, and eventually the fund became large enough to produce a respectable annual income on its own. Such comprehensive records attracted professional foresters during the 1930s and enabled calculation of an accurate timber yield per acre.[32]

Community attention to selection and management of ministry lands was undoubtedly a factor in the enduring ownership and productivity of public lots in Keene, Newington, and Danville. Physical proximity to village centers or important community buildings was also a contributing influence in Newington and to some degree in Danville. Both Newington and Danville also set a standard for town forests as that movement gathered energy, a factor that helped to further instill public interest. Lands adjoining the fifty-five-acre parsonage lot in Danville were acquired by the town and now comprise a three-hundred-acre town forest (fig. 2.11).

Preordained Ministry Lands

Not all parsonage and ministry lands were so closely tied to their communities, in either location or spirit. In fact, a great number were destined—by an almost predictable method of selection—for sites far removed from town centers. Absentee proprietors seeking maximum profit on their investment routinely assigned public lots to towns' least valuable regions. Parcels on distant mountain peaks or beneath bodies of water promised no more value for their communities than they did for the speculators who designated them. As often as not, location of public lots made their use for anything but logging impractical. Those not sold were occasionally just forgotten. Still, exceptions do exist, and a smattering of communities have managed to overcome proprietary cunning or avarice. Ironically, the same speculative interests that changed methods of town planning and contributed to the demise of common land also helped to preserve the communal character of many public lots because of their inaccessibility or poor soil quality.

Northwood, New Hampshire, located in the state's southeastern region, acquired its charter in 1773. A proprietor's map was drawn in 1848 by Samuel S. James (fig. 2.12), patriarch of a family of Northwood surveyors. James's plan shows a pair of two-hundred-acre public lots for parsonage and school set within a grid of land divisions. The parsonage lot is No. 11 in the seventh range, the town's most southerly range of lots, bordering Northfield and Deerfield. However, Lot 11 practically bestrides the peak of Saddleback Mountain, at 1,184 feet the town's only mountain to surpass the thousand-foot elevation mark, and part of the land is ledge. The school lot is No. 3, also in the seventh range, but it falls short of two hundred acres because a large segment is beneath a great pond. Ownership of both lots has remained with the town since its charter, although indifference among land speculators may have been the single most important factor. At one time or another, these meager tracts were probably rented for timber or pasture, but records offer little proof.[33]

New Hampshire's state foresters took notice of Northwood's public lands

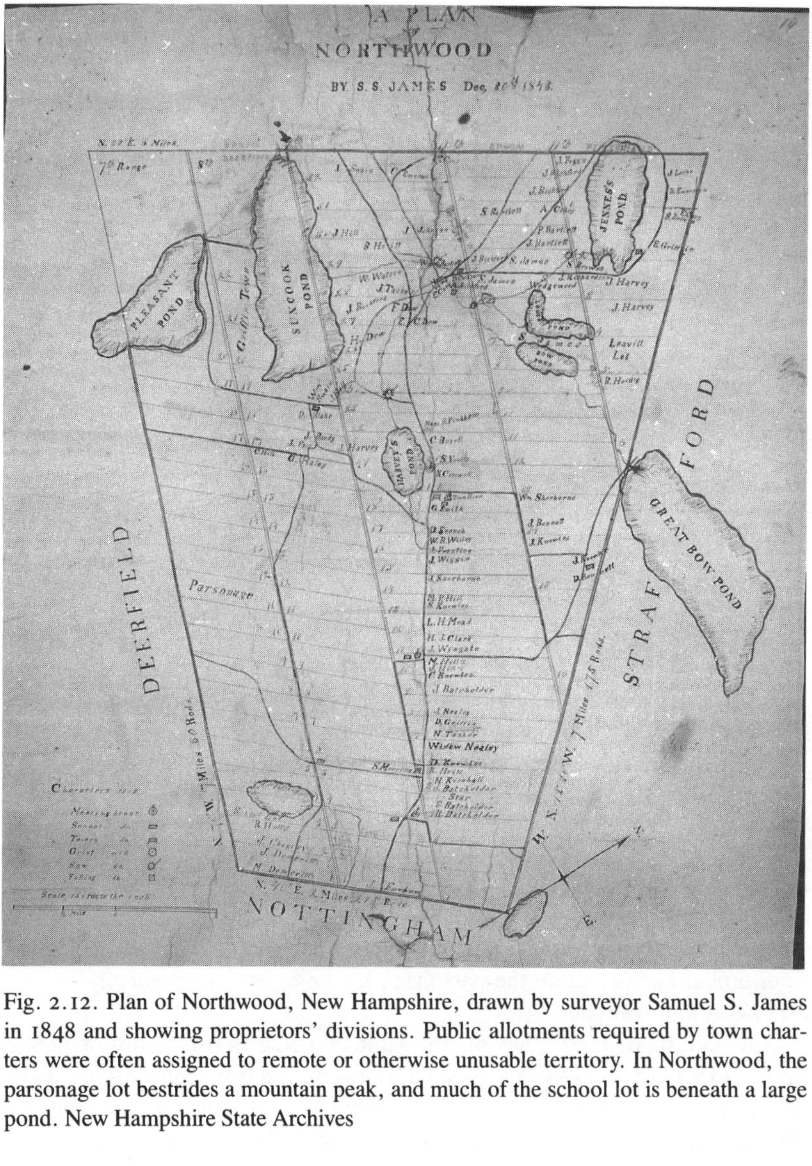

Fig. 2.12. Plan of Northwood, New Hampshire, drawn by surveyor Samuel S. James in 1848 and showing proprietors' divisions. Public allotments required by town charters were often assigned to remote or otherwise unusable territory. In Northwood, the parsonage lot bestrides a mountain peak, and much of the school lot is beneath a large pond. New Hampshire State Archives

as enthusiasm for municipal forests gathered momentum during the twentieth century. A town forest committee led by Orrin M. James, son of Samuel James, was appointed in 1924 and a year later recommended a management program of cutting and planting; a portion of the school lot continued to be rented for pasture. Surveys were plotted in 1933 by W. Harry Croke, a state forester, and acreage for each parcel computed: 235 acres for the parsonage lot

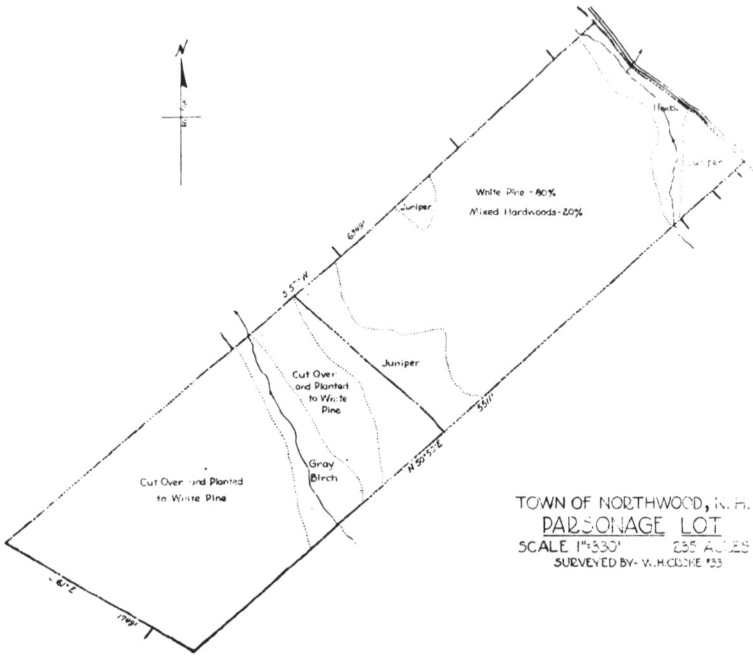

Fig. 2.13. Forest type map of parsonage lot in Northwood, New Hampshire, surveyed by W. Harry Croke for New Hampshire's state forestry department in 1933 and traced by Walter P. Knoettner in 1940. Northwood organized a town forest committee to manage its parsonage and school lots and appointed surveyor Orrin M. James, son of Samuel James, as chairman. New Hampshire Division of Forests and Lands—Concord

and 155 acres for the school lot (figs. 2.13, 2.14). Croke's type maps and notes indicate primary growth of white pine, with small percentages of hemlock, red pine, and mixed hardwoods, including gray birch during this period. Profits from timber harvesting on the parsonage lot were distributed equally among the town's four churches.[34]

Now and then, inertia has played a forceful role in ownership of public lands: faraway tracts of town woodland simply passed from collective memory and were never sold. Proprietors of Grantham, New Hampshire, not far from the Connecticut River in the state's central domain, assigned single, hundred-acre lots for minister and school to a hill region after the town received its charter in 1767. Boundaries for both were eventually lost, and their existence was overlooked until cutting by a local timber company in the 1920s pricked the recollection of several town members. A surveyor was hired to locate and reestablish original bounds, and a settlement of $1,100 was reached with the company.[35]

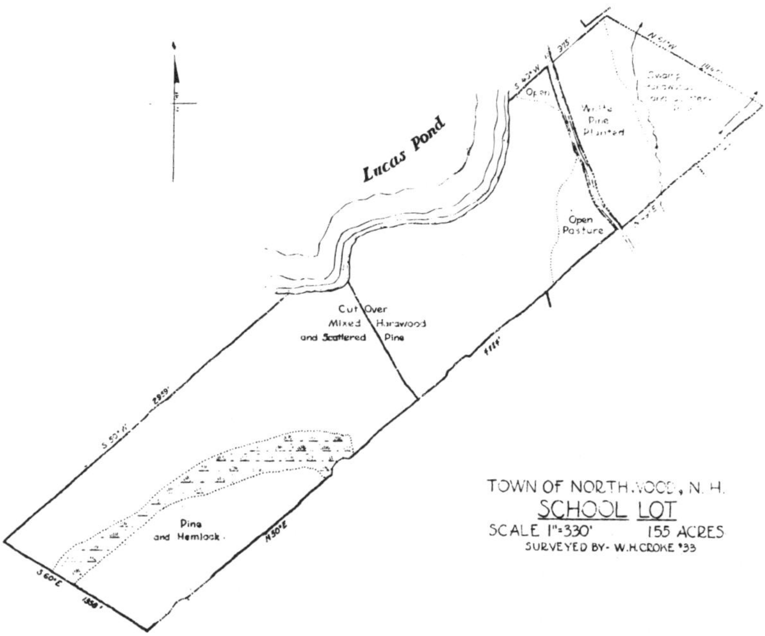

Fig. 2.14. Forest type map of school lot in Northwood, New Hampshire, also by W. Harry Croke and Walter P. Knoettner in 1933. Camp lots bordering Lucas Pond were rented under long-term leases, and income was applied toward the cost of operating town schools. New Hampshire Division of Forests and Lands—Concord

Serendipity has occasionally explained why certain towns continue to own their public lots and others do not. In Piermont, New Hampshire, also situated along the Connecticut River, a ministerial lot was originally allocated to the Church of England's missionary society, which later assigned its holdings to the state's Episcopal diocese. This institution finally relinquished its interest after a hurricane destroyed extensive timber stands in 1938. Steep slopes and ravines, rock outcrops, wet soil, and a remote, nearly landlocked site make the town's parcel suitable for forests and wildlife but little else. Networks of abandoned logging roads are the only means of access and leave few doubts about the area's historic use.[36]

Devious proprietors, distant or inaccessible mountains, and faint recollection among generations of town residents were not the only forces working against public ownership of ministry and church-related lands. During the seventeenth century, distinctions between town and church had been blurred. Dogmatic theology guided village life, taxes were levied to support minister and institution, and most communities were far from tolerant of challenges to accepted faith. Simmering tension between secular and ecclesiastical affairs surfaced every now and again in the form of conflict between minister and parishioners over salaries or use of ministry lands. Gradually, the institutional roles of church and town in village society became more clearly defined, and the two coexisted, sometimes with a measure of harmony. Nonetheless, confusion over ownership of public lands was inevitable where continuous use (or claim of use) by churches gave them color of title, and disputes between town and church were predictable.

Following the American Revolution, mandates of constitutional separation between church and state eventually made town ownership of ministry lands unnecessary or awkward. Ever greater numbers of religious denominations made equal division of limited town resources impractical, and many town-owned parcels were quietly sold. Towns that have retained title to parsonage or glebe lots have created separate trust funds from which modest income is apportioned equally to community churches. Religious societies that had successfully asserted ownership also felt the economic effects of this yawning gap between town and church, and petitions to state legislatures requesting permission to sell ministry lands were common. Still, some churches managed to cling to their woodlands, publicly assigned or otherwise acquired, well into the nineteenth century. In 1871, Dedham's First Church still owned two woodland parcels, identified as the Ledge Lot and the Pound Lot (fig. 2.15) on plans by Nathaniel Smith, a local surveyor who worked with Henry F. Walling in the preparation of numerous state and county atlases throughout New England.[37]

Legislation enacted during the nineteenth century sought to resolve lingering questions about the organization and authority of religious societies. Churches and parishes were treated as corporations and given all powers normally associated with corporate entities, including power to hold and sell real property. The Massachusetts legislature passed a comprehensive statute in 1836 that incorporated all previously established parishes and religious societies. Under this law, methods for governing the affairs of religious corporations closely followed those practiced by towns, including selection of officers at annual, properly warned meetings and levy of taxes. Only real and personal property belonging to members was subject to assessment, but methods of collection were similar. Ministers were authorized to take parsonage lands "granted to the minister and his successors, or to the use of the ministers, or granted by any words of the like import." However, conveyances by any ministers without consent of either town, parish, or religious society were valid for only as long as their ministry continued.[38]

Formidable obstacles to enduring ownership and management of church-related public lands have caused widespread disappearance of this once ordinary communal resource. Explanation can be attributed as much to changes in broad cultural patterns of New England communities as it can to any single

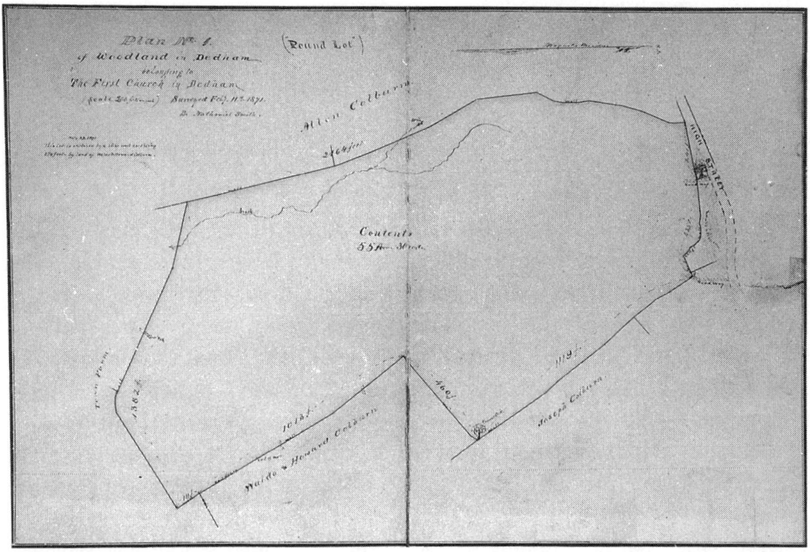

Fig. 2.15. Survey of the Pound Lot, a fifty-five-acre woodlot owned by the First Church of Dedham, Massachusetts, drawn by Nathaniel Smith in 1871. Town ownership of land utilized to support religious institutions eventually became awkward, and ministry lands were sometimes divided among local churches. In other cases, churches acquired their own woodlands to supply fuel. This parcel adjoined the town farm, and Smith employed an ancient oak as a witness tree, a practice—once standard—that now assists foresters who trace the evolution of tree species in a particular locale. Dedham Historical Society

factor. In this context, examples such as Keene, Newington, Danville, Northwood, and others are extraordinary for their tactile contributions to local history. More significantly, their current status as public lands is directly linked to management of their forest resources. The appearance of professional forestry in the early twentieth century no doubt reawakened community interest in these lands and quite possibly saved some from obscurity. Still, modern interest in forestry is simply part of the evolving character of collective stewardship in New England towns.

Occasionally, ministry lands long since given up to private ownership have returned to the public sector, reviving a sense of historical continuity. In Hopkinton, Massachusetts, a glebe lot was obtained by the state under a 1908 reforestation law and has been part of a state park for most of the twentieth century. In Plainfield, New Hampshire, the town's conservation commission recently acquired lands originally granted to the Society for the Propagation of the Gospel. Other communities have substituted less tangible alternatives. A monument in South Kingstown, Rhode Island, is the only reminder of a three-hundred-acre parcel of woods set aside for support of the ministry in 1668. The land was finally given to the town's Congregational church in 1752 but was sold and divided into house lots in 1821.[39]

School Farms and Woodlots

Public land and its resources also helped to subsidize costs of education, and towns selected rentable parcels of land to support their schools. These tracts were either chosen from common land or acquired outright from town members. In early villages, proprietors designated school lots by using methods similar to those employed for ministerial lands, and communal spirit again helped to ensure that parcels were thoughtfully selected. In addition, specific legislative grants of unappropriated territory often supplemented town-owned school lands. More often than not, these land endowments were quickly sold, generating value from timber or whatever other resources they happened to contain.

True to the barter economy of early villages, rented school lots typically yielded returns such as produce or grain. Forest resources undoubtedly contributed to these agrarian ventures, at the very least in the shape of fencing, building materials, and fuelwood. However, early farms were more properly characterized by number or location of open fields rather than by woodlots. Brief references to rented school lands in town records generally identify these parcels in agricultural nomenclature or simply mention location and amount of income. Unlike their support for ministers, towns did not consistently set aside specific woodlots for the benefit of their schools. The explanation probably lies in the fact that ministers were more prominent community members and commanded greater salaries, including wood dependably delivered.

Designation of school lands could be uncomplicated, as in Providence, where one hundred acres of upland and six acres of meadow were reserved in 1663 for maintenance of the town school.[40] In other cases, communities acquired a number of parcels in various locations over time. Ipswich voted to establish a free school in 1642, and several small tracts of land, including an orchard, had been given to the town for the school's use by 1650. A thousand-year lease in "fermeletten" was given to an optimistic John Cogswell that year for a neck of land beyond the Chebocke River for the yearly sum of fourteen pounds, to be paid in butter, cheese, beef, pork, and corn. A clause in the lease allowed reduction of yearly rent by any amounts levied against the land for town taxes. However, the agreement made no reference to use of any wood or timber on the parcel. Mr. Cogswell may have lived to a fine old age but not long enough to gain full benefit of his contract, and a similar, but shorter, sixty-year lease was signed in 1680 by a new tenant. The school farm at Chebacco,

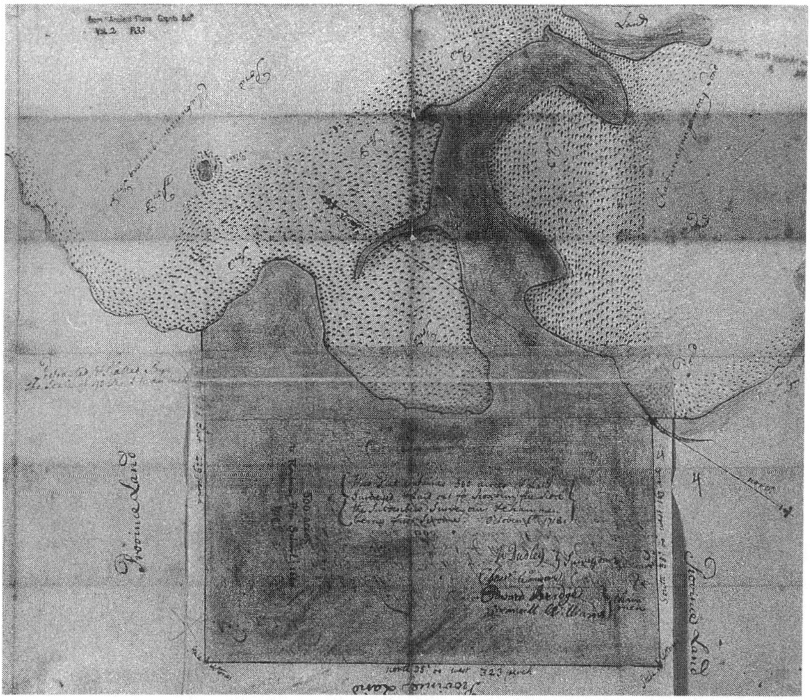

Fig. 2.16. Plan of a five-hundred acre land grant awarded to the Free School of Roxbury, Massachusetts, in 1660, drawn by surveyor Joseph Dudley in 1718. The Roxbury School Farm, as it became known, occupied a peninsula of land extending into Chabunagungamug Pond, a name that probably signifies a boundary water for Native American fishing territories. Massachusetts State Archives

as later records recite, was just one of a number of parcels that contributed to school coffers over the years.[41]

Farms also were established for schools in Roxbury, Duxbury, Walpole, and Charlestown. Five hundred acres of unappropriated province land bordering Chabunagungamug Pond were granted to Roxbury Free School in 1660 by the colony's general court, and a survey was prepared by Joseph Dudley in 1718 (fig. 2.16), followed by a second plan, limned by Isaac Learned in 1739. The curiously named pond may have represented an Indian boundary water, loosely translated: "You fish on your side, I fish on my side, no one fishes in the middle, no trouble."[42] A similar grant for school lands was made to Duxbury following a petition in 1733 by Col. John Alden, who pleaded that poverty had kept the town from meeting its obligation to maintain a grammar school. The Souhegan River provided a meandering southern boundary for

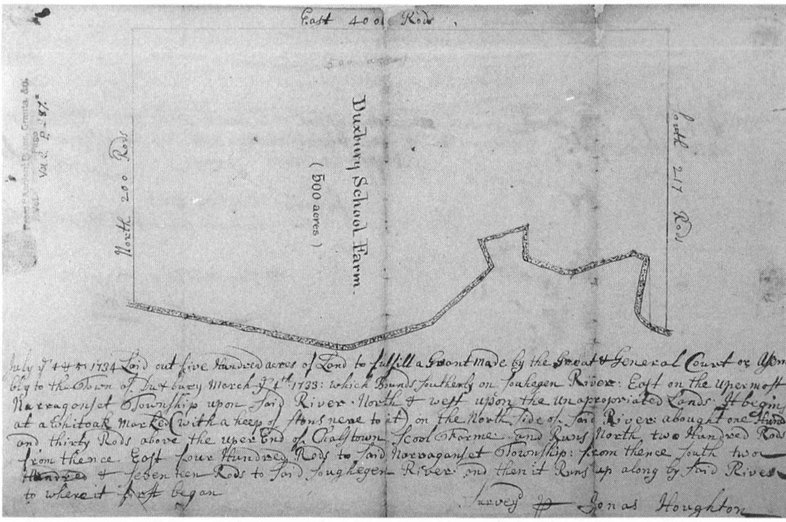

Fig. 2.17. Plan of a five-hundred-acre school farm in Duxbury, Massachusetts, mapped by surveyor Jonas Houghton in 1734. Bounded on the south by the Souhegen River, Duxbury's grant was located near Charlestown's School Farm. Although town records are generally silent on the subject, school farms from this period almost certainly included woodlots. The school farm in Walpole, Massachusetts, appears on John Butcher's plan (fig. 1.4). Massachusetts State Archives

Duxbury's school farm, and a survey conducted in 1734 by Jonas Houghton placed the tract roughly 130 rods north of a school farm given to Charlestown (fig. 2.17). Town records offer scant evidence about composition, volume, or duration of wooded areas on these large tracts. Consequently, it is difficult to conclude whether timber and wood were rapidly cleared or if the farms' woodlots steadily contributed to agricultural production.

As methods of town planning became more regimented, location of school lots was determined by proprietors who were less intimately involved in public concerns. As often as not, these lots were sold or became subject to long-term leases. In 1790, Hanover, New Hampshire, appointed a committee to sell its school lands and use the proceeds to benefit the town's school.[43] Dunbarton, New Hampshire, located just south of Concord, formed a similar committee in 1803 and leased their fifty-five-acre school lot at public auction for 999 years and the sum of $577.27.[44]

With expansion of new towns into more remote regions, unappropriated lands diminished and legislative grants to schools either filled gaps between towns already surveyed or were selected from increasingly distant territory. Cumbersome administration of inaccessible tracts fostered rapid sale, and

Fig. 2.18. Coe Academy in Northwood, New Hampshire, from a postcard by Henry Osgood, c. 1905. Built in 1866, the mansard-roofed school was privately chartered but charged no tuition for children of town-members. Currently Coe-Brown Northwood Academy, and much enlarged, the school continues to own woodland and has placed it under forestry management. Such school forests, as they are known, are fairly common in New England. Chesley Library, Northwood, New Hampshire

schools often converted land resources into more manageable endowments. Forest resources were culled to improve land or because they were useful for little else. Maine offered a large, untapped domain, and a number of Massachusetts academies acquired land in that region. Massachusetts schools in New Salem, Groton, and Westford, as well as Williams College, acquired large tracts of land in Houlton Plantation, bordering Canada's New Brunswick province, in 1818. A plan drawn two years later confirmed that Groton and Westford Academies had already sold their land.[45]

Public land endowments often proved inadequate or otherwise unavailable to support educational expenses, and many schools either purchased or were given nearby lands to produce rental income or wood for fuel. Royalton, Vermont, founded a public grammar school, which received a charter in 1807 as Royalton Academy. In addition to its legislatively assigned public school lot, located in adjoining Rochester, the academy received a generous donation of two large parcels in nearby Pomfret in 1817. Rental income was intended to pay for assistance and instruction of young men fitting for the ministry or, if

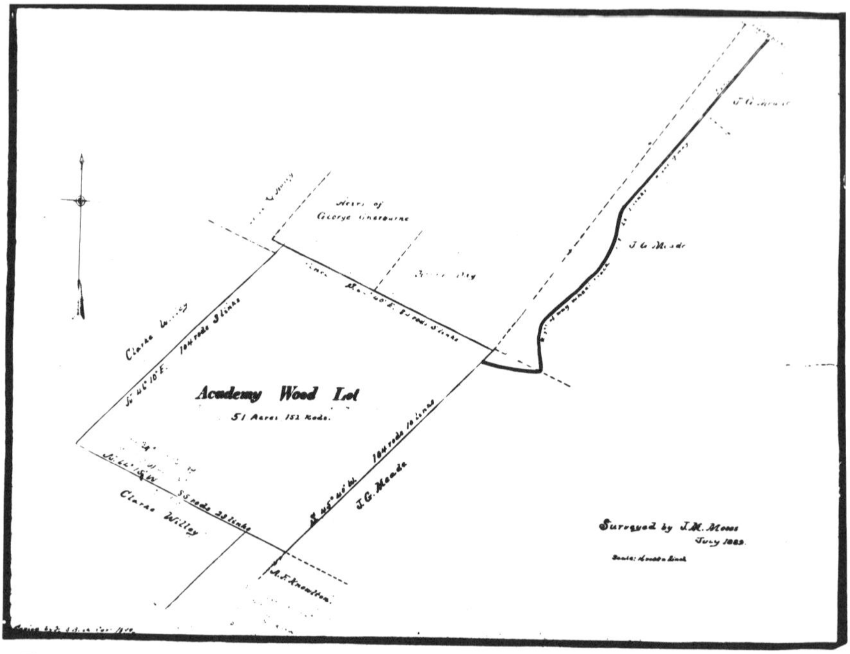

Fig. 2.19. Woodlot survey of land belonging to Coe Academy in Northwood, New Hampshire, drafted by J. M. Moses in 1889 and copied by H. B. Buck in 1899. For a period, the school was heated entirely by wood, and several woodlots were donated by the family of Ebenezer Coe, for whom the school was named. New Hampshire State Archives

worthy candidates were scarce, for the school's general benefit. These lands were leased in "fermeletten"—"for as long as grass grows and water runs"—but subject to several conditions regarding forest resources. Tenants were required to reserve thirty acres of woodland from every hundred acres cleared and were prohibited from committing waste or stripping forest growth. Land was to be cultivated according to rules of good husbandry, and every cleared five-acre parcel had to be groomed and well fenced before additional cutting was permitted. An allowance of timber and wood—enough to accommodate the needs of two families for fencing, building materials, and fuel—also was specified.[46]

Numerous institutions, both public and private, have retained their land grants, which continue to produce crops of trees; school forests, as they are called, are far from scarce. In southeastern New Hampshire, Northwood Academy (later Coe Academy and then Coe-Brown Northwood Academy) was established in 1866 as a private institution but served as a free public high school for its community (fig. 2.18). For a time, the school was heated entirely by wood, and one of its woodlots, a fifty-one-acre parcel, was surveyed by J. M. Moses in 1889 (fig. 2.19). The school currently owns woodland under forestry management.[47] As with other categories of public lands, school lots have occasionally become part of private forest reserves. A portion of Catlin Wood in Litchfield, Connecticut, now owned by a charitable trust, is a school lot that had passed into private ownership during the eighteenth century.[48]

Lease Lands in Vermont

As settlement penetrated remote regions of New England, colonial governments competed for lands and created a confusing battleground of title disputes. Massachusetts, in addition to its claims for territory in Maine, asserted rights to a large portion of New Hampshire and issued charters for towns along the Merrimac River and throughout southern New Hampshire. These charters routinely specified reservations for ministerial, parsonage, and/or school lots. New Hampshire's governor, Benning Wentworth, challenged these claims and often granted conflicting charters. He also maintained rights to land west of the Connecticut River Valley and in the Champlain Valley, where he awarded charters for a large number of towns in Vermont between 1749 and 1764. In addition to keeping a generous portion of land for himself, Wentworth's charters stipulated that four public lots be reserved, one each for glebe, the first settled minister, the town school, and the Society for the Propagation of the Gospel in Foreign Parts, often referred to as a Gospel Lot. In New Hampshire, public lots were regarded as direct grants to beneficiary institutions, and many passed into private ownership.[49]

Following the Revolutionary War, Vermont, first as a republic and later as a state, granted town charters requiring five categories of public lots: the first settled minister, town school, social worship of God (or gospel lot), county grammar school, and college or seminary. Unlike New Hampshire, Vermont courts prevented conveyance of fee title to public lots reserved by its own charters or those granted under earlier Wentworth charters. Generally, American courts have applied English common law to ownership of real property and have consequently discouraged perpetual or durable leases. Instead, such transactions are regarded as outright conveyances but are made subject to various conditions. However, Vermont's jurists devised an exception to this rule and permitted long-term leases, reasoning that public lots were intended as a trust to be held for future generations. As a result, public lands in Vermont became known as lease lands, and revenue was dispensed by towns to assigned beneficiaries.

Not surprisingly, a bewildering system of allocation developed. School lots reserved in Wentworth charters and gospel and school lots created by Vermont charters were placed under the control of town selectmen. After a series of legislative confiscations and court battles, income from glebe lots established in Wentworth charters also fell to town supervision. However, Wentworth reservations for the Church of England missionary society were eventually deeded to the state's Episcopal diocese. Vermont's grammar school lots were periodically granted by the state legislature to town schools or, alternatively, to local private academies, which qualified as grammar schools and burgeoned during the nineteenth century. Lots assigned for the first settled minister in both Wentworth and Vermont charters were occasionally acquired by ministers but just as often remained under town supervision. Revenue from college and seminary lots created by Vermont charters was distributed to the state university. Legislation passed in 1937 finally permitted sale of Vermont lease lands subject to any outstanding rights of lessees, and proceeds were placed in trust for designated recipients. Nevertheless, a large number of public lots still remain under the control of town selectboards.[50]

Management of lease lands in Vermont has been haphazard at best. Beneficiaries have not taken an active interest in supervising their lands, town administrators have lost or not kept careful records, and inflexible rents have yielded a much poorer long-term return than have taxes as land values have increased. Information from town records, if available, is usually limited to simple income tallies, which offer little help in sketching patterns of land use. Leases, frequently assigned, were often not recorded, and towns sometimes failed to keep track of their tenants. In other cases, lease lots were mistakenly conveyed outright by deed when individuals who owned adjoining tracts became confused about title to their property.

A great many lease lots were intentionally placed on distant, inaccessible sites, such as mountaintops, ravines, steep grades, or swamps, and names of more than a few reflect the wry humor of their towns' founders: Glebe Mountain in Londonderry and Governor's Mountain in Guilford, for example. In such cases, lots were often difficult to locate and survey, and many towns simply forgot or ignored their lands for long periods. A glebe lot atop the westerly peak and face of Mt. Ascutney in West Windsor, for instance, has yet to be delineated on the ground. It is rare to find public lots in close proximity to village centers, and active community involvement with a large percentage of these lands has not occurred.[51]

Sizes of individual lease lots varied, and proprietors were allowed to balance quantity with quality as well as assign more than one parcel to a particular beneficiary. Absent even rudimentary records, general conclusions about use of woodland resources are difficult. Nevertheless, certainly a large portion of Vermont's public lands were either cleared for pasture or left wild for a succession of timber harvests. Many of these lands were, and still are, valuable primarily for their forests, and leases amounted to little more than sale of timber rights. For instance, Lemington, located in the far northeast corner of the state, leased its lot reserved for the town's first settled minister for "as long as wood grows and water runs" in 1870. However, the lease was made subject to a condition that ample security against damage to the lot be provided prior to any logging.[52]

Despite considerable shortcomings in methods of managing Vermont's lease lands, a great many of these public lots still exist. As interest in municipal forests increased after 1915, towns began looking for available land to place under woodland management. Delinquency in lease rentals allowed towns to reclaim some, and inaccessibility helped to preserve others. The towns of Arlington, Bloomfield, Essex, Huntington, Lemington, Morrisville, Pomfret, Reading, and West Windsor have all either designated lease lands as municipal forests or committed them to forestry management. Bestriding the peak of Saxon Hill, for example, the school lot in Essex bordered land acquired for the Essex Junction Village Forest, at one time one of Vermont's most actively managed town forests. School lots in Arlington, Bloomfield, Lemington, and Pomfret, are similar examples, although Pomfret's land later became part of the Appalachian Trail.

Maine's Public Lands

While Vermont was carving up its lands into towns after the Revolution, Massachusetts embarked on a similar but more complex undertaking for its vast and empty territories in Maine. Wildland townships of six square miles were surveyed, identified by township and range numbers, and held for sale, with hopes that incorporated towns would develop. A law passed in 1784 required designation of public lots for the first settled minister (200 acres), for the public grammar school (280 acres), and for future appropriation by the general court (200 acres) in every township sold. Four years later, an amendment required a fourth public lot for ministry use, and sizes were increased to 320 acres for each category. Lots set aside for future appropriation by the general court became known as state lots and were eventually sold through special legislation, occasionally as grants to schools and colleges in Massachusetts. As a result, townships established after 1788 typically contained 960 acres of public lands.[53]

Maine became a state in 1820 and, with Massachusetts, assumed joint ownership of unsold, unsurveyed territory in its northern regions, identified simply as the "public domain." For a short period, reservations of public lots for school and ministry in newly developed townships followed the well-established practices created in 1788. However, legislation adopted in 1824 consolidated future reservations of public lots to a single thousand-acre parcel intended for general public use, as determined by Maine's legislature. After incorporation, each town received the benefits of its thousand-acre public lot. An office of land agent was created to administer lands remaining in the public domain, and officials were given authority to sell timber and grass rights, grant small tracts of land to settlers, and prevent timber trespass. Concurrently, town inhabitants were given title to previously created ministry lots except when ownership of those lands had already been acquired by an individual or church. Power to sell these lots was granted to a board of trustees composed of municipal officers. Later, statutes enacted in 1831 and 1832 required that proceeds arising from sale of any ministry lots be applied to support primary schools in that same town.[54]

Enormous portions of Maine's public domain were unsuitable for settlement, and thousand-acre public lots began to accumulate as townships were sold without plans for forming any towns. Control of these lots was chaotic, and timber trespass continued to deplete the only resource available to attract settlement. In response, the land agent was authorized in 1850 to sell timber and grazing rights on all reserved lands until towns became incorporated or organized for plantation purposes. Another obstacle to settlement was removed three years later, when Massachusetts relinquished its entire interest. By 1878, incorporated towns had acquired title to reserved public lots within their boundaries, and only supervision of timber and grassland sales in unincorporated townships remained. In 1891, the land agent became state forest commissioner, and the latter assumed control of all lands. Vast tracts are still part of the public domain.[55]

Town Poor Farms

During the nineteenth century, well after most land in New England had been organized into towns, an entirely different category of public lands emerged, as communities resorted to agricultural subsidy of indigent populations. Although almshouses were common in New England villages much earlier, poor farms not only housed the resident, ill, destitute, or aged, they also provided opportunities for work. By the middle of the nineteenth century, such farms had become a widely practiced method of local welfare throughout New England. In addition to crop and pasture land, poor farms often included woodlots from which timber and fuelwood were culled. As responsibility for welfare gradually shifted from towns to county or state governments toward the close of the nineteenth century, many communities retained ownership of farm woodlots and converted them to town forests. Woodlots on or near these farms supplied fuelwood for resident paupers, building materials for farming operations, and lumber or other wood products for sale to the general public. As such, they represent a category of communal woodlands worthy of consideration.

As did so many aspects of New England's society, care for the poor was transplanted from England, where a substantial body of welfare law had been assembled prior to the Great Migration. Emancipation from feudal tenure in England encouraged greater freedom of movement and helped to create an increasingly mobile indigent population. In response to widespread begging, a series of laws was enacted that culminated in passage of comprehensive legislation during the second half of the sixteenth century under the reign of Elizabeth I. Together, these laws implied a public obligation to offer relief for the destitute. Ordinarily, responsibility for the support of needy persons was assigned to local parishes, where overseers of the poor were appointed. Parish farms were well known and offered a source of employment when farmers sometimes turned out their hired men during lean winter months. Funds were raised by taxation, and a primitive scheme for employment was developed that gradually led to organization of workhouses.[56]

Members of New England's early villages assumed responsibility for their poor, much as local parishioners had in England. At first, care was administered informally, and common pasturing of cows or swine included a few animals that were made available to paupers. Timber for building materials and cordwood were provided from common land when necessary. At a town meeting in Marshfield in 1645, inhabitants agreed that everyone should contribute a peck of wheat and twelve pence for maintenance of their indigent.[57] Watertown selectmen ordered a survey in 1660 to assess wants and needs of impoverished families and six years later assigned a portion of the town rate for their support.[58]

Colonies quickly enacted legislation converting custom into law. In 1647, Rhode Island required each town to provide carefully for relief of their poor, to maintain the impotent, to employ the able, and to appoint an overseer to enforce these obligations.[59] A New Hampshire law of 1687 ordered that justices of the peace provide for relief and maintenance of paupers according to the laws of England.[60] A year later, New Hampshire town selectmen were appointed overseers and given authority to levy taxes and arrange for employment.[61] While town members may have accepted an obligation to care for resident paupers, duty clearly did not extend to strangers, who were usually warned from town before they could be considered residents.[62] In many villages, the helpless resided in homes of relatives or were given shelter in almshouses, which often appear among the buildings shown on late-eighteenth- and early-nineteenth-century town maps. Paupers, especially children, also were bound at auction to low bidders and became indentured servants until they reached maturity.[63]

Acquisition of poor farms began to occur early in the nineteenth century as growing populations increased the burdens of charity. New England lawmakers again seemed to take their cue from well-established practice. Several states enacted laws between 1790 and 1800 that gave communities liberal authority to establish town-owned almshouses, workhouses, or tenements and to obtain necessary land.[64] Undoubtedly, authority to acquire farms rested on this legislation. By 1853, however, only Maine's laws made reference to an agrarian-based welfare system and specifically allowed towns to purchase farms to support their poor.[65] In any event, poor farms became a conventional and occasionally efficient means for public assistance in New England communities well before midcentury. Although ancient failures to distinguish between criminals and the destitute had been corrected by this time, the cultural imprint nevertheless survived in language, and residents on many New England farms were referred to as inmates.

Ordinarily, town farms were far from desirable places to live, a distance measured by both emotional and physical journey. Shame and humiliation accompanied "going to the poor farm," and towns selected many farms for their remote locations, far removed from village centers. This forced paupers to travel, in symbolic penitence, along "poor farm road" or "alms road." In communities such as Foster, Rhode Island, the stigma was made more emphatic by the designation "town asylum farm."[66]

Despite a prevailing hardship associated with public assistance, many poor farms were managed competently and provided adequate, even comfortable refuge. Agricultural production reached substantial levels, and yearly reports regularly itemized produce sold. With a measure of irony, town members of Henniker, New Hampshire, applied proceeds from the sale of the town's ministry and school lots to purchase and stock a poor farm in 1836. Debate on the matter had surfaced several years before when a committee was appointed to examine methods of caring for village poor. Their report, presented at a November town meeting in 1833, suggests a trend:

> We have endeavored to obtain correct information as to the result of this experiment in various places, where it has been tried, and have ascertained that it has been invariably successful in a great savings of public expense. The belief is now so generally prevalent that this mode of providing for the poor is attended with greater advantages than any other that has been practised, that very many towns are preparing to adopt it. In some places, the income of the farm has exceeded the pauper expenses, and, in all, the taxes for this object, have been reduced nearly or quite one half.[67]

Henniker's poor farm had periods of respectable productivity, and an overseer's report from 1856 itemized articles sold, including peg timber, butter, beef, calves, pigs, poultry, neat stock, hides, apples, pears, plums, and knitting.[68] Poor farms such as Henniker's represented an important part of nineteenth-century New England villages, and many farmhouses remain, as do the overgrown pasture and crop lands that supplied hay and ensilage. Town farm roads may persist in communal memory although buildings have long since been demolished or converted to different use. Although agricultural production served as the backbone of poor farms, woodlots contributed significantly to farming operations and intermittently supplemented revenue through sales of lumber, fuelwood, and a variety of wood products, including shingles, posts, rails, clapboards, railroad ties, bark, and even ship's timber. In Massachusetts, Groton's poor farm churned out a steady supply of bridge planking and pine lumber between 1859 and 1891; production peaked in 1889, when 10,000 feet of boards yielded a return of $180.[69] Petersham purchased a poor farm in 1825 and periodically sold wood and lumber,[70] while Pepperell's farm marketed cord wood annually between 1883 and 1892.[71] Occasionally, income from poor-farm woodlots went directly into general coffers, as in Carlisle, where voters authorized overseers of the poor to sell wood on the town farm in 1880 and apply the proceeds toward reduction of the town debt.[72]

New Hampshire's poor farms have an equally productive history of woodlot management. After considering adequacy of lands, water supplies, and stands of wood and timber on several farms, a town committee in Weare, located in Hillsborough County in the state's south-central region, finally made a selection in 1838. Management proved as scrupulous as selection, perhaps too much so, and town voters curiously decided to tax the farm in 1840. Records consistently show supplies of wood and lumber kept on hand, as well as recurring sales of timber, hemlock bark, joists, clapboards, and shingles. In 1900, for example, 50,000 feet of hemlock plank were milled locally, and a large portion was used to repair town bridges.[73]

Concord's two-hundred-acre city farm, acquired in 1827 and located north of the city, was tucked between Penacook Lake and the Merrimack River and bordered by Rattlesnake Brook, which flowed between the two (fig. 2.20). In 1858, lumber, cordwood, and walnut timber netted a modest return; and a year later more lumber, along with shingles and sixteen cords of wood, produced $183.44. In 1863 and 1864, income soared as sales of ships' timber, cordwood, and lumber produced nearly $1,400. Frequent sales of lumber, fuelwood, or special items such as derrick timber continued until the farm was closed and partly sold at auction during the winter of 1883. Large areas of pasture and woodland were retained and became part of the city's watershed forest and park surrounding Penacook Lake.[74]

Doubtless, not all poor farms contained woodlots, and among those that did, not all were fully utilized. Some, such as that owned by Newmarket, New Hampshire, located in the state's southeastern corner, generated revenue almost exclusively from dairy products, crops, or animal husbandry. However, when Newmarket's town farm was surveyed by Orrin M. James prior to its sale

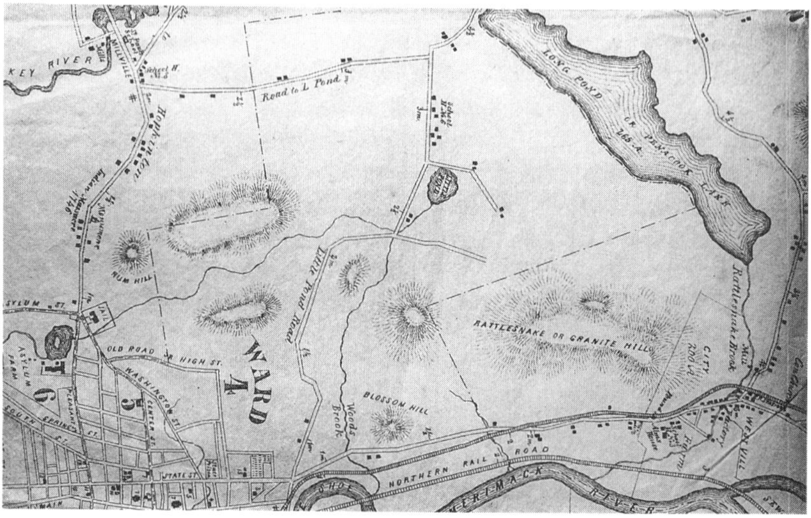

Fig. 2.20. Portion of a plan of Concord, New Hampshire, drawn by S. C. Badger in 1855 and printed by Chandler and Bro. Lithograph, Boston. The city farm is located between the Merrimack River and Long Pond (also called Penacook Lake), which later became a reservoir. Land surrounding the lake, including part of the poor farm, was converted to a watershed forest. Concord Public Library

in 1909 (fig. 2.21), the town paid two men to examine and value its woodlot. Nor were all poor farms always located entirely in towns they served, as James's plan shows.[75] In other cases, towns recognized that woodlots were an indispensable part of farming operations and actively sought separate parcels of woodland to supplement inadequate reserves. Fitchburg, Massachusetts, obtained its city farm in 1827, along with a separate, fifty-acre woodlot on the town's boundary with Leominster (fig. 2.22). Known as the Wanoosnoc Lot, this tract of woodland later became one of four parcels converted to a municipal forest in 1914.[76] Bernardston, situated in western Massachusetts near the state's border with Vermont, received its hundred-acre charity farm as a gift from the estate of Job Goodale, along with a separate woodlot that was managed by town selectmen and later named after its benefactor.[77] Rochester, Vermont, has maintained a large sixty-acre woodlot separately from its town farm and continues to own the land.

Throughout their histories, New England towns have probably secured the odd parcel of timberland whenever opportunity and need became aligned. Whether to supply poor farms with pasture or fuelwood or simply to reduce town debt, woodland has always been a readily available commodity. In Warner, New Hampshire, a nineteen-and-one-half acre parcel of woods and

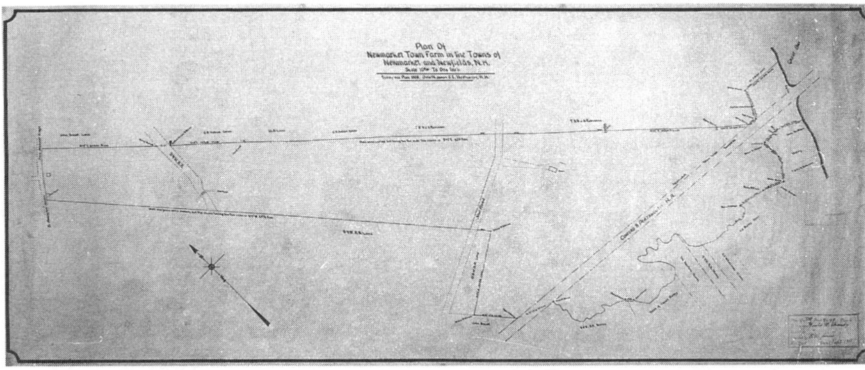

Fig. 2.21. A survey of the town farm owned by Newmarket, New Hampshire, mapped by Orrin M. James in 1908. When towns owned land in adjoining communities, as did Newmarket, property taxes were often an unavoidable expense. However, tax rates among New England towns have often varied substantially, and frugal selectboards have sometimes been able to scrimp by purchasing land in communities with lower rates. New Hampshire State Archives

pasture owned by the town and located a short distance from their poor farm was plotted in 1847 by Harrison H. Harriman, a local surveyor, and a plan was drawn in 1852 (fig. 2.23). By then, the town had also acquired a four-acre parcel to the south, also shown on Harriman's plan. Warner's poor farm had been acquired years before, 1834 to be precise, and it produced a dependable supply of planks and boards, mill logs, timber, bark, and shingles until its sale in 1890. In addition to income from forest resources, Warner's farm records consistently mention pasture rentals, and it is likely that at least a portion of both belonged to this mysterious lot.[78]

Active town farms declined in number toward the close of the nineteenth century as state and county governments gradually assumed greater control over public welfare. A great many towns marketed their farms, but others withstood pressures to sell. With routines of woodlot management in place, some communities diligently continued harvesting timber long after farms had been closed. As the municipal forest movement gathered momentum during the twentieth century's first decades, many towns simply converted woodlots and abandoned pasture into town forests. In Massachusetts, more than twenty-five poor farms were transformed in this manner, and a number became very successful municipal forests, including Fitchburg, Groton, Norton, Petersham, and New Salem.[79] Town Farm Road in Groton became, instead, Town Forest Road. Athol continued to operate its poor farm well into the 1920s and at the same time embarked on a campaign of forestry management on a portion of its woodlot. A three-year contract was sealed with the Franklin Forestry Company

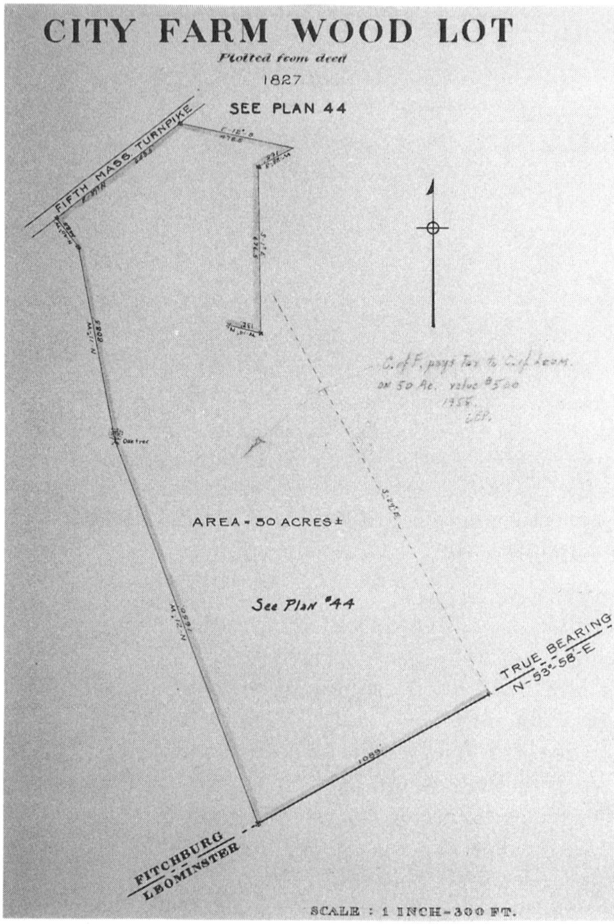

Fig. 2.22. Undated map of the poor-farm woodlot belonging to Fitchburg, Massachusetts, drawn by Herman S. Cushing, a city engineer. At about the same time that responsibility for the care of towns' indigent members shifted to county and state governments, interest in local forestry began to develop. Many poor-farm woodlots, Fitchburg's among them, were conveniently transformed into town forests. Office of the Fitchburg City Engineer

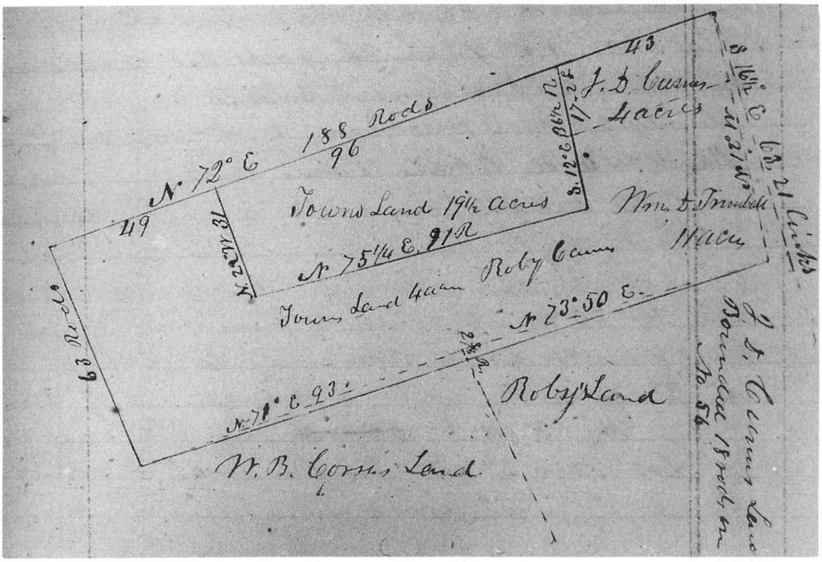

Fig. 2.23. A survey of town pasture and woodland belonging to Warner, New Hampshire, mapped by local surveyor Harrison H. Harriman in 1847. Harriman's carefully detailed drawings and notes, penned in a bound volume, place this parcel about a mile from the town's poor farm. Receipts from Warner's farm show income generated through sale of timber and rental of pasture, quite possibly involving this property. Pillsbury Free Library and Frederick A. Brofos, Warner, New Hampshire

to plant 10,000 white pine and spruce seedlings yearly. Clean-up from timber cutting and clearing for planting yielded cordwood, and a log cabin was constructed at a municipal campground on part of the land. Petersham employed similar forestry management. Its town farm was acquired in 1826, and sporadic sales of wood occurred until 1910. In 1921, the year town farm buildings were sold, a sale of stumpage netted almost $5,200 and paid for reforestation of open land. Remaining land was not designated as a town forest until 1925.[80]

In New Hampshire, town farms at Weare, Franklin, Sullivan, and Milton were among those converted to town forests. The Vermont towns of Calais, Danville, St. Johnsbury, and Rockingham adopted similar decisions. Old Town, Maine, and Colchester, Salisbury, and Suffield, Connecticut, are other examples. Suffield's plan is a fairly rare example, showing town farm and town forest together (fig. 2.24). In some cases, land from town farms became part of municipal watersheds. Concord and Hanover, New Hampshire (see fig. 6.6), are both good examples, and the poor farm in Westfield, Massachusetts, cultivated a town nursery begun in 1914 and used for both street planting and for stocking the community's watershed forests.[81] Parcels from several Massa-

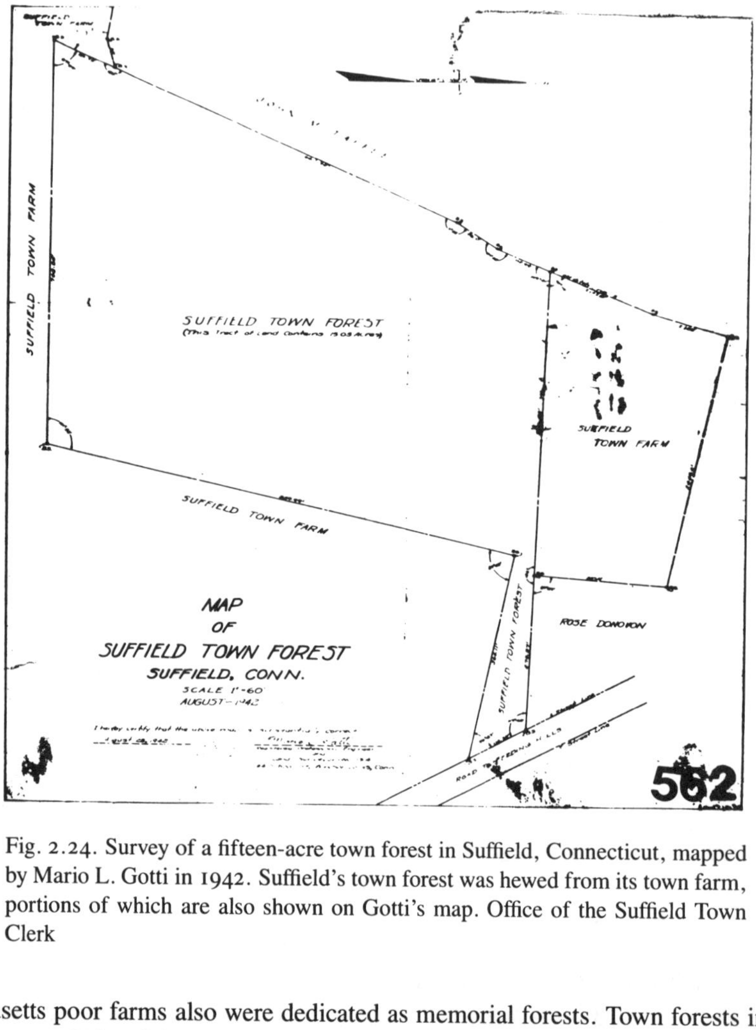

Fig. 2.24. Survey of a fifteen-acre town forest in Suffield, Connecticut, mapped by Mario L. Gotti in 1942. Suffield's town forest was hewed from its town farm, portions of which are also shown on Gotti's map. Office of the Suffield Town Clerk

chusetts poor farms also were dedicated as memorial forests. Town forests in Groton and New Salem were dedicated to the memory of World War I veterans, and Duxbury named its town woodlot in honor of forester Frederick B. Knapp (see Epilogue, fig. 9.1). Methuen converted neglected land from its poor farm to a woodland and beach recreation area suitable for hiking, camping, scouting, and swimming, as well as for commercial production of timber.[83]

. . .

Public lands in New England towns present a complex and unwieldy subject riddled with subtle characteristics peculiar not only to individual states but to regions within each state as well. A multitude of campaigns to establish new towns and variations in proprietors' methods for town planning are in large part responsible. However, glebe, ministry, and other public woodlots represent a remarkably expansive link between seventeenth-century common woodlands and twentieth-century town forests. Collective stewardship of forest resources evident in many communities reinforces that connection, and places such as Keene, Newington, and Danville, New Hampshire, assume a special significance in New England's urban and forest history. Moreover, frequent and easy transition from poor farm woodlot to town forest is indicative of community participation in development of forest conservation policy at the turn of the century.

Chapter 3

Community of Interest in Forest Policy

The nature of collective stewardship in New England's town woods, their character of use, and their function in community structure all began to change as the nineteenth century waned. Forces propelling that change materialized along a number of fronts. Public concern for America's depleted forests and damaged ecosystems, quietly voiced by the close of the eighteenth century, steadily swelled in unison (see fig. 3.1). Timely, influential writing helped to foment opinion and innovation, both public and private. Forestry gradually developed into a profession, and policy slowly matured. Conservation, a movement commanding foresight and restraint during a time of diminishing resources, became a rallying cry. Government forestry programs created national and state forest reserves; the seeds for wilderness preservation were also sown. Nurtured by these events, New England's communities sprouted new growth shaped as town forests.

Connections between dynamic trends in America's forests during this period and developing interest in community woodlands are pivotal. The arrival of European-trained foresters on the national scene, Bernhard Fernow in particular, marked a beginning of serious attempts to introduce community forestry. Swiss and German city forests, by then well established and managed, became much-publicized models for New England towns. National trends also influenced state programs, which embraced town forests as part of comprehensive plans to reverse profligate timber management. Formative state and federal policy-making also prompted creation of private, nonprofit forestry associations that actively encouraged local participation. Awareness of a turning point seeped into public consciousness and reappeared in the form of sylvan community initiative.

THE RETURN OF THE PRODIGAL

Fig. 3.1. Cartoon from the *Twenty-Sixth Annual Report of the Massachusetts Forestry Association*, Bulletin 138, issued in 1923. New England's town forest movement developed from concern about spent timberlands and abandoned, idle farmland. Emphasis on cultivation of trees for commercial sale was the driving force behind this movement, but appeals for local forestry were couched in the broadest possible way and included promotion of recreational, educational, and aesthetic benefits.

A Quest for Policy

When Thoreau journeyed to northern Maine between 1846 and 1857, he found a dense forest of fir and spruce intermixed with hardwoods, primarily paper or yellow birch and sugar or red maples. Northern cedar occupied swamplands; elm and ash bordered stream valleys and lake shores. Absent were extensive stands of white pine, once the largest and most common tree in the northern pine forest and the most widely sought by the lumber industry. Red cherry and aspen sprouted in cut-over lands or in the exposed mineral soils of fire-scorched earth. Thoreau's account, published as *The Maine Woods* in 1864, gives an important description of New England's northern forests at midnineteenth century. Throughout that region, few stands of virgin timber still existed. Measureless forests in other regions—the Great Lakes, the South, and the Pacific Northwest—faced imminent exploitation.[1]

Depletion of America's timberlands was caused by the irrepressible growth of a society dependent on wood for practically all of its needs and on an agriculture that routinely, even single-mindedly, cleared land. Ever larger and increasingly mobile populations—and the roads, canals, and railroads that transported them—devoured forest products. The opening of canal and rail corridors from the Atlantic seaboard to the Great Lakes regions and eventually to the Pacific Northwest caused an even greater impact on forest consumption. Unbounded pursuit of newly cleared land, a fertile commodity, continued. Industrial expansion, much of it erected with wood, fueled the economy of a nation. Large commercialized lumber companies, cheap and ready labor, improved saw and mill technology, and development of wholesale centers in such places as Buffalo and Chicago trumpeted the arrival of industrial capitalism in the timber industry. Cuthroat competition, overproduction, unstable markets, and unrestrained clear-cutting were the results.[2]

Inattention to sensible timber management had far-reaching ecological consequences. Only valuable trees were cut, leaving cull timber, or "weeds," the poorest of the species, to regenerate forests. New growth was usually narrow in diameter and generated poor yields per acre. Forest composition also changed as only those trees capable of sprouting poles from stumps produced new growth. Earlier practices of removing timber by handsaw and horses left the productivity of the forest floor relatively undisturbed; regeneration through natural seeding and sprout growth occurred steadily. Portable steam-powered sawmills attacked the next generation of forest cover with far less patience. Capable of being assembled on isolated hilltops, these mills made it possible to transport dimension lumber directly from forests. Remaining stands of virgin timber, once inaccessible, were quickly plundered. Waste of otherwise usable portions of trees was the price; high stumps and long tops were left to rot. Sparks from machinery and slash left uncleared created fire hazards, further disrupting ecological order.[3]

Throughout much of the nineteenth century, Washington pursued a policy of transferring lands from public domain to private ownership, with little regard for forest resources. Special grants, particularly those to railroad companies, were greatly abused and often seen as little more than opportunities to strip land of their trees. Unmanaged private reserves were mined, not cropped, with little thought to the future. Lands left in public domain were not protected by legislation, and timber trespass occurred freely at the hands of mining prospectors, settlers, and railroad contractors. Forests were endless, timber supplies inexhaustible, and laissez-faire a formal strategy.[4]

Harbingers of change appeared late in the eighteenth century. From his travels throughout New England in 1796, Timothy Dwight was able to envision the region's ecology as a whole, a view unrestricted by artificial political boundaries. He recognized that second-growth trees were dominant, "almost all the original forests of this country having been long since cut down." If agriculture was at fault for ceaseless clearing of the land, farmers nevertheless may have been the first to recognize a loss and to move to restore forest ecosystems. In 1804, the Massachusetts Society for the Promotion of Agriculture and Arts offered prizes for the best plantations of certain hardwoods.[5]

Attention also began to turn abroad. Between 1785 and 1805, two French botanists, Andre Micheaux and his son Andre Francois Micheaux, explored American forests, documenting characteristics and distribution of tree species. Their treatise, published in 1819 under the title *North American Sylva*, anticipated a period during the second half of the nineteenth century when Americans borrowed regularly from forestry knowledge and practice in Europe. Although English translations of *North American Sylva* were expensive and not widely circulated, several summaries were made available to New England readers. In four consecutive weekly publications during the spring of 1837, *New England Farmer* published a lengthy series under the heading "American Forest Trees." The articles were based on a book by Daniel Browne, *Sylva Americana*, which reproduced Micheaux's work. The final installment encouraged forest plantations and pleaded for "the necessity of economizing what yet remains of these rich national treasures, and of replacing what has been so carelessly wasted." Another article, appearing four years earlier in the same journal, had tolled a similar call of alarm and had cited the European example:

> Already has the supply become so small throughout Germany and France and England, that laws have been enacted in these countries for the very purpose of preserving forest trees, and for the additional purpose of promoting their cultivation. In Germany this cultivation has been made a science, schools have been established for the purpose, and a forester's education requires a long period of study and of close application?"[6]

Not until 1864, however, did writing of momentous future import appear. That year, while America was preoccupied with the Civil War, George Perkins Marsh published *Man and Nature; Or, Physical Geography as Modified by Human Action*. Marsh, whose boyhood days were spent in Woodstock, Vermont, wandered from one career to another, sometimes unsuccessfully, as lawyer, farmer, engineer, businessman, newspaper editor, politician, and diplomat. His book, described by Lewis Mumford as the conservation movement's fountainhead, offered a startling assessment of America's misused environment. An extensive chapter examined forests and their relationship to ecosystems, concluding that a natural balance was being drastically altered by humankind: "It is certain that a desolation, like that which has overwhelmed many once beautiful and fertile regions of Europe, awaits an important part of the territory of the United States . . . unless prompt measures are taken to check the action of destructive causes already in operation."[7] Marsh also contributed to America's sharpening focus on European forestry practice, discussing at length progress made on that continent. He briefly mentioned the existence of woodlands owned in common by members of a community or owned outright by municipalities: "In countries where there exist municipalities endowed with an intelligent public spirit, the purchase and control of forests by such corporations would often prove advantageous." [8]

An important body of writing discussing the plight of American forests soon materialized, helping to raise the level of governmental and public concern. Granted, federal interest in small town-owned woodlands was anything but a priority during this period. Nonetheless, most European systems received scrutiny, communal forests among them, and both national and local interest in forest conservation began to take root. An 1876 article by Dr. John Warder, U.S. commissioner to the International Exhibition in Vienna in 1873, exposed America's blindness to timber management by recounting, in careful detail, extensive forestry exhibits displayed by European countries. Warder, a retired physician, had established what amounted to an experimental station for forestry on his Ohio farm in 1855 and had convened a forestry congress in Chicago in 1875, leading to creation of the American Forestry Association.

A significant portion of Warder's report investigated division of European forests among state, communal, and private owners. He praised German methods, writing: "In this part of the exposition were found some of the richest and most complete representations of the forest-wealth of Europe. This is what we might have expected in a country that has for so long a time occupied the first rank among nations by her forethought in bringing skill and science to bear upon the protection and improvement of her forest-lands." [9] Chiding Americans for their wasteful practices, he nevertheless hedged at the prospect of a broad governmental role in American forestry similar to that of European sovereigns: "In this matter, it may not be wise to imitate the practices of European states. We have no crown-lands, no royal preserves, no imperial forests nor any church and communal reservations of woodlands. We desire no such tenures, but we do need more forests in many parts of the country than the individual greed of private interests has left us." [10] Warder was incorrect in his conclusions about church and communal forests, but he unerringly assessed the power of individual property rights in America and the resulting unchecked pillage of timberlands. His words portended an emerging battle between government intervention and individual freedom, one whose fires have never been extinguished in American forests.

While the world displayed its wares at Vienna's exposition, the American Association for the Advancement of Science embarked on a quest of its own. At an 1873 meeting in Portland, Maine, Franklin B. Hough presented a paper titled "On the Duty of Governments in the Preservation of Forests." Hough, like Warder a physician, was also an amateur botanist, geologist, and historian. He called for public education as a critical first step in giving state and federal governments power to protect forests. The association promptly endorsed legislation for a federal forestry commission accountable to Congress on matters affecting the extent and consumption of forests, influence of forests on climate, and European planting and management methods. Hough supplied a comprehensive report concerning estimates of timber supply, rates of use and destruction, the effect of forests on rivers and water courses, and forestry in Europe. He emphasized well-established French codes that organized a central forest administration for oversight of public and communal woodlands. He also observed that communal forests and those owned by quasi-public entities such as hospitals exceeded crown and state forests in total acreage.[11]

During succeeding decades, America began to move more rapidly toward forest conservation. In August 1876, Congress appropriated funds allowing the commissioner of agriculture to employ an expert to study forest conditions. Franklin Hough was appointed and was sent to Europe to study forestry. He also inaugurated publication of the *American Journal of Forestry*, which he devoted to tree planting, formation and care of woodlands, and ornamental planting. Short-lived, the journal began circulation in October 1882 and ended in September 1883. However, that year was an important one. The American Forestry Congress met in Cincinnati in April 1882, attended by delegates from across the country who presented more than eighty papers on a wide range of topics related to forestry, many extensively researched. Four months later, at a joint congress held in Montreal by the Society for the Promotion of Agricultural Science, the American Association for the Advancement of Science, and the American Forestry Congress, the latter absorbed John Warder's American Forestry Association. It waited until 1889 to assume that group's name and then began a well-organized movement dedicated to the interests of American forests.

Meetings of the American Forestry Congress at Cincinnati and Montreal were a turning point for reform. Several papers by key figures addressed broad policy issues, such as the role of governments and the need for formal forestry education. Advanced forestry practices and administration in Germany were stressed, and, significantly, a number of papers discussed regional and local forest conditions. At the Montreal conference, Matthew C. Read presented a plan titled "The Coppice for the Village and the Farm" and encouraged cultivation of miniature forests—ten to forty square rods—in the middle of village blocks adjoining ends of individual lots. Read, a professor of zoology and practical geology at Western Reserve University in Hudson, Ohio, devised his plan primarily to improve sanitation and protect wildlife, and he included a thick undergrowth of creepers and vines to accompany the coppice. Although it is unclear whether he preferred ownership of these plots by towns or in common by adjoining lot owners, his paper is one of the earliest plans for village forestry.[12]

In 1886, forestry became a formal division within the Department of Agriculture. Bernhard E. Fernow, a professionally trained forester from Prussia, was selected to manage the newly established office, and three years later the Department of Agriculture achieved cabinet rank. In 1891, Congress took a giant step and passed the Forest Reserve Act, which authorized American presidents to set aside designated areas of public land. Within a year, seventeen reserves had been selected. Still, the law was silent on the secretary of the interior's authority to regulate cutting of timber in these domains, and much work remained. Read's plan excepted, contributions to local forestry by Marsh, Warder, Hough, and others from this period are far from direct. In general, references to European community forests were offered primarily as good examples of forestry practice rather than as a forest type suitable for America. Nevertheless, a context in which municipal forestry could be cultivated had taken form, and local attitudes were clearly a part of broader national trends. Appointment of Fernow as chief of the Forestry Division signaled a dramatic shift in America's pursuit of sane forest policy. It also precipitated a much closer inspection of European forestry methods and community forests, topics entirely familiar to Fernow.

Community Forests of Germany and Switzerland

When Americans finally awakened to the plight of their forest legacy, policy in Germany was already well established. Diverse topics, among them silviculture, mensuration, timbering methods, protection of watersheds, science and literature, education, and administration, were all firmly ensconced as features of timber management. In Germany, too, local forestry had advanced to proficient levels. Community forests in Germanic countries developed by two very distinct methods. Those owned by many small towns embodied ancient common rights that had evolved from the Mark Society, a form of settlement similar to England's nucleated villages in origin and constitution. In contrast, city forests connected to major urban centers developed as a direct result of strong, independent political states that had emerged in central Europe during the Middle Ages. Such city forests are wholly absent from Anglican urban or forest history, and an explanation lies primarily with England's historically strong monarchy.

There is considerable irony in the contributions of England and Germany to New England's history of communal forests. Common lands in New England represented an evolutionary segment of village structure that can be traced to both English and Germanic civilizations. However, common rights to woodland developed differently in the two countries, managing to withstand subjugation of feudal tenure better in Germany than in England. In the latter, wooded commons were reduced to fractured portions of the landscape as nobility and an emerging landed class methodically sought enclosure. In New England, too, outlying common lands were unable to withstand individual pursuit of property and soon vanished. Almost two centuries later, New England rediscovered communal forests by turning to Germany, where such lands had persistently evolved into a well-defined community resource.[13]

The Mark Society and Imperial Forests

During the fifth century, Germanic tribes pressed southward—the Huns in what is now eastern Germany, the Franks and Alemans in Gaul, and the Angles, Saxons, and Jutes in Britain. In central Europe, urban legacies of vanquished Romans amounted only to scattered sites that, because of their locations along principal roads or waterways, attracted settlement. These communities developed into compact agrarian units with dwellings clustered for easy access to surrounding arable land. As did England's nucleated villages, *Mark* communities encompassed both buildings and cultivated landscape. The *Mark* forest (*Markwald*) delineated district perimeters; in the *Markwald*, a boundary forest (*Grenzwald*) consisting of woods, lakes, rivers, and swamps identified specific precincts. Finally, common land (*Allmende*) surrounding a tiny nucleus of dwellings belonged to village inhabitants. Members possessed rights to cut timber, pasture livestock, clear land for cultivation, and hunt or fish in surrounding common lands. Individually owned plots were confined to sites for dwellings and adjacent farmyards. By the close of the sixth century, however, principles of land ownership had emerged. Communal fields assigned to private ownership (*Feldmark*) were distinguished from undivided communal woods. Dominant community members acquired large holdings and began to control rural civilizations.[14]

During the eighth, ninth, and tenth centuries, the Carolingian empire consolidated and unified Germanic territories. Conquered lands and unoccupied regions were gathered into the imperial realm. Forests became the emperor's private domain, to be cleared only with permission. Large land grants were awarded to the loyal; churches and monasteries received endowments as well. Royal privileges in imperial forests gradually devolved to a growing class of nobles. As in England, lands held by royalty (Ban Forest) extended over fields and waters as well as woodlands and could envelop entire villages. Still, boundless weald was valueless without populations to extract wealth from the land, and new settlements were encouraged to strengthen the territories of feudal lords. Enticements such as forest privileges were granted to supply domestic needs. As generation succeeded generation, privileges ripened into prescriptive rights and gradually became part of village land tenure.[15]

With disintegration of the Carolingian empire during the tenth century, strong independent feudal states began to emerge, a consequence of several factors. Many grants of land by Carolingian kings had included all rights to revenue, and nobles were completely free from tribute to the Crown. Land tenure had gradually become hereditary and eventually included rights of sovereignty and jurisdiction over populations. Manorial economy had been perfected and networks of new settlements created, thus furnishing independent sources of income. Absence of a strong ruler capable of unifying these states, coupled with civil wars and papal strife, served to further goals of ambitious princes. During this period, too, fortifications at monasteries or castles provided security from attack and beckoned rural populations. For many, such centers lay beyond the lawful reach of feudal lords and offered escape from servitude. Growing cities attracted craftsmen and merchants who contributed to a stronger economy, which in turn encouraged even greater migration from surrounding regions. Large urban centers also buttressed the political power of independent states, helping to undermine an already weakened central monarchy.[16]

City Forests of the Middle Ages

From the time of the Carolingian empire's dissolution until the fifteenth century, powers of independent feudal states became stronger and control by kings weaker. Civilization spread eastward, fueled by population increases and an economy growing increasingly dependent on trade. Colonization of unsettled lands expanded feudal realms and established greater independence from kings. Reclamation of wasteland became an organized movement controlled by territorial lords. Colonies for villagers engaged in industries such as mining were founded near towns. Forest communities for woodcutters and charcoal burners appeared in the midst of clearings to facilitate transport of their products. Village members were granted tracts of woodland and pasture outright or received rights to use the domain of lords. Clergy frequently acquired timber rights in nearby forests. Relentless clearing of forests continued unabated, balanced in part by primitive forestry management in scattered regions. Regulations typically controlled amounts and methods of cutting, required immediate removal of felled timber, and prohibited owners of livestock from allowing cattle and pigs to graze on new growth.[17]

Colonization policies also were furthered by planning new towns, which became outlets for overflowing populations and promised defense from threat of attack. Moreover, if located on favorable sites near rivers or roads, these towns regularly proved to be successful commercial ventures. Those who migrated became freemen, often adding to the ranks of artisans and helping to build a system of trade guilds. Peasants plying surrounding land changed from bondsmen to suppliers. In so doing, they acquired an ability to pay taxes to their lords in the form of money rather than services and helped to undermine feudal tenure. During the twelfth century, a series of towns was planned or improved in what is now Switzerland and southern Germany by rulers of the Zahringen dynasty in order to secure regional domain. These fortified towns were carefully designed with a grid pattern of streets, a market thoroughfare, public buildings, a sewage system, and homestead lots representing units for taxation. In all, fifteen towns were controlled by the dukes of Zahringen, including Zurich and Villingen, both of which possess important city forests.[18]

During the Middle Ages, the creation and growth of urban centers in central and western Europe were followed by the appearance of city forests. Woodlands often were acquired by grant from founders or by direct purchase. In other cases, emperors donated or mortgaged imperial forests in return for funds needed to underwrite unceasing conflict with feudal barons. Remarkably, nearly all imperial lands eventually were frittered away through sale or mortgage, hereditary feudal tenure, or political prizes to favored districts. Cities and free village communitics generally adopted and administered their own laws, including those affecting forest use. Management of municipal and communal forests did not fall under strict control of political states until nearly the end of the Middle Ages. By then, more sophisticated methods of forest management had begun to appear as emphasis on regulated clearing shifted to concern for recuperation. Wealthier cities continued self-government, including management of their forests.[19]

The significance of independent political states in Germany and their many prosperous urban centers can be underscored by distinguishing feudal land tenure in England during the same period. While Germanic kingdoms were struggling to gather political power, William the Conqueror sought to check the influence of landed barons in England. The Domesday Book not only served as a record of estate lands owing tribute to the Crown but also compelled formal recognition of William's authority. While German emperors were squandering ownership of their imperial forests, successive Norman kings fashioned a comprehensive body of law for enlargement and control of royal forests. Ironically, large portions of these lands eventually became privately owned. Moreover, the planting of new towns in England was accompanied only by grants of timber rights necessary for the construction of settlements. Ownership of forests remained with the Crown or, alternatively, with the lord whose property had been requisitioned. In sharp contrast, many forests acquired by cities and towns in Germany and Switzerland remained in municipal ownership and continued to serve their communities for centuries.

Management of Community Forests

Between the fifteenth and eighteenth centuries, political tensions continued to shape and reshape Germanic kingdoms. Wars resulted in the destruction of villages, and lands were seized by conquering states. Extensive forests and fields succumbed to feudal domain following secularization of churches and monasteries. As the Mark Society disintegrated, ownership of communal lands became muddled. In some respects, uncertainty of land tenure contributed to forest decline; after timber was cut, lands grew to brush for want of dominion. Yet woodlands also embodied wealth and filled an increasingly important economic role as states struggled for autonomy during this period of upheaval. During the sixteenth century, state authority was extended over private forests. A rudimentary forest policy gradually emerged, focusing on conservation and improvement of forest conditions as well as wasteful cutting or use of wood materials. As secular authority solidified, methods of fiscal administration and resource management were improved. By the eighteenth century, the study of state finances had developed into a discipline known as cameral science. Because of their size and distinct product, forests required new methods of administration—the dawn of forestry science. Quantitative approaches to regulation and measurement of forest products led to advanced resource management.[20]

The course of Germany's empire led to formation of clearly defined states as the eighteenth century waned. By that time, control over state-managed economies was well established, and forestry science was rapidly improving. Private forestry schools had been established and important books and journals published. Wilhelm Gottfried von Moser's *Principles of Forest-Economy*, published in 1757, corrected archaic methods of timber harvesting based on equal land divisions. Instead, Moser focused on older trees, which produced a better cash crop, pointing the way toward forest mensuration and proximate sustained yield. Forest economics based on mass volume of wood quickly displaced methods of measurement by area and led to identification of wood mass as the crucial variable in forestry management. Mathematical procedures for calculating mass were refined, as were modes of tree cultivation. Monocultural, evenly aged forest compartments resulted from a quest for minimum diversity and sustained yield to simplify mathematical calculation. Sustained yield, however, hinged on a variety of factors, such as growth rates, which in turn depended on weather and soil conditions. Forestry science subsequently expanded into these areas as well.[21]

Throughout this period, cities were annexed to particular states but were permitted to retain their forests. Control and management increasingly fell to state supervision by trained foresters, chiefly to avoid mismanagement. The manner of oversight varied from strict to tolerant depending on the type of forest and the size of the community. At the start of the nineteenth century, about half of Germany's municipal forests traced their origins to common land, and revenue from forest yields was apportioned among community members in states such as Baden, Hesse, and Brunswick. In other regions, Westphalia, Weimar, and Wuerttemberg, for instance, states provided supervisory management, insisting only that forestry operations should pursue a goal of sustained yield and follow a working plan. Plans were subject to approval by state personnel. Communities also were required to employ trained foresters, some of whom were shared by several towns. Where forests grew on land owned entirely by cities, income flowed directly into local treasuries, and state management was less rigid, nominally intended to achieve forest maintenance. Communities were not obliged to adopt a management plan nor to employ foresters. Sale, clearing, or changes in utilization, however, required approval.[22]

Germany and Switzerland were not the only countries in Europe to establish a tradition of community forests. Austria and eastern European countries, notably Bulgaria and Czechoslovakia, maintained a large number of city forests; Vienna has managed a municipal forest for centuries. France also has a large number of community forests and has adopted a comprehensive system for their regulation. However, community forests in Germany and Switzerland were most influential in New England's towns before World War II and were more closely observed than any others, particularly Zurich's Sihlwald.

The Sihlwald

Among Zurich's many forest holdings, the Sihlwald is the greatest in size and the most widely known. Its more than 2,500 acres cling to the mountainous eastern slopes of the Albishorn as it descends into a narrow valley scoured by the Sihl River. Apart from being a model of successful management, Sihlwald's origins typify those of many city and communal forests in central Europe. Its shaded, well-traveled paths quietly wind their way through an intriguing history in which common villagers, barons, kings, clergy, and descendants of Charlemagne vied for land rights. Ulrich Meister, for many years city forest master, amassed a treatise titled *Die Stadtwaldungen von Zurich*, first published in 1883.[23]

Written accounts of the Sihlwald begin in 853, when King Ludwig, grandson of Charlemagne, bestowed certain property rights from his royal court at Zurich to a daughter, Hildegard, abbess at a convent church founded that year. Ludwig's grant included interest in a "forst" on Albis Mountain, possibly a reserve for royal hunting, with privileges of taxation. Underlying title to the "forst" and surrounding territory probably remained with royalty. As primitive settlements developed, villagers acquired communal rights to gather fuelwood and graze livestock. In exchange, they furnished annual payment of a *Fastnactshuhn*, a chicken due on the weekend before Lent, as well as convent wood.

Over time, nobility—the barons von Schnabelburg—also began to assert privileges on Albishorn's woodlands. Zurich became part of the Carolingian empire and was governed by one of its royal families until 1172, when the last of its members died. Duke Berchtold IV of the house of Zahringen was installed as ruler, and Zurich became one of the Zahringen towns. Prosperity ensued, and the city claimed freedom when this dynasty ended in 1218. In addition to gaining control over municipal affairs, it acquired title to property once belonging to royalty. This included "forst" as well as other woodlands that together comprise the Sihlwald. By 1280, city burghers asserted rights to all wood culled from the land, while convent and local baronage disputed rights to the "ban forest" and the tax owing from it.[24]

Political intrigue and wars with Austrian royalty in 1309 resulted in the defeat of the von Schnabelburg family. For the price of neutrality and as reimbursement for damage to city property, the victors relinquished all interest in the Sihlwald to Zurich. Thereafter, only city, convent, and villagers competed for a share of forest wealth. Populations grew, new villages sprouted, tenants' communal use increased, and scattered plots were sold as wood products became more valuable in an expanding economy. City income from the forest amounted to little more than a tax paid by tenants in the form of fuelwood, which was then distributed to citizens. Zurich's burghers eventually reversed their policies, and in 1506 the city began acquiring available parcels to enlarge its rights of use. In 1524, the convent dissolved, and its remaining claims were extinguished by conveyance. Appointment of city foresters had been authorized as early as 1460, and improvements in management continued. An administrative court was established, yearly accounts by tenant users were ordered, and an annual report of conditions by forest masters was required. Oversight became so sophisticated that tenants were permitted to secure loans by pledging their forest rights as collateral.[25]

By 1798, the Sihlwald had grown to encompass more than eighteen hundred acres. That same year, Zurich's status as an independent city-state changed when it merged with the larger Swiss state of Helvetia. In 1805, the city was given complete possession of nearly 3,144 acres of forest. In return, officials agreed to supply fuelwood to citizens, as well as lumber and wood for city buildings. Throughout the centuries, communal rights to graze, gather dead wood, glean leaves and twigs for livestock bedding, and dig up tree roots had been continuously observed. Charitable privileges—supplies of wood to widows, clergy, schoolmasters, and the poor—also had been customary. With increasing populations, some of these rights eventually accrued to separate communities, and Zurich was able to extinguish many through purchase. In other cases, individual interests were sometimes acquired by purchase. This benevolent policy, albeit more organized, continued during the nineteenth century, and permits were given to the wood-poor in exchange for one day's annual labor. In 1884, that system was abandoned in favor of wood-gathering cards that could be purchased for a nominal fee, with repayment guaranteed after labor was performed.[26]

Technical working plans for the Sihlwald date from as early as 1697. How-

Fig. 3.2. An ox-drawn log car in the Sihlwald, Zurich's ancient city forest. American foresters traveled frequently to Germany and Switzerland to observe management of community forests, and the Sihlwald became a highly praised model for town forests in New England. Terence J. Hoverter College Archives, SUNY College of Environmental Science and Forestry, Syracuse, New York

ever, modern forestry practice originated with a series of nineteenth-century management plans first issued in 1834 with major revisions in 1860, 1880, 1900, and 1920. Forestry practice fell under the supervision of the canton of Zurich, equivalent to an American state, and also became subject to certain federal laws. Forest ecology was historically dominated by mixed conifers and hardwoods. By 1900, however, beech had become prevalent to accommodate fuel demand; ash, maple, and elm trees filled out compartments. Early-twentieth-century plans called for a mixed forest, 40 percent coniferous, to capitalize on greater commercial value of softwoods. Shelterwood, a silvicultural practice that calls for removal of crops in a series of cuttings to induce natural reproduction of young trees, was in place well before 1900. Compartments of equal productive capacity contained stands of every age, and age gradations succeeded one another in a regular series. Prepatory cuttings opened the crown canopy for younger growth, and thinning occurred between the ages of thirty and eighty years. Compartments were managed for sustained yield on a rotation of 100–110 years.[27]

Steep, inaccessible slopes were overcome by construction of a narrow-gauge railroad with lightweight, movable tracks (fig. 3.2); a network of log chutes was also built for fuelwood (fig. 3.3). When American foresters began visiting the Sihlwald during the late nineteenth and early twentieth centuries, a city-

Fig. 3.3. A log chute in the Sihlwald. Efficient wood products utilization and local marketing were key factors in the profitability of many German and Swiss community forests. In addition to a network of log chutes and a narrow-gauge railroad with movable tracks, the Sihlwald's improvements included a mill that produced finished products as well as lumber. German community forests often contained nurseries, and crops such as potatoes were sometimes grown between rows of newly transplanted seedlings. Terence J. Hoverter College Archives, SUNY College of Environmental Science and Forestry, Syracuse, New York

operated sawmill was churning out lumber, railroad ties, and ash tool handles. An excelsior machine spun "woodwool," which was then shipped to Italy as packing material for fruit. Fuelwood and kindling were transported to a city woodyard, where local citizens purchased their supplies. Together, mill and forest employed a modest community of more than a hundred, and workers were eligible for a retirement pension after thirty years of labor. With good reason, the Sihlwald's productivity offered ample enticement to America's nascent forestry profession.[28]

Elsewhere in Germany and Switzerland, a number of other important city forests were inspected by visiting Americans. Not far from Zurich, Winterthur managed a productive coniferous forest on roughly 2,800 acres, including a 160-acre nursery that offered seedlings for public sale. Located on hills immediately behind the town, the forest benefited from working plans adopted as early as 1800. In the canton of Bern, the commune of Grindelwald consisted of seven mountain villages, each of which owned a forest. Supplies of wood and timber were allotted according to priorities based on intended use. Most important were small barns for cattle in high pastures, followed by mountain fences, valley barns, repairs to valley houses, new houses, and finally fuelwood.[29]

A number of communities in the Black Forest in southern Germany actively managed their forests. Villingen, in the state of South Baden, worked more than 9,000 acres of spruce, pine and fir and first adopted a management plan in 1837. Baden-Baden's 11,250-acre forest completely surrounded this medieval city and was regarded by many Americans as one of the most beautiful. The small town of Forbach owned a forest of 1,482 acres, which produced enough income to pay all taxes for the townspeople, many of whom worked in the forest. A portion of the land was set aside as a reserve to meet unusual expenses, such as new town buildings. At Freiburg, records of the communal forest date to the late thirteenth century, and a forestation plan was prepared early in the fifteenth century. City forests in Heidelberg, Frankfurt, and Munich also attracted considerable attention. In Frankfurt, crops of oats and potatoes were planted between rows of white pine seedlings, and a bird sanctuary was designed. Munich's forest, owned by a monastery during the thirteenth century, furnished income from timber, fuelwood, hunting, and house rentals. All revenues were applied to the expense of running a municipal hospital.[30]

Knowledge of German management practices in all categories of forests, as well as full cognizance of the ancient history of communal forests, arrived in America in the person of Bernhard Fernow. As inaugural chief of the forestry division, his principal mission called for installation of policy based on systematic management and supported by scientific principles. Although his primary accomplishments were national in scope, he also can be credited with formal introduction of communal forests to the American public. Use and stewardship of New England's local woods assumed new dimensions as a result.

Bernhard Fernow: Forestry in Every Town

With passage of America's Forest Reserve Act, the focus of debate on forests in this country switched from elementary questions about preserving public timberlands to more perplexing quandaries affecting management. Large-scale scientific forestry was a new idea to Americans, one that faced immense obstacles, and Fernow was well qualified for his assignment. Born to an aristocratic family, he had acquired an early interest in forestry while caring for his uncle's estate. Enrollment in Muenden's forest academy in western Prussia followed, and, upon completion of his formal education, Fernow worked as a forester in Silesia and Brandenburg. He subsequently married an American woman, Olivia Reynolds, and established a home in Brooklyn in 1879. Fernow quickly joined the American Institute of Mining Engineers, forming a friendship with its secretary, Rossiter W. Raymond. Through Raymond, he obtained employment with an iron manufacturing firm, Cooper Hewitt and Company, and managed its forest at Lehigh Furnace in Pennsylvania. This gave Fernow a chance to study American forest conditions, and he began to write. At the American Forestry Congress in Cincinnati, he presented a paper titled "Historical Sketches of the Development of the Forest Policy in Germany" before a section chaired by Franklin Hough.

Immediately after the Cincinnati and Montreal conferences, Fernow's status as America's only professionally educated and trained forester boosted his prominence. He continued writing and speaking in an attempt to guide the forestry movement toward practical objectives suited to American conditions, political as well as climatic. In 1884, he began publishing the *Forestry Bulletin*, intended as a voice for shaping American forest policy, but abandoned the project after its third issue. A short two years later, but by then a leading figure in American forestry, he was appointed by President Grover Cleveland to head the forestry division in the Department of Agriculture.[31]

In his new post, Fernow worked assiduously to introduce sound principles of forestry management. He set a policy permitting supervised cutting and sale of timber on public reserves but one that protected forest cover against a multitude of antagonists, chiefly fire; proper valuation of timber also was urged. He stressed the importance of conditions created by soil, temperature, air, atmospheric carbon, moisture and evaporation, humidity, and light or shade. Americans' lack of knowledge about these matters was all too apparent, a circumstance aggravated by vast regional variations. Fernow viewed forestry as a utilitarian enterprise, one whose primary objective was efficient use of soil for production of wood crops intended to command maximum revenue and one only incidentally concerned with aesthetics. Emphasis on several new fields of science, timber physics and wood technology, which promised to yield profitable information about uses of different wood types, was entirely consistent with his practical plan. By making technical information available to owners, he pursued a strategy intended to implement sustained-yield management during a period when economic conditions favored only rapid clearing from one forest to the next.[32]

Despite his understandable emphasis on national reform, Fernow did not neglect local conditions. His presentation at Cincinnati's 1882 congress explained evolution of state, private, and communal forests to an interested and surely inquisitive audience. That paper anticipated his lengthy "Report upon the Forestry Investigations of the United States Department of Agriculture," published in 1898, which included a discussion about origins of the Mark Society and communal ownership of forests associated with these early villages. With considerable insight, he remarked: "It is interesting to note that through all the changes of centuries, these so-called servitudes (certain rights in the substance of the forests) have lasted until our own times, much changed, to be sure, in character, and extending by new grants especially to churches, charitable institutions, cities, villages, and colonists."[33]

In 1890, Fernow contributed an editorial letter titled "Communal Forests" to *Garden and Forest*, a journal published by Charles Sprague Sargent, who headed Harvard University's Arnold Arboretum. In his letter, Fernow urged inauguration of a movement to establish community forests in America and referred to several German towns where profitable forests were managed. He wrote: "In Germany I know of communities where not only all taxes are paid by the revenue from the communal forests, but every citizen receives a dividend in addition." This statement, borrowed repeatedly in almost identical language, reappeared in countless journal publications throughout the first quarter of the twentieth century and became a key argument for those advocating creation of town forests.[34]

In that same article, Fernow selected the Sihlwald to demonstrate the long-term profitability of forestry management. Fernow was enthusiastic but practical in his evaluation of America's need for local forests:

> If every community will concern itself in the rational use of the land within its borders, if every town and every county will give profitable occupation to its waste lands by utilizing them for forest-growth, the movement would not only increase the financial prosperity of each community, the efforts of those who work for a rational forest-policy in the country at large would be subserved by every communal forest established. In fact, no better method of forest reform could be suggested than by beginning forestry in each town, which as a part of the country at large, will influence the movement of the whole.

In more cautious terms, he then warned: "It would be folly for the townships or counties to engage in forest planting, except with a full realization that they build for the future, and that no returns can be expected within a decade or more."[35]

Garden and Forest was an ambitious periodical devoted to horticulture, landscape gardening, and forestry, and it began circulation in February 1888. Published weekly from offices in New York's Tribune Building, the journal maintained freedom from trade influence and offered articles of international scope by individuals prominent in their fields. A limited number of subscribers, coupled with the absence of advertising revenue, caused its demise in 1897. Fernow and Gifford Pinchot were frequent contributors, and a broad range of topics related to forestry were presented, including several articles about community forests. In October 1889, nine months before Fernow's letter, an article about Lynn Woods in Lynn, Massachusetts, appeared. Although no longer common land, large areas remained undeveloped and were being returned to town ownership, primarily as a watershed forest and secondarily as a park. Fernow referred to Lynn Woods in his 1889 "Report of Chief Forester," noting that other communities could benefit from the city's example.[36]

Small Policy Hewed from Large Timber

Focus on proper management of the nation's timberlands—and an attending increase in public knowledge—inevitably influenced local events. Forestry science and study of regional ecosystems was pursued at forestry experiment stations and in private forests. Washington Vanderbilt's summer estate, Biltmore, near Asheville, North Carolina, became a model farm, game preserve, and arboretum. Gifford Pinchot, educated as a forester in Europe, was retained by Frederick Law Olmsted to manage the arboretum on recommendation of Pinchot's father, a wealthy lumberman and close friend of Olmsted. Pinchot's work at Biltmore was exhibited at the World's Columbian Exposition in Chicago in 1893, where forestry science was put on show for all America to see, and much of the country's populace did so. Fernow devised the exposition's program and with William Buchanan, chief of the fair's agriculture department, arranged a series of short, thirty-minute lectures—"practical talks by practical men upon practical topics," as Buchanan put it. Buchanan's list of seventeen topics was international in scope; two, "Woods of Europe" and "Government Legislation and Control of Forests," might logically have referred to communal woodlands.[37]

Competent management of forests necessitated formal education. The first forestry professionals in America—Fernow, Pinchot, Carl A. Schenck, Henry Graves, Filibert Roth, and others were trained in Europe. However, programs in forestry began to surface in America toward the close of the nineteenth century, first at Biltmore in 1897, headed by Schenck, and at Cornell in 1898, where the New York State College of Forestry was directed by Fernow. Other programs soon appeared at Yale in 1900, headed by Graves; at the University of Maine in 1903; and at Harvard in 1904. Graduates of these schools allegiantly disseminated forest gospel to state and local governments. Need for management plans establishing rules for cutting, protecting young growth, and preventing fires became widely recognized.

Fernow's goal of a three-tiered policy based on federal reservations, forestry science, and regulatory management was finally realized in 1897. A year earlier, the National Academy of Sciences, a counseling body for the government on matters of scientific import, issued a report calling for a rational forest policy. Addressing the government's inability to manage public reserves, their report recommended a national forest service patterned after that of Germany. The secretary of the interior would be given the power to regulate certain areas of concern, such as forest growth, methods of protection and improvement, cutting, timber sales, and use for water power, irrigation, or mining. The report also called for thirteen new forest reservations. During the final days of his administration, President Grover Cleveland complied by proclamation, and a bill assigning regulatory powers and appropriating funds was later enacted. Fernow's departure from Washington for academic responsibilities in Ithaca, New York, paved the way for Gifford Pinchot's appointment as chief forester in 1898.[38]

Pinchot's intuitive political abilities transformed the fledgling forestry division into a well-organized, better-staffed, and adequately funded body. He became a friend and key advisor to Theodore Roosevelt, elevating the forestry division's status to cabinet equivalency during Roosevelt's administration. The Division of Forestry became the Bureau of Forestry in 1901, and in 1905 the U.S. Forest Service was born. Under Pinchot, policy was built on Fernow's carefully placed foundation. Sustained-yield management, forestry science, public education, and utilitarian goals were its cornerblocks. Many of Pinchot's exploits unfolded around conservation, a password during Roosevelt's years in office. Economic interest groups competing for a share of natural resources pushed from one direction. From another pressed those who argued for more chaste policies of wilderness preservation and recreational use. In conservation, Roosevelt and Pinchot wielded a political sword in a forest battleground where tactics demanded wise and productive use of the nation's resources. The weapon was a convincing one in a progressive era fueled by antitrust sentiment.[39]

Although Pinchot, too, was familiar with European communal forests, he was not as staunch an advocate for municipal forestry as Fernow. He had studied in France at L'École Nationale Forestière in 1889 and had observed forest management in the Sihlwald for a month during 1890. Results of his observations in Switzerland were published in a series of three articles entitled "The Sihlwald," which appeared in *Garden and Forest* during the summer of 1890. He emphasized its management and working plan based on a compartment system and described its transportation network; the final installment addressed origins and productivity. In a brief concluding paragraph, Pinchot questioned whether similar city forests would be successful in America. Noting their justification as parks and sources of revenue as well as for their educational benefits, Pinchot was nevertheless cautious about their prospects in this country: "The value of the Sihlwald to the City of Zurich is great and unquestioned, but whether the system might be advantageously transplanted to America, local circumstances must go far to decide."[40]

Shortly after returning from his studies in Europe, Pinchot presented a paper at a meeting of the American Forestry Association in Washington in December 1890 on the topic of European forest policy, and an abridged text of his paper was reprinted in *Garden and Forest*. He addressed communal forests of Germany and France but emphasized their extensive management by state governments. In certain cases, town activity was limited to selling timber cut under supervision and paying for plantings and other improvements. Pinchot anticipated American reaction to this type of governmental interference, noting:

> But however galling so extensive an interference with the rights of property may appear, it is none the less unquestionably true that in France, as well as in Germany, the state management of communal forests lies at the root of the prosperity of a very large proportion of the peasant population, and the evils which have attended its withdrawal in individual cases are notorious.[41]

As these writings suggest, Pinchot's concerns and policies lay well outside the boundaries of small public forests. Unlike's Fernow's annual reports to the secretary of agriculture, Pinchot's were completely silent on the subject of communal forestry. Pinchot's concern for national issues, his focus on western forests, and his attitudes about federal or state intrusion in local affairs—and the political arrows predictably loosed—steered the Forest Service away from municipal forestry. That course did not alter until a community forestry program was begun during Franklin Roosevelt's administration. However, other events during Pinchot's years in Washington portended significant activity in town forests.

The National Conservation Commission

In the spring of 1908, concern for formal policies affecting the nation's natural resources prompted Theodore Roosevelt to convene the nation's governors.

The National Conservation Commission sprang from this gathering, and Pinchot was appointed chairman. Within a year the group outlined a plan to inventory the country's natural resources, called for preservation of public forests through practical forestry methods, and recommended creation of state forests. In preparing its report, the commission heard papers covering a variety of topics, two of which considered community forests. Raphael Zon, a Russian-born forester in Pinchot's ranks, discussed timber sources in foreign countries and reviewed communal forests in France, Switzerland, Germany, and Austria.

A second paper, by forester Samuel T. Dana, considered European methods of forestry administration and their suitability for America. Dana focused on the broad role of Germanic states in management of their community forests and observed similarities to America's federal government. Noting the high yields of many communal forests, a fact he attributed to state management, Dana concluded that a program for community forests in America could be just as successful:

> Communal forests are a natural result of the development of property conditions in Germany and are a European institution which has no counterpart in this country. They are an important source of revenue to the community, sufficient in some cases not only to make taxation unnecessary, but even to yield a surplus to be divided among the citizens. Although circumstances have not led to the establishment of communal forests in the United States, there is apparently no good reason why they should not be established with equally satisfactory results.[42]

Although no formal policies materialized in Washington as a result of Dana's recommendation, similar optimism appeared in many state programs shortly after.

New England Woods

Undeniably, the mixture of federal, state, and local influence on growth of municipal forestry in New England was an uneven blend. Although national events were important, state and local strides were gaining better ground. In Massachusetts, enabling legislation permitting towns to acquire and manage woodlands had been passed in 1882, and similar laws followed in New Hampshire, Vermont, and Maine by 1915. The region's long and intimate relationship with its forests was in good measure responsible for the vigorous commitment to timber conservation that emerged, and municipal forestry sprouted as part of that movement. Support materialized from various sectors. Effects of poor forestry management became apparent to increasingly larger population segments as timber shortages hurt local economies dependent on wood-products industries. In an opposite corner, regional interest in cultivation of trees matured from a desire to use forests for recreation. Others found sufficient reason between.[43]

Forestry Commissions and Legislation

Forest conservation evolved through a great variety of initiatives in New England, both public and private. In 1869, five years after Thoreau's *Maine Woods* appeared in print, Maine's legislature appointed a commission to study the state's forestry practices. New England's remaining states fell quickly into line: Connecticut in 1877, New Hampshire in 1881, Vermont a year later, and Massachusetts by 1890. Early commissions were temporary ones, created primarily to gather information about the extent and condition of forests or the influence of forest destruction on rainfall and, in turn, on ponds and streams. Most were asked to consider legislation protecting woodlands from fire or indiscriminate cutting, and their reports spawned laws of assorted shapes and sizes. Commissions were soon a permanent part of state government and received expanded assignments; reports gradually became dependable biennial publications.[44]

Remarkably, Maine's first report in 1891 contained an article encouraging town forests through legislation requiring tax-forfeited lands to vest permanently in towns. All such lands would then be devoted to growth of timber and wood. Authored by George F. Talbot, a Portland judge and historian, the paper focused on questions about Maine's expanding waste lands, methodically stripped of their forest growth. Talbot included a draft bill that prohibited any expenditures for care or planting of forfeited lands in excess of income generated, or 10 percent of tax income produced. Management was to be controlled by town selectboards, who would report to the state's forest commissioner once every two years. Success in one village would lead others to follow, and Talbot foresaw an awards program for the best-managed forest, with prizes going to enlarge or improve winning examples. Although his bill was not enacted, Talbot's article represented the first formal proposal for a town forest program in New England.[45]

Other commissions approached the subject less directly. In 1893, New Hampshire's commission was charged with investigating original and secondary forests, ascertaining removal and disposition of timber (both domestic and exported), calculating revenue derived from forest products, and evaluating "different methods of lumbering pursued, and the effects thereof upon the timber-supply, water power, scenery and climate of the state." Significantly, the commission was authorized to acquire forest lands for public parks and given power of eminent domain if agreement concerning price could not be reached. Unfortunately, the effectiveness of this bill was negated by the fact that only private, not public, funds were permitted for the acquisition of public forests.[46]

The creation of permanent forestry commissions was followed by appointment of state foresters, first in Maine in 1891 and in remaining states between 1903 and 1909. State nurseries, desperate campaigns against forest fires, publications urging prudent management, and other trappings of sound forest policy steadily surfaced from organized forestry agencies within state government after the turn of the century. State ownership of forest lands became part of this policy, and states began to acquire large tracts of land as public reserves and parks.[47]

Private Initiatives

Individual experiments also helped to advance conservation and forestry science in New England. Although direct connections to municipal forestry are rare, the results of these demonstrations were often publicized in government reports or received local attention. As such, they contributed to a general public awakening on matters of forestry. Seeking to increase knowledge of tree planting among working farmers, the Massachusetts Society for the Promotion of Agriculture awarded a series of prizes in 1876 and again in 1878 for private plantations of trees. Competitions focused on experiments for improved arboriculture, not necessarily reforestation. However, examples inevitably contributed to the latter, as where an award was given for successful and inexpensive plantation from seed on land too sterile for agriculture or profitable pasturage.[48]

An increasing number of small, private plantations by farsighted individuals appeared after the Civil War. Many were begun simply to put wasteland to commercial advantage and were started with seedlings dug from surrounding fields. Some were begun by wealthy individuals who simply desired extensive forests on their estates. A few reflected genuine interest in conservation. In Vermont, Frederick Billings returned to his childhood town of Woodstock as a successful lawyer and president of the Northern Pacific Railroad. At the former home of George Perkins Marsh, acquired by Billings in 1869, he established a model farm and began planting an experimental forest of white pine, Norway spruce, white ash, European larch, and several other species. Bending to the task with his men, he reforested a large portion of Mt. Tom overlooking the Ottauquechee River in central Vermont. In 1884, he hired George Aitken, a Long Island breeder of Jerseys, to manage the project, and both farm and forest prospered as a result. Billings's personal interest in forestry carried him into the public sector, and he served on Vermont's first forestry commission in 1893.[49]

Private nurseries sometimes capitalized on reforestation projects or made planting stock available to village improvement societies intent on furnishing shade trees for streets or parks. Jacob Manning's nursery in Reading, Massachusetts (fig. 3.4), was established in 1854 and supplied a number of plantations with hemlock or white pine seedlings, including Seward Webb's farm at Shel-

Fig. 3.4. An 1899 view of Jacob Manning's commercial nursery in Reading, Massachusetts, from a printed advertisement. Manning, who established his landscaping business in 1854, worked on a number of important commissions for wealthy patrons, including a plantation of pine and hemlock at Seward Webb's farm in Shelburne, Vermont. U.S. Forest Service Collection, National Archives

burne, Vermont. Manning acquired more than a passing interest in forestry and attended a number of important conferences, including the American Forest Congress in Montreal in 1882.[50]

Forestry Associations

The formation of nonprofit forestry organizations served as a catalyst for policy-making at the turn of the century and helped to launch town forest programs. The Connecticut Forestry Association, chartered in 1895, was the first, followed by the Massachusetts Forestry Association (MFA) in 1898, the Society for the Protection of New Hampshire Forests (SPNHF) in 1901, and the Forestry Association of Vermont in 1904. Both the MFA and the SPNHF were very influential in New England's town forest movement, particularly the former through efforts of Harris A. Reynolds, association secretary from 1911 to 1953.

Founders of the Massachusetts body were a visionary group of nine men, among them Allen Chamberlain, its first secretary, and Frederick Law Olmsted, Jr. Chamberlain, a journalist and writer who began his career with the

Boston Herald in 1889, became active in a number of conservation organizations in New England and regularly contributed articles to their bulletins and reports. The association was created to foster careful woodland management, educate the public on matters affecting forests, promote afforestation of unproductive lands, and encourage the planting and care of shade trees. Many of these goals were pursued through vigorous lobbying before the Massachusetts legislature. Regularly published bulletins addressing a wide range of topics related to forestry augmented the group's efforts.[51]

Town forests and appointment of town foresters became principal objectives for the association after Harris Reynolds was hired (fig. 3.5). Reynolds, a native of West Newton, Pennsylvania, acquired a civil engineering degree from West Virginia University and then completed Harvard's graduate program in landscape architecture in 1911. Almost immediately he began producing a steady flow of articles about town forests; the first, written in 1913, was titled "A Citizens Movement to Reduce the Tax Rate." In it he argued that towns could substantially reduce their taxes by planting wastelands with trees. In August of that same year, Reynolds and his bride of German descent, Alice Hecker, traveled to Europe, where he combined, apparently with success, a honeymoon and a study of European community forests. William P. Wharton, a member of the association's executive committee and later its president, joined Reynolds in Europe for that study.[52]

The SPNHF, established to counter the effects of unrestricted logging on private lands, became a counterpart to its Massachusetts neighbor. It was founded by a group of influential political figures who advocated public acquisition of forests and important scenic areas. Many of its early members and officers (Allen Chamberlain, for example), were affiliated with the MFA. Particularly alarmed at the devastation wrought by clear-cutting on steeply sloped lands in the White Mountains, the society immediately became instrumental in garnering political support for the Weeks Act, passed in 1911, which authorized federal acquisition of the White Mountain Forest Reserve.[53]

The society also encouraged public acquisition of forest lands by New Hampshire's timber-wealthy communities. In November 1901, a year after its inauguration, it appointed Philip Ayres as its forester. Ayres was a remarkable individual, whose studies in philosophy and history, first at Cornell and then at Johns Hopkins, led him to charitable and social work. In 1900, at age forty and plagued by ill-health, he returned to Ithaca to spend a year at the New York State College of Forestry at Cornell under the tutelage of Bernhard Fernow. As the SPNHF forester, he actively assisted towns in the acquisition and management of forest lands, and the group's annual reports monitored the progress of town forests from an early date. In a 1909 report, Ayres advocated public ownership of land, predicting that New Hampshire's plan for state aid to town forests, similar to their practice of aid for town roads, offered an incentive. He

Fig. 3.5. Harris Reynolds, secretary of the Massachusetts Forestry Association (later the Massachusetts Forest and Park Association) from 1911 to 1953. Reynolds generated a prolific amount of writing about town forests and, through his work with the MFA, was a singularly important figure in the movement for town forests. He also founded the New England Forestry Foundation in 1944. Private Collection, Clinton Reynolds

remarked, too, that town forests in Germany, France, and Switzerland were sources of considerable profit. A year later, the society's annual report listed eighteen New Hampshire towns as owners of forest lands, most associated with watersheds.[54]

Both the MFA and the SPNHF encouraged formation of local forestry organizations. Aware that public education was critical to the success of much-needed forestry legislation, both groups focused on local forestry instruction, and the Massachusetts group successfully founded numerous branch associations. Property taxes, levied by towns and paid yearly based on land values, were a singular disincentive to long-term forest management as timber value matured. Acquisition followed by rapid clear-cutting proved most economically efficient, and local support for corrective legislation was needed. The SPNHF devoted its greatest energies to that task and to state acquisition of critical forest areas. It played a less active role in promoting town forests than did the MFA, partly because of the intense lobbying and fund-raising required to fulfill its policy goals and partly because of Harris Reynolds's strong commitment to town forests. Ironically, an SPNHF pamphlet titled "Forests for Towns and Villages" was more influential in Massachusetts than in New Hampshire and was later credited by Reynolds for inaugurating the town forest movement in his state. Reprinted by Reynolds's organization in 1912, it contained eight short articles by George H. Maxwell, a California lawyer and water specialist who formed the National Irrigation Association in 1899 and the National Reclamation Association in 1912.[55]

Maxwell was an enigmatic figure who strenuously promoted government regulation of rivers and waters to achieve reclamation of arid or wastelands through irrigation and flood control. He also served on a number of flood commissions and supported legislation to prevent speculative concerns from obtaining extensive water rights in western states. The core of his philosophy seemed to be an idealistic, pastoral image of man toiling to extract earth's bounty, a movement identified during that period as homecrofting. Gardening, husbandry, forestry, and irrigation all contributed to a broad scheme for agrarian self-reliance summarized by his rough-edged catchphrase: "Get an Acre and Live on It." Not surprisingly, municipal forests fit neatly into his vision for America. From Chicago in 1905, he began publishing *Maxwell's Talisman*, an obscure monthly journal described as the "Homecroft Magazine and Journal of Social Economy." [56]

In "Forests for Towns and Villages," Maxwell extolled potential benefits of municipal forests for the entire country and commended Pennsylvania's 1909 municipal forest law permitting its towns to acquire and manage forests for production of revenue. His fifth and sixth articles referred to German cities (Baden, Freiburg, Heidelberg, and Villingen) identified as communities of sizable populations whose forests produced a substantial dollar yield per acre. Smaller villages also were considered. With a population of 220, Aufen owned a forest of 163 acres, gave each of its citizens a yearly quota of fuelwood, and sold a sufficient amount of timber to meet all annual expenses. Although Maxwell possessed an obvious flair for promoting ideas, he remained a very minor figure in the movement for town forests. Nonetheless, "Forests for Towns and Villages" served as an important writing. Its wide circulation in New England by the region's two most influential private forestry groups occurred at a time when little else had captured a similar audience.[57]

Municipal Forests outside New England

Maxwell's contribution is noteworthy in another respect. Ironically, he wanted to address a midwestern and western audience and specifically congratulated the town of Grand Forks, North Dakota, owner of a small plantation used as a city park. However, his was only one of sundry articles about municipal forests that had begun to appear around the country by that time. Municipal forestry developed amid a national clamor for reform in forest use and became popular in midwestern and western regions as well as in eastern states. For example, in 1910, San Diego set aside three thousand acres of unproductive, steeply graded land for forestry and appointed a city forester to manage the lands.[58]

Pennsylvania and New York were particularly active in the national movement. Both states were quick to enact enabling legislation—Pennsylvania in 1909 and New York in 1912—and their state forestry associations were among the first to encourage community forests. Pennsylvania's forestry association began publication of *Forest Leaves* in 1886, and numerous articles about municipal forests have appeared among that bulletin's pages. In New York, an 1890 article in *The Rural New Yorker* offered community forests as a solution to problems caused by unused pasture or arable and suggested practical means for acquisition, planting, and management. Its author concluded by arguing that communal forests were a thousand times better than wild blueberries, unpaid taxes, and wretched moaning about abandoned farms.[59]

At Cornell University, New York's State College of Agriculture and State College of Forestry also played a significant role in publicizing municipal forests. In 1909, Raphael Zon endorsed communal forests in a monograph for the exhaustive *Cyclopedia of American Agriculture*, compiled by Liberty Hyde Bailey. Zon's writing continued themes he had voiced before the National Conservation Commission that same year. He accurately traced communal and city forests in Europe to their diverse origins, including the Mark Society, evolution of German city-states, feudal land tenure, and formal town planning. He also evaluated their extent, distribution, administration, forest yield, and financial returns, remarking, as had Fernow and Pinchot, that Germany's success hinged on state management. Several years later, Arthur B. Reckgnagel, a professor at the State College of Agriculture, prepared a reading course titled "County, Town and Village Forests" as part of that school's farm study program. The innovative roles of Pennsylvania and New York in a national movement for town forests are exemplary. However, New England's politically autonomous towns, with ancient traditions of communal and town-owned forests, seized the idea of municipal forestry in a way that eventually outlasted similar efforts in other regions.[60]

Independent but Not Isolated

Predictably, forestry reform in New England followed an independent and often resourceful course. Self-reliance notwithstanding, individuals and fledgling associations energetically sought forestry advice wherever it was available. For their part, professional foresters realized that science would remain little more than "book forestry" without demonstrations showing profit as well as practicality. On Pinchot's directive in 1898, the Division of Forestry established an assistance program for lumbermen, farmers, and other forest users. Initial inspections were free to woodland owners; management recommendations were given on small parcels, also at no charge; fees for larger tracts requiring plans were set at cost. A program summary, *Circular 21*, was distributed to those who inquired. Although aimed at private woodland owners, the campaign by chance sometimes found its way to public lands and thus embodies Pinchot's most noteworthy contribution to municipal forestry. For instance, the

Pennichuck Water Works in Nashua, New Hampshire, received assistance, and Maine's Bar Harbor Village Improvement Association obtained advice about uncontrolled cutting of firewood on Mt. Desert Island.[61]

By far, the program's most laudable brush with communal forests occurred on Brunswick Commons. In 1900, town members voted to improve eight hundred acres remaining from the Pejepscot Proprietors' legacy. For years, young pine growth and scrub had been burned over regularly to accommodate an annual pilgrimage by squatters in search of blueberries. Heeding a request by the village improvement society, James W. Toumey, supervising forester for tree planting under Pinchot, examined the soil and concluded that it would support a valuable growth of white pine as long as fire could be controlled. The common was divided into three parts (one indulging the squatters) and seedlings were set in bunches, four feet apart. Pitch pine was allowed to seed naturally, to be overgrown eventually by white pine. Nurseries for pine and oak were begun during 1901. By end of the second year, the forest contained 5,500 white pine seedlings, most obtained from old fields or roadsides; beds of a thousand red oak seedlings were ready for transplanting; and a nursery of 75,000 yearling pines, some of which the town planned to sell, was in place.[62]

Planting on Brunswick Commons was accomplished under the supervision of Austin Cary, a rough-hewn and extremely capable forester. Cary's self-acquired knowledge of silviculture served northern New England's forests well at a time when timber companies scoffed at suggestions of forestry management. Born in East Machias, Maine, in 1865, Cary developed an interest in the natural sciences and received degrees from Bowdoin College in 1887 and 1890. While on a fishing expedition in Maine in 1892, he met Bernhard Fernow, who persuaded him to become a professional forester. A year later, he accepted a position with the forestry division and in 1896 traveled to Europe to observe German forestry practice. Soon after his return, Cary began work for Berlin Mills Company in New Hampshire.[63]

Probably through his association with Fernow or his study in Europe, Cary became aware of German communal forests. In his report delivered to Brunswick voters at their town meeting in 1901, he described planting progress during the preceding year. Commenting on the novelty of town forests in America, he observed that such forests were customary in Germany, where they often yielded revenue sufficient to eliminate taxes. Cary also predicted that Brunswick's wasteland would soon be converted from eyesore to forest park. In a second report, issued in 1902, he recommended selection of a special committee (partly from the farming community) to protect against fire, plan improvements, and supervise harvesting of wood and valuable blueberries. Considered together, Cary's reports are important because they sketch, in rough form, essential parts of town forest planning. Moreover, they were prepared at a time when little else had been written about the subject in this country.[64]

Brunswick Commons is also significant as one of the earliest forests planted entirely from open land unrelated to watershed protection. Credit belongs to Cary and to town members, some of whom were associated with Bowdoin College. Important characteristics of Brunswick's forest—planting of wasteland with primarily coniferous types, encouragement by reference to successful German community forests, a promise of future revenue from wood harvests, appointment of a management committee, and attention to aesthetic and parklike qualities—appeared in a majority of town forests created during the first quarter of the twentieth century.

Two states away and by quirk of fate, Gifford Pinchot endowed the founding of a large church forest in Westmore, Vermont, during the same year that Brunswick Commons was being planted. Westmore is situated near the northern end of Lake Willoughby, a long, narrow, and especially scenic finger of water that pokes its way between two mountain peaks in the state's northeastern corner. The village's first pastor and founder of its Congregational church, Charles O. Gill, had been a classmate of Pinchot's at Yale. During a visit to Westmore by Pinchot, Gill persuaded his friend to contribute toward the purchase of a church woodlot. Gill subsequently acquired almost four hundred acres and then donated it in 1900. His conveyance included all of Lot 29, together with part of Lot 44, in the town's first division of lands (fig. 3.6) and was conditioned upon management according to forestry division regulations, with income to the church treasury. The Gill/Pinchot grant was combined with adjoining Lot 43, the town-owned Minister's Settlement Lot, to create a church forest of nearly seven hundred acres.[65]

Perplexed by fickle timber markets, scarcity of loggers, and deep snows, church members resolved in 1904 to place all lands under management by Vermont's Domestic Missionary Society but still subject to Gill's qualification. Proceeds were held in trust to sustain an annual grant of two hundred dollars for the church or its pastor, and rights were reserved to cut fuelwood or lumber for building repairs. Church officers also wrote to the Department of Agriculture and received a copy of *Circular 21*. Despite their reluctance to delve into forest management, church trustees nevertheless approved periodic timber sales and used lumber from their forest to construct a parsonage and community house. In 1934, birch, maple, and white ash were sold, and provision was made to mill additional stands at least ten inches in diameter with precaution against injury to spruce or young timber. Management authority was finally returned to the church in 1979 after interest in preserving the forest was renewed and the resolution of 1904 rediscovered. Town ownership of Lot 43 was conceded, but lumber remains available to the church upon request to a town committee.[66]

Prominent foresters also were frequent speakers at gatherings sponsored by concerned groups in New England. In 1904 and 1905, the Forestry Association of Vermont conducted a series of four public meetings and invited Fernow

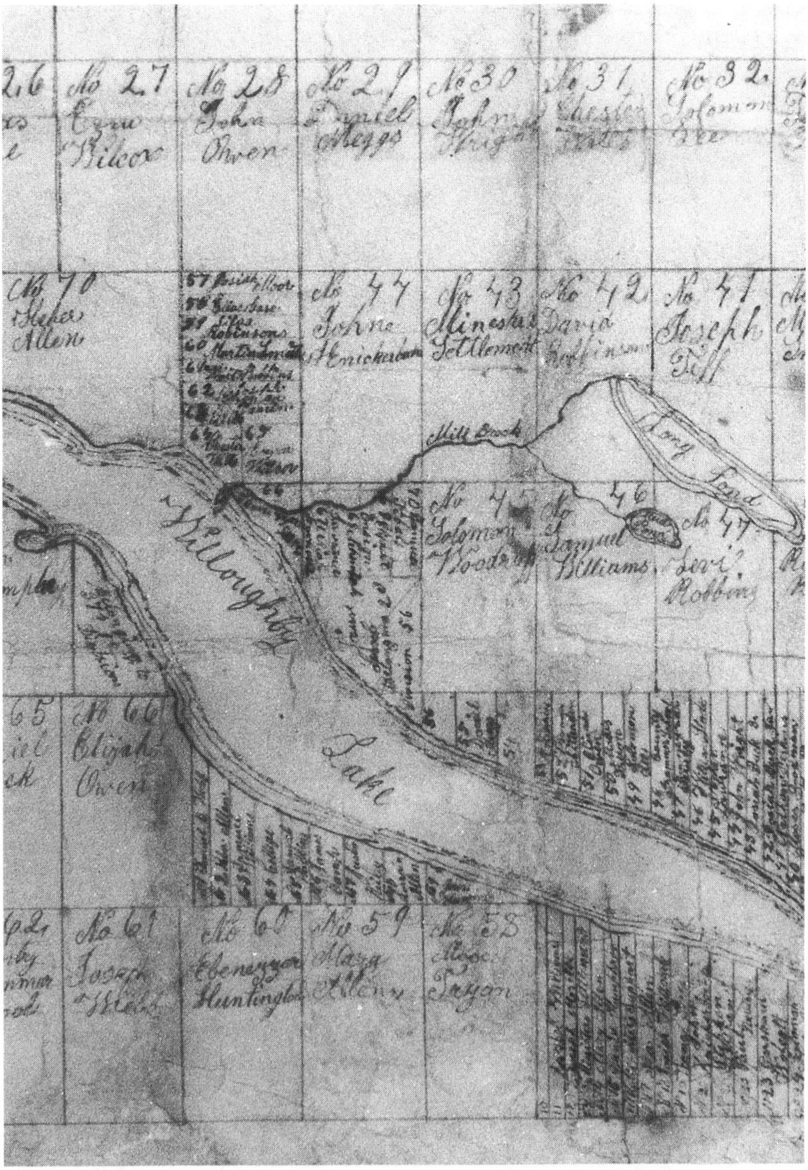

Fig. 3.6. Portion of an undated proprietor's plan of Westmore, Vermont, showing Lots 29 and 44 as well as the Minister's Settlement Lot, No. 43. Assisted by a financial donation from Gifford Pinchot, Westmore's Congregational Church was able to acquire four hundred acres of woodland. When combined with privileges to cut timber from the Minister's Lot, now owned by the town, the church forest encompasses nearly seven hundred acres. Office of the Westmore Town Clerk

Fig. 3.7. Bernhard Fernow at the Red Deer Ranger Station in Bow River Forest, Alberta, Canada, c. 1915. As chief of the Department of Agriculture's Forestry Division between 1886 and 1898, Fernow introduced community forestry to America. He later developed forestry programs at Cornell University and the University of Toronto. Forest History Society, Durham, North Carolina

(twice), Pinchot, and Henry Graves to speak. Fernow, more outspoken since leaving his job in Washington, addressed a Montpelier audience on the topic "Fundamentals of Forestry." He enumerated eight principles essential to effective policy and suggested necessary steps. A year earlier, he had given a similar speech to the Massachusetts Board of Agriculture. Fernow argued strenuously that "the forest, like all life, is preserved not by non-use, but by regeneration, a succession of generations," and he regarded attempts to prohibit cutting in the name of forest preservation as puerile and ignorant. His other axioms addressed the importance of local conditions, production of quality wood crops, recognition of the length of time needed to produce harvests, and the need for present expenditure to obtain future income. He also emphasized use of land unfit for agriculture, the inability of the private sector to sustain management of forests, and protection from fire. As part of state forestry policy, he recommended improved fire laws, appointment of state foresters, financial aid to forestry organizations and educational institutions, and state acquisition of wastelands to be held as forest reserves under proper management. In addition, he encouraged assistance to private woodlot owners, improved tax laws, and, finally, support of town forests through state loans and management assistance.[67]

Although no longer a government policymaker, Fernow remained an important figure in American forestry (fig. 3.7). Many of the principles and policy goals outlined in his speaking engagements and expansive writing were closely observed by state forestry programs. Town forests were an integral part of his broad plan to regenerate the country's woodlands and were promoted as revenue-producing ventures. In 1913, by then founder and director of the University of Toronto's forestry school and president of the Society of American Foresters, Fernow addressed the SPNHF annual meeting in Sunapee, New Hampshire. He called for a large-scale planting program to meet future demand for wood and timber and assigned special obligations to federal, state, and municipal governments. States, he argued, needed to develop assistance plans for towns. Plant materials, expert advice, and credit to pay for planting of wastelands, were all critical. The federal government, in turn, would provide state loans. Fernow remarked that in New England, municipal enterprise offered the most promising course for regeneration but he cautioned that state control might be preferable.[68]

Enabling Legislation for Municipal Forests

Proponents of municipal forestry also faced questions about the legal authority of communities to acquire lands for production of wood crops and revenue. Generally, American towns and cities obtain their authority to exist, legislate, govern, and tax from state constitutions or laws. The extent and limits of their power are thus defined, and authority, unless specifically granted, remains with the state. True, New England villages had owned and maintained woodlots for centuries. Nevertheless, these lands often had been donated or acquired pursuant to a charter for the support of school, church, or minister. Rents or harvests of fuelwood and timber were generally devoted to these ends. In odd cases, towns had acquired lands by gift or through tax forfeiture and had little difficulty putting them to public use. Still, purchase of real property to gain income from the sale of timber, fuelwood, or other products was, absent a specific state law, beyond the scope of municipal authority.

Following an appeal by Governor John Long in his annual address in 1882, Massachusetts became the first state to confront this problem. That year, legislation enabled towns to purchase land and place it in the public domain to preserve, reproduce, or promote culture of forest trees for wood and timber or to protect water supplies. Acquisition was permitted by direct purchase or through eminent domain, but title was vested in the state as trustee for towns. Administrative chores were given to the State Board of Agriculture, which had power to adopt management regulations, appoint a keeper, sell forest products, lease any existing buildings, construct new buildings for recreation, and balance books. Funds cleared from wood and timber sales were paid to towns. For their part, towns were required to appropriate an amount sufficient to meet acquisition expenses and were allowed to issue bonds to cover debt.[69]

Backing for this legislation stemmed in large part from interest in Boston's plans for a metropolitan park system. Surrounding towns were quick to recognize the benefits of preserving woodland or planting lands for park purposes. However, the law implicitly recognized the potential for timber-crop management by towns and acknowledged an important relationship between forests and protection of water supplies. It also received public attention from the forestry community outside New England when an article applauding its enactment and quoting its full text appeared in Franklin Hough's *American Journal of Forestry* in 1883.[70]

A number of towns promptly responded and began successful acquisition projects. In Middlesex County, several towns joined resources to purchase four thousand acres of barren, open land known as Middlesex Fells, which contained large reservoirs. A group of trustees was appointed to accept donations, with the hope of returning the Fells to pine forest. Lynn Woods also was acquired under authority of this legislation, sometimes referred to as the Public Park Act of Massachusetts. Any ambiguity about distinctions between park and forest or about the ability of towns to hold unfettered title to revenue-producing woodlands was finally resolved in 1913. That year, Massachusetts amended its law and permitted towns to own forest lands directly and to administer them without supervision by a state agency. Communities were given the choice of appointing a city forester or assigning that task to the state forester. A month earlier, New Hampshire had adopted legislation permitting towns to purchase, manage, and improve lands for forestry purposes but placed responsibility for management under direction of the state forester. Proceeds from the sale of wood, however, belonged to town treasuries.[71]

In 1915, Vermont joined the group with a law that established legal authority for towns to acquire, manage, and improve lands for wood and timber. It also created a special category of forests intended to support local schools. State foresters were required, upon request by a town, to examine parcels of land larger than forty acres and to designate suitable tracts as school-endowment forests. Management of school forests remained beyond town control, and only trees bearing the state mark could be cut. Sale among towns of up to 150,000 seedlings from the state nursery was authorized annually, and towns were required to file yearly reports of income to the state treasurer. The task of protecting school forests, however, was given to local forest fire wardens. With enactment of town forest enabling laws in New Hampshire, Massachusetts, and Vermont, New England's movement to establish municipal forests had taken a beginning step. Simultaneously, categorization of town forests as units of woodland devoted to production of timber received a strict statutory definition. The classification survives today and distinguishes town forests from other forms of communal woodland.[72]

A Blueprint for Town Forests

Although only two years separated Vermont's enabling law from those of Massachusetts and New Hampshire, by 1915 the town forest movement was well underway in the latter two states. Preparations to launch this movement had already begun before enactment of town forest laws, and both the SPNHF and the MFA had encouraged formation of branch associations as a likely means for generating community interest in forestry. The SPNHF's annual report for 1911 discussed objectives of local organizations, including plantations on wastelands and acquisition of town forests through gift or purchase. The biennial report of the New Hampshire Forestry Commission for 1913–14 considered the state's new municipal forest law but noted that extensive forest lands already were owned by Newington, Manchester, Concord, and Franklin. The report also recommended that the state forestry department provide free consultation to towns and offer trees for planting town-owned land at no cost.[73]

In 1913, Harris Reynolds revealed the MFA's carefully prepared blueprint for municipal forests in an article for *American City*. He reasoned that forestry education for the general public offered optimal solutions for conservation problems, and he outlined a plan to achieve that end. Several years earlier, the association had divided the state into twelve districts and had placed a trained forester in each to solicit members and generate formation of branch associations. Foresters received commissions based on membership dues and provided forestry consultation at no charge to town residents. Now, branch associations were encouraged to champion proper forestry management at every opportunity and to promote reforestation of water supply areas, reclamation of wasteland, and planting or protection of shade trees. The plan's critical part, however, required associations to prod local officials into creating an office of town or city forester. Reynolds maintained that appointment of trained foresters in each town would cause municipal forests to multiply rapidly in number.[74]

The office of city forester was an ambiguous one from birth, due in part to its considerable responsibilities outside strict forestry practice and in part to some difficulty in distinguishing town forests from town parks. Reynolds contributed to this ambiguity in an article titled "An Opportunity for the Young Landscape Architect," also written in 1913, which appeared in *Landscape Architecture*. Conceivably recalling his own quest for employment, Reynolds contended that the term "city forester" was a misnomer and argued that landscape architects were more qualified than commercial foresters to fill the office. City foresters needed proper training in the planting and care of shade trees, nursery and greenhouse management, road location, and aesthetic qualities of trees and park design, and this education was generally unavailable from forestry schools. He acknowledged that the curriculum in landscape architecture schools was also weak in certain areas but could be adjusted more easily than in forestry programs. Finally, he called for both working forestry plans and landscape designs for all municipal forests.[75]

Reynolds, trained as a landscape architect, was obviously writing to a specific audience and may have been somewhat biased. His later articles did not repeat these arguments, and the matter of landscape designs for municipal forests surfaced only occasionally. If one ignores an inclination to focus on his comparison of forestry and landscape architecture schools, the core of his argument reveals the origins of parallel development between commercial and recreational use of town forests. Reynolds explicitly predicted that cities and villages would soon establish forests, by name implying a commercial use, but would consider them to be outlying parks as well. He also predicted that recreational use would ultimately play a very significant role in maintenance of these forests. To Reynolds, the value of city foresters lay "in the certainty that these outlying forests are from the start wild parks, which, with the growth of the community must become real parks."[76]

Reynolds's predictions have proved to be more accurate in regions where metropolitan growth has been substantial. In many other parts of New England, distinctions between town forest and town park remain clear. In such cases, commercial logging continues to be a primary use, often to the exclusion of other activity where lands are inaccessible. The significance of Reynolds's article for *Landscape Architecture* lies in its clear identification of these two primary uses on the eve of New England's movement for town forests. Commercial and recreational use have vied for center stage throughout this movement and continue to do so.

• • •

Events that shaped policy reforms in America's forests during the late nineteenth century were also precursors of important community trends. Town forests, enabled by state legislation, soon became New England's dominant category of local woodland. The term frequently included watershed lands and, just as often, was applied broadly enough to encompass forest parks. Whether for commercial harvest, protection of water supplies, recreation, or some combination of all three uses, the town forest movement pursued a single objective: cultivation of trees for production of wood and timber to common benefit.

The tenor of the period also guided the character of stewardship in New England's town woods toward a different bearing. Heightened levels of public responsibility, professional management standards, and levels of productivity displayed by German and Swiss communal forests, all fostered a different mood. In local forests, issues such as sustained yield management and multiple use began to reflect, albeit on a smaller scale, problems that preoccupied the nation's conservation policymakers. Relationships between forests and ecosystems received greater emphasis. Perception of loss now extended far beyond the simple economic or material need that had long influenced collective stewardship in New England towns, and communities sensed the transformation. Finally, as towns gradually acquired more land to accommodate different forest uses, community structure began to change as well.

Chapter 4

Forestry for the People: The Town Forest Movement in New England

In a paper titled "Forest Revival in New England," which appeared in a 1916 issue of *Forest Leaves*, Harris Reynolds concisely portrayed the mission of the Massachusetts Forestry Association (MFA) as that of helping forestry become part of people's everyday lives. Alert for a mechanism to inaugurate this plan, Reynolds boldly predicted that municipal forests were destined to become the single greatest factor in America's move toward woodland conservation. Such optimism, a bit overstated, is nevertheless forgivable if one places his comments into context, particularly the broad spectrum of local interests to which municipal forestry could appeal in New England.

In town forests, Reynolds and others envisioned an impressive list of beneficial uses: a chance to alter the course of deforestation, reclaim idle lands, increase property values and related tax rates, reduce timber shortages, eliminate reliance on imported lumber, encourage local wood-using industries, provide employment, protect water supplies, and, not to be overlooked, simultaneously generate town revenue. Advocacy for commercial forestry was frequently balanced by appeals to recreational purposes—wild parks or bird and game sanctuaries—as a way of enticing local participation. Town forests also provided ideal memorials to war veterans or family members. Underlying each appeal was a subtle prod to advance public education on matters of forestry. If these blessings were not sufficient to entice a skeptical or uninterested public, successful examples of city forests in Germany and Switzerland were available to point the way. Little wonder that Reynolds keenly sensed a golden opportunity.[1]

As a movement, town forests grew from escalating interest in forest conservation and pursued the goal of timber cultivation through sustained yield management. Yet motives for establishing forests varied from town to town and often evinced a collaborative effort between town committees and forestry associations or state foresters. Community initiatives were sometimes prompted by economic concerns—a desire to generate income from tax-forfeited lands, for instance; in other cases, altruism swayed opinion. Watershed protection was a primary incentive in a number of key locales. Between 1915 and 1930, proponents directed most of their energies toward helping towns create forests according to guidelines set by state enabling laws. Articles published in an assortment of forestry, conservation, civic, recreational, and planning journals poured forth advantages and emphasized commendable examples, such as Newington and Danville or several early watershed forests that by then had begun to produce a crop of trees. Generous offers of pine and spruce seedlings from state forestry agencies, usually either free or at reduced cost, encouraged planting campaigns. Schoolchildren, scouting groups, and local clubs were enlisted to turn community soil. State-sponsored municipal forestry programs in New Hampshire and Vermont soon developed. World War I stifled civic improvement activities, and the movement sputtered until the decade between 1920 and 1930, when scores of forests were started and multitudes of seedlings planted.

Although forest conservation mapped a general course for town forests, New England's culturally diverse communities interpreted trends in subtle and intriguing ways. These local translations invite a journey from place to place. Disparate use and stewardship of municipal forests tended to reflect the region's cultural diversity, as did size and location of lands selected. Parcels physically or visually accessible to community residents, particularly lands near village centers, helped to reinforce community structure (see fig. 4.1). Regional variations unfold as well, revealed by the comparative strengths and weaknesses of state programs for municipal forestry. Although emphasis on timber productivity guided the movement and defined its character, recreational forest use steadily gained popularity; town forests and forest parks sometimes merged as a result. Full consideration of emerging recreational themes, however, belongs to the latter. Still, public education benefited in either case and served as a principal objective for those engaged in advancing the movement.

Having a Forest and Cutting It Too

The campaign for town forests in Massachusetts began in earnest with contests held in 1915 and 1916 by the MFA. The first contest was announced by the association's Bulletin 112 issued in June 1914, with judging to occur during autumn of the following year. The rules of the game were simple. Towns were required to obtain one hundred acres of land by either gift or purchase, officially designate that land as a town forest, and plant at least fifty acres with white pine seedlings spaced at six-foot intervals. Points were awarded under carefully arranged criteria, including land area, quality of planting, advantages

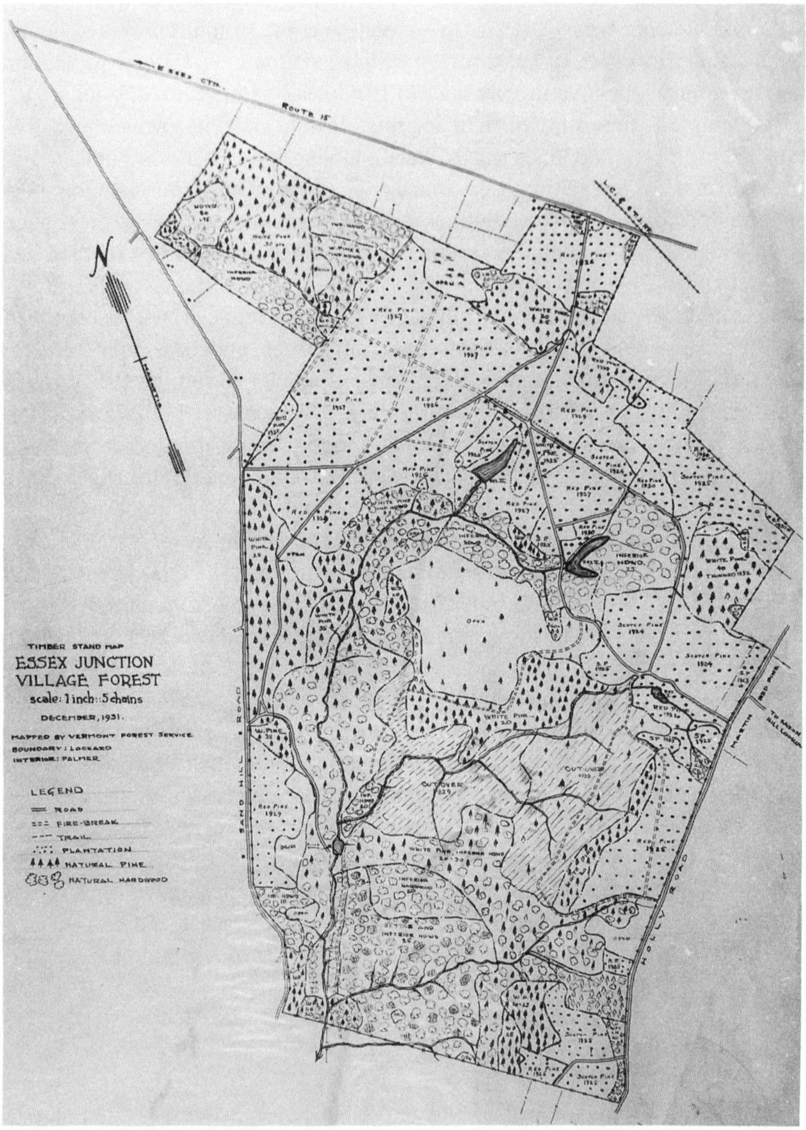

Fig. 4.1. Timber stand map of the Essex Junction Village Forest in Essex, Vermont, drawn by Charles Lockard and Huntley Palmer for the state's forestry department in 1931. One of the state's most capably managed, the municipal forest was planted to protect water supplies. After the city shifted to a regional water utility, a large portion of the land was sold in the early 1970s and converted to an office park. Restrictions, however, require that part of the land remain undeveloped. Vermont Department of Forests, Parks and Recreation—Waterbury

for lumbering, water and soil protection, recreation and aesthetics, potential for enlargement, fire prevention, and general improvements (such as roads). Highest points could be scored in fire protection and recreation categories, with logging a close third. The winner's forest would receive fifty acres of free planting, 1,200 white pine seedlings to the acre, with a 75 percent guarantee. Unless ten or more towns entered the contest, all bets were off. The 1916 contest was similar, but the prize was reduced to 1,000 trees and the rating of criteria was rearranged to give general improvements top billing, followed by fire protection, logging, and recreation. Both contests were publicized through the associations's bulletins and annual reports distributed to members, by then numbering more than two thousand.[2]

Town forest contests may have stirred excitement, but alas no winners emerged from among the fewer than ten entries each year. Only a smattering of Massachusetts towns officially established forests during the years preceding America's entry into World War I, but those that did became important examples. In the face of joint urging by the MFA's local branch association and city park commission, Fitchburg designated four parcels of unused, city-owned land as a municipal forest late in 1914. Included were the poor-farm woodlot (see fig. 2.22) and two worn-out homesteads known as the Taylor Farm and the Raymond Farm. Most of the land consisted of overgrown pasture with isolated stands of white pine and mixed hardwoods. Page Bunker, city forester and superintendent of parks, who later became state forester of Alabama, prepared working plans and supervised planting, primarily with white and scotch pine. Portions of the land were devoted to watershed protection. Three months after the town forest was created, Bunker described the project in "A Town Forest in America," which appeared in *Forest Quarterly*, the journal of the New York State College of Forestry at Cornell and predecessor to the Society of American Foresters' *Journal of Forestry*. After World War I, Guy Hubbard replaced Bunker as city forester and carried on the city's planting efforts. Hubbard, too, was capable of literary endeavor and penned an article for *The American City* in 1921, one of many town forest endorsements to appear in that journal.[3]

A much more elaborate town forest was devised in Walpole, located roughly twenty miles southwest of Boston. Chaired by Charles S. Bird, Jr., a wealthy paper manufacturer and MFA member, Walpole's planning committee had launched a commendable town-improvement venture. In 1912, the committee retained John Nolen, landscape architect and city planner from Cambridge, to assist. At the time, Nolen was just emerging as an important figure in America's nascent planning profession. Born in Philadelphia in 1869, he had faced poverty through much of his childhood but struggled diligently to gain an education, first at Girard College, a school for fatherless boys, and later at University of Pennsylvania's Wharton School. His interest in landscape architecture was sharpened at Onteora Park, a New York resort where he supervised

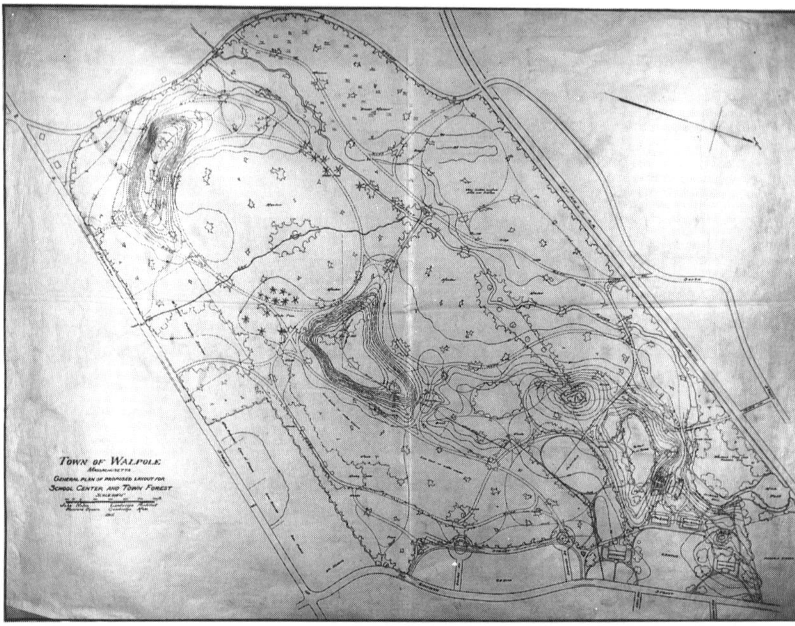

Fig. 4.2. Plan of the town forest and school center in Walpole, Massachusetts, designed in 1915 by John Nolen, a prominent urban planner. Conception of Walpole's forest, more elaborately devised than most, coincided with contests launched by the Massachusetts Forestry Association encouraging towns to plant local forests. Trails still follow Nolen's original plan, but entrances are not well marked and cultural interpretation of the forest is lacking. Cornell University Archives

landscaping during summer recess from Wharton. In 1905, six years before Harris Reynolds, Nolen completed Harvard's graduate school of landscape architecture. Many of his early plans emphasized features popularized by the City Beautiful Movement, such as parks; wide, tree-lined boulevards; and public plazas framed by classical architecture. However, his mature work helped to push city planning toward confrontation with more significant problems facing America's growing urban class.[4]

At Walpole, Nolen had completed a comprehensive town plan by 1914 but was soon asked to prepare additional designs for a high school park and town forest following a donation of 150 acres by George A. Plimpton. In October 1914, Bird solicited Nolen's opinion, and Nolen wrote to Harris Reynolds inquiring about the contest and asking for copies of *MFA Bulletin* 112. Nolen advised Bird that the town forest plan should be simple, less elaborate than the high school park, but the intriguing design unveiled in 1915 was at least as much park as forest (fig. 4.2). Replete with open meadows, circling roads, sheltered vistas over the Neponset River valley, and trails climbing to elevated points, Walpole's Town Forest was cast in the romantic image of America's forest parks. Its location near the town's center simplified connection of forest roads and paths to the community's general park system as well as to the adjoining high school campus. To his credit, Nolen recommended a report on the property's development as a town forest with suggestions for tree planting. A consulting forester from Boston, George Carlisle, was hired in 1916 to supervise planting, which began with white and red pine. Carlisle suggested white ash or red maple for swamp and moist lands but cautioned against the use of oak, which would provide food for gypsy moths. A dedication ceremony was held on Arbor Day, 1916, and Lieutenant Governor Calvin Coolidge planted a symbolic first tree while hundreds of Walpole's schoolchildren waited for their chance to assist (fig. 4.3). That summer, Walpole's town forest was featured in an issue of *American City*, whose advisory board included Nolen and well-known planner Charles Mulford Robinson.[5]

Although contests and planting projects dominated town forest activity in Massachusetts before World War I, ownership and management concerns were not overlooked. William W. Colton, a member of the MFA's Fitchburg branch association, addressed forestry management topics at the MFA annual meeting in 1914. Colton first summarized available methods for acquiring lands and emphasized workable agreements with town water departments as well as procurement of tax-forfeited lands. Thousand-acre tracts were recommended as an ideal size, but Colton acknowledged that much smaller forests would also be acceptable. Once boundaries had been established by survey and topography charted, forestry work would commence with division of maps into compartments based on tree species. Estimates of acreage as well as age of timber growth would then be supplied for each compartment. Geographic features such as roads, bodies of water, swamps or wasteland, as well as soil character, also would be plotted. A working plan prepared from these data followed, and strategies for planting, thinning, and harvesting were to be devised to attain maximum market value from the forest. Cutting would occur in rotation, based on yearly rate of growth, in order to sustain growing stock. Improvements, including fire lanes and stations, roads, and a small nursery, were recommended during beginning years. Permanent caretaker's quarters were encouraged if the forest was to be opened to public recreation.[6]

In addition to Fitchburg and Walpole, the Massachusetts towns of Brookline, Fall River, North Adams, Pittsfield, and Westfield all began planting activity on watershed lands prior to or shortly after World War I. Promotion of these early watershed projects, including Fitchburg's, helped to emphasize timber cultivation as a key goal of town forests. Walpole's meticulous aesthetic and recreational elements were atypical and can be attributed to Nolen's involve-

Fig. 4.3. Dedication ceremonies for the Walpole Town Forest on May 6, 1916; *left to right:* George Plimpton; Calvin Coolidge, then lieutenant governor of Massachusetts; and Charles S. Bird, Jr. As president in 1924, Coolidge convened the National Converence on Outdoor Recreation, and presentations by key figures, including Harris Reynolds and Benton Mackaye, stressed the recreational value of forests. Walpole Historical Society

ment. Still, publicity surrounding Walpole's project generally helped to advance primary themes. Although the MFA quickly became a vanguard for New England's movement, New Hampshire and Massachusetts towns established approximately the same number of forests during this period. The Society for Protection of New Hampshire Forests (SPNHF) had played a premiere role in advancing local forestry in its state and undoubtedly sparked community interest. As the SPNHF adjusted its aim toward legislative lobbying, however, New Hampshire's forestry commission became the state's most energetic advocate, a role sustained by the state's enabling law.

Unlike Massachusetts, New Hampshire's 1913 enabling act obliged towns to relinquish management of their forest lands to the state forester. In practice, however, supervision was shared with town committees. Beginning in that year, the commission's biennial reports regularly devoted a section to municipal forests and identified towns that had begun forestry programs. Although these government publications never achieved the breadth in scope and quantity of Harris Reynolds's work for the nonprofit MFA, they nevertheless included review of European community forests as well as activity in other New England states. In addition, the state forestry department's quarterly journal, *New Hampshire Forests*, provided a forum for a number of articles about town forests, and Warren F. Hale, assistant state forester, was a frequent contributor.[7]

A number of town forests were established in New Hampshire at about the same time that contests were being introduced in Massachusetts. Franklin, Gilsum, Richmond, and Jaffrey all began modest forestry work on town-owned land during this period. Jaffrey initiated planting of fifty thousand trees on five hundred acres of open land at the top of Mt. Monadnock. In Franklin, a demonstration forest was begun in cooperation with the U.S. Bureau of Entomology to develop moth-resistant trees. Hollis, located in Hillsborough County on the state's southern border, accepted a gift of 189 acres in 1916 from one of its residents, Silas Spalding, and named its forest Spalding Park. Woodland management was assigned to the state forester, and under terms of the donor's will the town was required to make specified improvements for ten years. Planting of fifteen thousand pine seedlings began that fall, and culling of mature timber soon followed. Harvesting of cordwood and lumber over the next several years helped to keep the town in winter fuel and furnish revenue for improvements to the land.[8]

After considerable debate at their 1919 town meeting, residents of Warner accepted a valuable gift of eight hundred acres with extensive stands of merchantable timber and nearly limitless cordwood. The donation came from Lloyd H. Chandler, then a navy captain who later rose to the rank of admiral, on behalf of his father, former U.S. Senator William E. Chandler. The elder Chandler had spent summers in Waterloo, a small village within the town, and had encouraged public use of his forest. Captain Chandler's correspondence from aboard the battleship USS *Minnesota* conditioned the gift upon townmembers' willingness to forgo taxes on mature timber. Chandler pointed out that he could easily profit by clearing the land but had no wish to destroy the resource. Opponents feared the consequences of permanently removing such a large parcel of land from town tax lists. Resistance, however, was overcome and a town forest committee appointed to assist the state forester. Cordwood heated the town hall, school, and library over the coming years, and surplus was sold to residents at moderate prices. Following preparation of a type map by the state forestry office in 1920 (fig. 4.4), a logging operation was begun. Revenue from lumber and topped-off cordwood was placed in a town forest fund, which helped to pay for planting and blister rust control. A nearby chemical company that produced charcoal, wood alcohol, and a product sold under the dubious label "Black Iron Liquor" provided a ready market for low-grade hardwoods.[9]

Town forest advocates repeatedly faced concerns about depleted tax revenues, an undying contention often raised in opposition to public lands. When

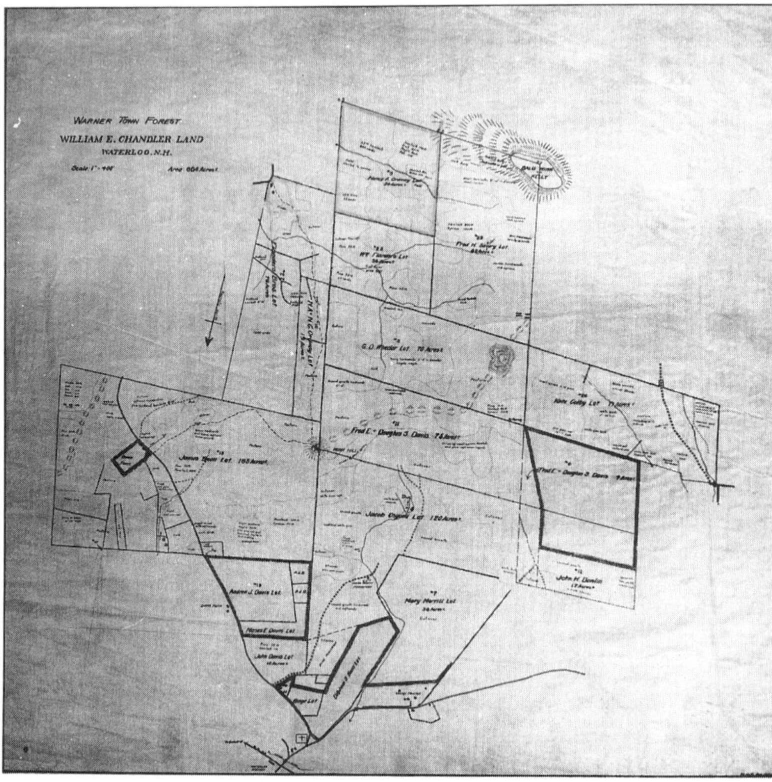

Fig. 4.4 Survey and forest type map of the town forest in Warner, New Hampshire, drawn by the state's forestry department in 1920. Known as the Chandler Reservation, the eight-hundred-acre tract with extensive stands of mature timber was given to the town a year earlier. Today, leaflets for a self-guided woods walk identify the commercial value and uses of various tree species as well as old "scoot" or logging-sled trails. New Hampshire Division of Forests and Lands—Concord

opportunities arose, foresters were quick to point out the ability of properly managed timberlands to generate revenue capable of balancing tax lists. Their arguments worked best in places such as Warner, where recently established forests could produce immediate revenue, either through thinning or harvest of mature timber. Newly planted lands, however, offered no immediate hope for economic return, and loss of tax income became an issue that sometimes influenced local votes. Site selection, too, was occasionally affected by this issue. Landlocked or remote, logged-off parcels unsuitable for anything but growing trees generated negligible tax receipts and became likely candidates for town forests. Unfortunately, physical distance from community centers also could be gauged by the extent of public involvement.

Municipal forestry was stronger in Massachusetts and New Hampshire than in other New England states during the movement's early years. Comparisons between the two, however, are probably unfair at this stage. Undoubtedly, many towns in both states had quietly owned and utilized forest land to community benefit for ages. The status of town-owned land sometimes changed quickly, and records are not always clear. Enactment of town forest laws introduced formal management, but not all towns willingly participated at first. Reports from this period identify those towns that had actively encouraged management assistance from either nonprofit or state forestry groups or had met technical requirements for creating town forests. As a result, they may not be complete. Assembling accurate and comprehensive information about forestry activity in the region's independent towns has always proved difficult. This weakness was first observed in a 1916 article by James W. Toumey, who by then had replaced Henry Graves as director of Yale's School of Forestry. Titled "County, City, Town and School Forests," Toumey's article appeared in *American Forestry* and contained a list of municipal forests in ten, mostly eastern states as well as brief descriptions of important examples. By far, the largest portion of Toumey's examples were watershed forests.[10]

Comparisons aside, certainly the most significant municipal forestry work in both New Hampshire and Massachusetts, as well as in other New England states, initially occurred on watershed lands during the period before World War I. In New Hampshire, Concord, Hanover, Keene, Manchester, and Nashua had all begun forestry work on old pastures surrounding public water supplies well before enactment of the state's town forest law in 1913. Most had already solicited assistance from the state forester or, in the case of Nashua's privately chartered utility, Pennichuck Water Works, from the U.S. Forest Service. A substantial number of Connecticut towns, among them Hartford, Middletown, New Britain, New Haven, and Waterbury, also had planted lands bordering city reservoirs. The Vermont cities of Essex Junction, Rutland, Bellows Falls, and Montpelier were not far behind. Forestry activity on watershed lands provided an effective source of information about municipal forestry and heralded the movement's period of greatest popularity in New England during the Roaring Twenties.

Public Need—Public Duty

Numbers of town forests began a rapid climb after 1920, pushed upward by a steady advance of literature, encouragement and assistance from state foresters, zealous promotion by the MFA, and increasing community interest.

Within the forestry profession, concern for reclamation of idle lands and fear of regional timber scarcity—and a desire to acquaint the public with these problems—propelled the movement more directly than any other single factor. Advocates sought to prick a civic conscience by arguing that private industry had pathetically failed to conserve an important natural resource and that society had a duty to supply a public need for future generations.

Still, the appeal of town forests touched a broad range of local interests, all of which contributed to the movement's popularity. Public approval of forest management and concern for abandoned land unsuitable for agriculture plainly influenced voters at town meetings. Water companies, owners of land for which uses were very limited, had everything to gain by growing crops of trees. Interest in wild parks as well as the expanding role of city planning also were instrumental. More than a few opinions, too, were impelled by the prospect of increased town revenue and opportunities for employment. Forestry groups capitalized on the wide attraction of town forests and routinely appealed to community sentiment in the broadest possible way. Bulletin 132, issued by the MFA in October 1921 and titled "Why Massachusetts Needs Town Forests," is typical, and it wasted little prose in its matter-of-fact enumeration of the reasons that every town could find for establishing a town forest.[11]

Public education continued to be a keystone for the municipal forest movement, and periodical literature took aim at a sizable segment of America's population. While forestry journals carried many of these articles, the MFA made a calculated appeal to public support for recreational use of forests. Addressing the National Conference on Outdoor Recreation held in Washington in May 1924, Harris Reynolds recommended that a special congressional committee be formed to devote full-time educational work around the country on the subject of town forests. That same year, the text of Reynolds's address appeared in *Playground*, a journal with obvious recreational leanings.[12]

The following year, Reynolds and the MFA collaborated to publish a monograph titled *Town Forests: Their Recreational and Economic Value and How to Establish and Maintain Them*. Reynolds's short book, published with a foreword by Charles Lathrop Pack, president of the American Tree Association, was a major publication for the town forest movement and achieved wide circulation. The prominence of recreation in the book's title was clearly intentional (fig. 4.5). However, Pack's foreword stressed forestry and sounded a clarion alarm for America's timber shortage: "We need more forests and we must look upon timber as a crop. We must plant trees, mature them and cut them when they are ripe, always looking out for next year's crop."[13] As if to balance his choice of title, Reynolds began the book with a discussion of the economic benefits of town forests but followed it with a chapter devoted to recreational merits, in which he stated: "The recreational value of the town forest, however, is one of the chief, if not the most important, benefits to be obtained from such

Recreation and Nature Study Are Assets of the Forests

Fig. 4.5. Vignette from Harris Reynolds's 1925 monograph, *Town Forests: Their Recreational and Economic Value and How to Establish and Maintain Them.* The need for public education about prudent management of timberlands steadily propelled advocacy for town forests. Recreation and nature study were ideal ways to generate public involvement with forestry, and schoolchildren often transplanted seedlings on Arbor Day.

an area. From the purely investment standpoint the recreation feature is only a by-product of the forest which costs the town nothing." [14]

Vacillation between the importance of commercial and recreational use of town forests, typified by Reynolds's remark, occasionally appears in literature from this period. A 1916 editorial in *American Forestry* had candidly suggested that recreation was the most fundamental use for town forests. Reynolds and the MFA were not always consistent in their appeals, and results in a number of towns (Walpole and Hollis, for example,) were equally ambivalent and may have reflected this attitude. Eventually, as the movement progressed, the accent on recreational benefits betrayed a concern about the ability of towns to capably manage timber production. At this stage, however, appeals to recreational users were a tactical, simple acknowledgment that such benefits held wider local appeal than commercial forestry and facilitated means to an end, namely, reforestation. Reynolds admitted as much in his address to the National Conference on Outdoor Recreation, although he hedged on the willingness of towns to pay for forest parks:

> . . . but in dealing with public opinion, rapid progress is not always made by direct action. To secure public support in these matters, every argument that will apply to the case in hand must be advanced. I believe that the greatest good that may come to a community from the creation of a town forest will be recreation which such an area will afford, but it is not the argument that wins with the average town committee on finance.[15]

Emerging tension between recreational and commercial use of local forests mirrored similar trends in state and national forests. While enabling laws made it clear that town forests were created for cultivation of trees, a practical balance was struck between the goals of recreational and commercial benefits. Multiple use thus surfaced as a preliminary issue, one of expectation resolved by a town's decision whether or not to establish a forest. In contrast, debate over multiple use in national and state forests has typically been addressed through management policy, not always a consistent or effective process.[16]

In 1925, the same year that Reynolds's book appeared, the Connecticut Forestry Association sponsored a similar monograph, this one by Phillip L. Buttrick and titled *Town Forests and Parks for Connecticut*. Buttrick emphasized the public's role in forestry and repeated Bernhard Fernow's arguments that only governments, whether national, state, or town, could manage the long-term investment necessary to produce high-grade lumber; only governments could afford to grow timber on land too poor for any other use. Buttrick deserves credit for briefly sketching a connection between twentieth century town forests and America's first community woodlots sequestered from common land. In addition, he distinctly classified three types of town forests: commercial, watershed, and forest parks. Buttrick provided the compulsory list of advantages, placing recreation toward the end, but also responded to arguments that town forests were ultimately unprofitable because they removed land from tax lists. He pointed out that cut-over and wastelands produced small or no gain from any source, taxes included, other than trees. Loss in taxes from forest parks, on the other hand, was balanced by overall increase in community welfare as well as long-term revenue from timber management.[17]

Most literature also stressed benefits from employment opportunities, public involvement with forestry practice, and subsidy for the poor. Town forests provided ideal training grounds for young foresters. Furthermore, much forestry work required unskilled labor for tasks such as logging, thinning, culling weed trees, road building, and clearing fire lines. These jobs were close to home, required little travel, and could be accomplished during winter, when unemployment was most serious and other work often not possible. In some cases, forestry chores could be delayed until workers were available. Local forests also were regarded as a home outlet for swelling labor ranks during the postwar years and in the midst of the Depression. Employment advantages extended beyond forest boundaries to reach local wood-using industries. Municipal forestry thus became part of a broad plan to stabilize local economies through forestry education and example.

Charles Lathrop Pack, writing for *Nature Magazine* in 1925, emphasized public education as the paramount issue facing forest conservation and considered town forests to be community textbooks. Increased public awareness of good forestry practice presumably led to better stewardship of private forests, particularly farm woodlots. With demonstration forests close at hand, farmers and other landowners were given an opportunity to observe the application of silvicultural practices to local conditions, a crucial factor in generating quality timber and reasonable profit. Accessibility to superior timber encouraged local industry, which in turn provided a market for tree growers. In an article appearing in *World Agriculture* in 1926, Harris Reynolds remarked that any assistance to farmers in improving their woodlots would ultimately be a good community investment. To Reynolds, town forests were the best way to offer a practical demonstration in forest management.[18]

Dissemination of literature continued to be an important part of the town forest movement, but pragmatic incentives were not overlooked. In June 1920, the Massachusetts legislature enacted a bill that encouraged the purchase and development of state and town forests and authorized the state's conservation commissioner to provide free seedlings to towns for planting their forests. About a year later, the MFA made a standing offer to plant five thousand trees (roughly five acres) without charge to any town that established a forest of more than one hundred acres, and the offer was repeated yearly. Soon, several Massachusetts organizations, the Berkshire Forestry Association, the Springfield Chamber of Commerce, and the Hampden County Improvement League, offered similar inducements for towns in their respective counties. The New England Box Company, a conglomerate timber industry founded by Ansel Dickinson, offered to plant five thousand trees for any municipal forest in Franklin County.[19]

Communities also were encouraged to place articles in their town meeting warrants requesting a vote on appointment of a town forest committee. Once formed, committees then consulted tax rolls or solicited gifts of land and reported about available parcels. Following a committee's recommendation, towns could then decide whether to begin forest cultivation under the state's enabling law. Occasionally, state foresters also participated in the preparation of surveys and management plans.

Massachusetts towns promptly responded to this flurry of literature, nudges, and incentives. By 1924, forty-one towns had voted to establish town forests;

nearly all of these had already set aside land, and twenty-eight had begun planting activity. By 1930, the number had increased to ninety forests covering more than 25,500 acres on which roughly three million trees had been planted. Petersham, in 1921, began a reforestation plan on a section of their poor farm and helped to start a trend heartily approved by the MFA. Silvicultural work did not begin until 1926, when students from Harvard's forestry school weeded young growth and a type map was prepared (fig. 4.6). In addition to providing a useful management tool at the time, such type maps now help to document the history of forest cover. Petersham's town forest committee was appointed in 1927, staffed by Richard T. Fisher, Albert C. Cline (director and assistant director, respectively, of Harvard Forest), and Daniel Broderick. Two other parcels were acquired, and Petersham's forest became one of the state's most efficiently managed.[20]

Groton, too, elected to convert a portion of its poor farm to town forest in 1922, based on advice from resident William P. Wharton, who was also a member of their town forest committee. That group had recommended that a memorial forest be established on land near the confluence of the Nashua and Squannacook rivers. Wharton prepared a special minority report urging voters to add fifty-four acres of adjoining pasture and woodlot from the town farm. Over the years, Groton's forest has quietly illustrated the unpretentious character of town forests envisioned by Harris Reynolds. Planting was accomplished with free seedlings provided by the MFA; weeding, thinning, and improvements provided employment and fuelwood for the poor; timber stands generated revenue from lumber; forestry management by the community's 4-H club proved educational; and a wildflower sanctuary was carefully planned by a local garden club.[21]

In southwestern Massachusetts, the sparsely populated village of Russell began a one-hundred-acre forest on a small portion of town-owned land in 1923 and persistently increased its size to roughly three thousand acres. Long a center for logging and paper manufacturing, Russell had acquired its extensive lands to protect water supplies and reduce costs of furnishing utilities to a few meager hill farms. Large-scale forest management was begun following preparation of a working plan by C. A. Galarneau, a state forester. Old pastures and cut-over lots, many of them located in the adjoining town of Blandford, were planted with thirty thousand trees annually for ten years. Beginning in 1929, Russell's forest became the site of a number of annual conferences for the state's town forest committees. A second management plan and type map (fig. 4.7) was prepared in 1945 by Philip T. Coolidge, consulting forester for the New England Forestry Foundation. Heeding Coolidge's advice, the community hired a full-time forester, Elmer R. Foster, a year later.[22]

Town forests in Massachusetts developed around the carefully focused leadership of the MFA or its branch associations and were buoyed by a steady

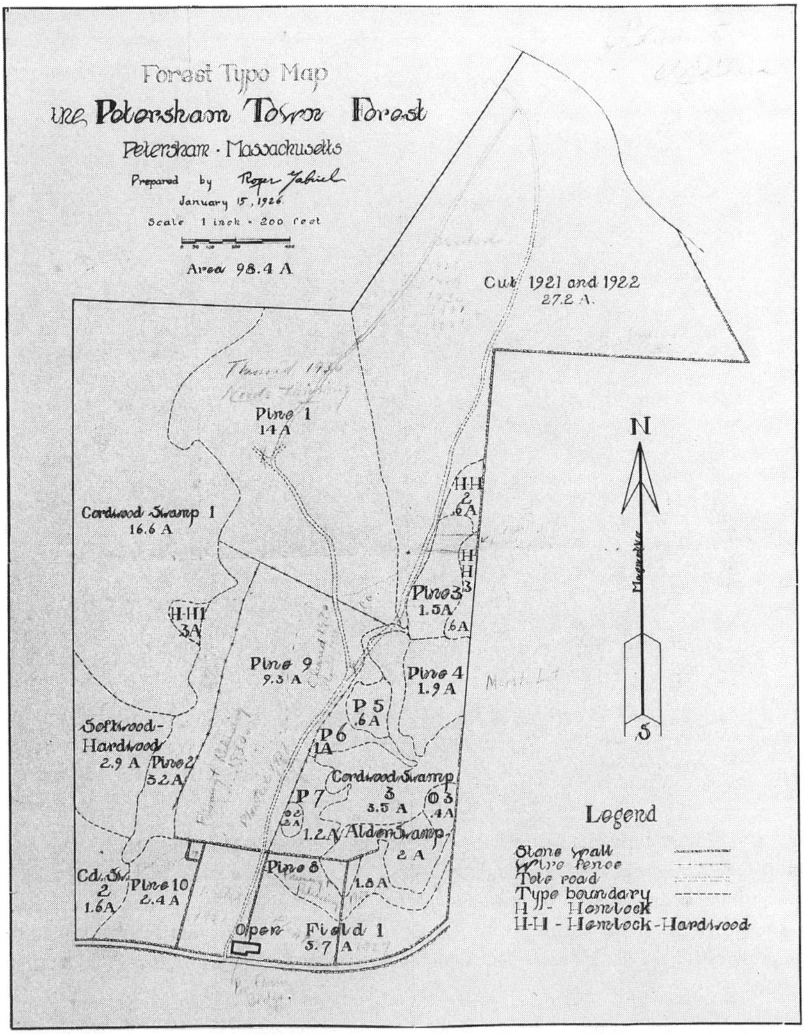

Fig. 4.6. Forest type map of the town forest in Petersham, Massachusetts, prepared by Roger Gabriel in 1926. Members of Petersham's town forest committee included the director and assistant director of Harvard Forest, and the town's example became a model for other communities. Two additional parcels, the Shaw Tract and the Monroe Lot, were also placed under forestry management. Harvard Forest, Petersham, Massachusetts

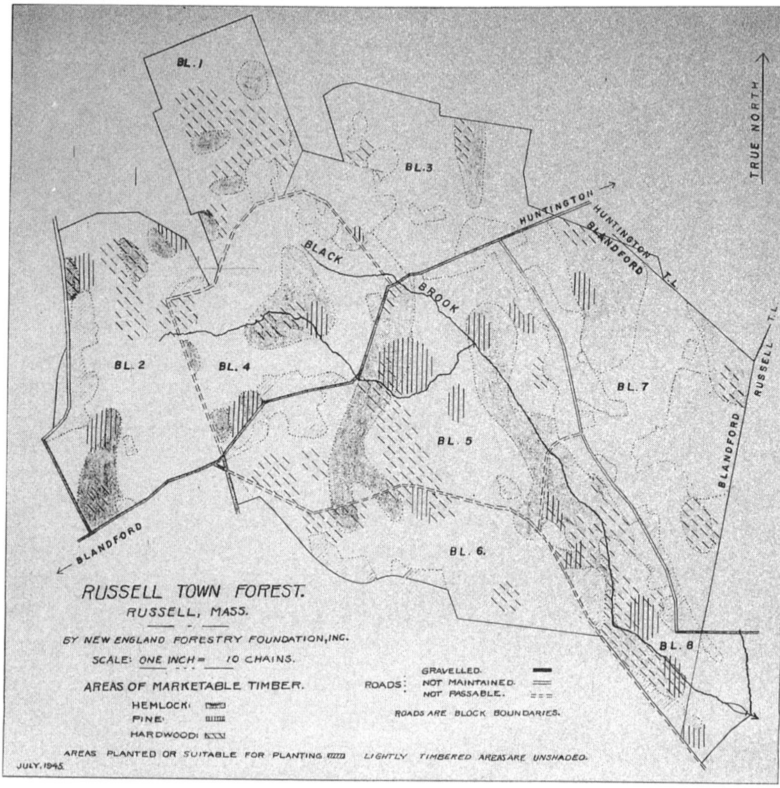

Fig. 4.7. Forest type map of the town forest in Russell, Massachusetts, drawn by Philip T. Coolidge for the New England Forestry Foundation in 1945. Russell acquired its woodland primarily to protect water supplies and to reduce the cost of town services to a few struggling hill farms. Both Petersham and Russell frequently hosted annual conferences for the state's town forest committees. New England Forestry Foundation, Boston

flow of published literature. Initiatives by state foresters were generally confined to donation of seedlings, preparation of an occasional management plan, and the chance presentation of a paper. In contrast, New Hampshire's program was shaped by a small core of government foresters with myriad duties. Yet town forests multiplied just as rapidly as they had in Massachusetts. In 1925, the state forestry commission reported forests in sixty-one communities, and by 1930 that number had climbed to seventy-nine on more than sixteen thousand acres, with plantations exceeding two million trees. Generally, both the commission's biennial reports and the state's quarterly forestry bulletin,

New Hampshire Forests, lacked the sweeping public appeals that characterized town forest literature in Massachusetts. Instead, many articles simply described worthy examples of town forests throughout the state. While New Hampshire towns may have lacked the volume of literature available in Massachusetts, strong involvement by state foresters more than compensated for any shortcomings. Unlike Massachusetts, woodland records for New Hampshire's town forests are relatively complete, centrally located, and easily accessible.[23]

In other aspects, New Hampshire's town forest movement followed patterns similar to those in Massachusetts. Local selectmen were encouraged to include articles about municipal forests in their warrants for town meetings. Letters outlining a comprehensive plan for creating forests were sent to each town that formed a committee. Questionnaires also were mailed periodically to local officials in an effort to assemble data about community-owned woodland. In 1913, the state forestry commission recommended that trees be supplied from the state nursery at no charge to towns. To some extent, this policy was applied inconsistently, and a law specifically authorizing the state forester to provide free seedlings was not enacted until 1929. Activity by New Hampshire's government foresters also extended beyond state boundaries. During the spring of 1925, Warren Hale obtained a leave of absence and journeyed to Europe, where he studied community forests in France, Germany, and Switzerland. John H. Foster, who succeeded Edgar C. Hirst as state forester, addressed the same topic but stayed a little closer to home when he ventured to Boston in 1928 as one of three principal speakers for the first meeting of Massachusetts town forest committees.[24]

New Hampshire's public woodlands in Newington, Danville, and Northwood received particular attention during this period and in turn helped the state to keep pace with Massachusetts. The state forestry commission's biennial report for 1923–24 devoted several pages to a history of Newington's church and parsonage lots, and in 1927 the town participated in white pine blister rust control. Elsewhere in New Hampshire, gifts, tax forfeitures, and an occasional purchase supplied land needed for town forests. Winchester, tucked in the state's southwestern corner, received a sixty-six-acre gift of land on Meetinghouse Hill, which rises above the village center. The donor, Lucy Jennings Dickinson, managed her father's local lumber business for a short time before marrying the president of New England Box Company, LaFell Dickinson. Named in memory of her father, the Willard H. Jennings Forest was dedicated in April 1927 and later mapped by her cousin, Dwight Jennings, a company surveyor and timber cruiser (fig. 4.8). A thousand pine seedlings donated by the Keene Forestry Association were planted by schoolchildren. Neighboring Richmond acquired one hundred acres of tax-defaulted land and received a contribution of seventeen thousand pine seedlings from Ralph L. Morgan, chairman of Fitzwilliam's town forest committee. Morgan, who spoke at the

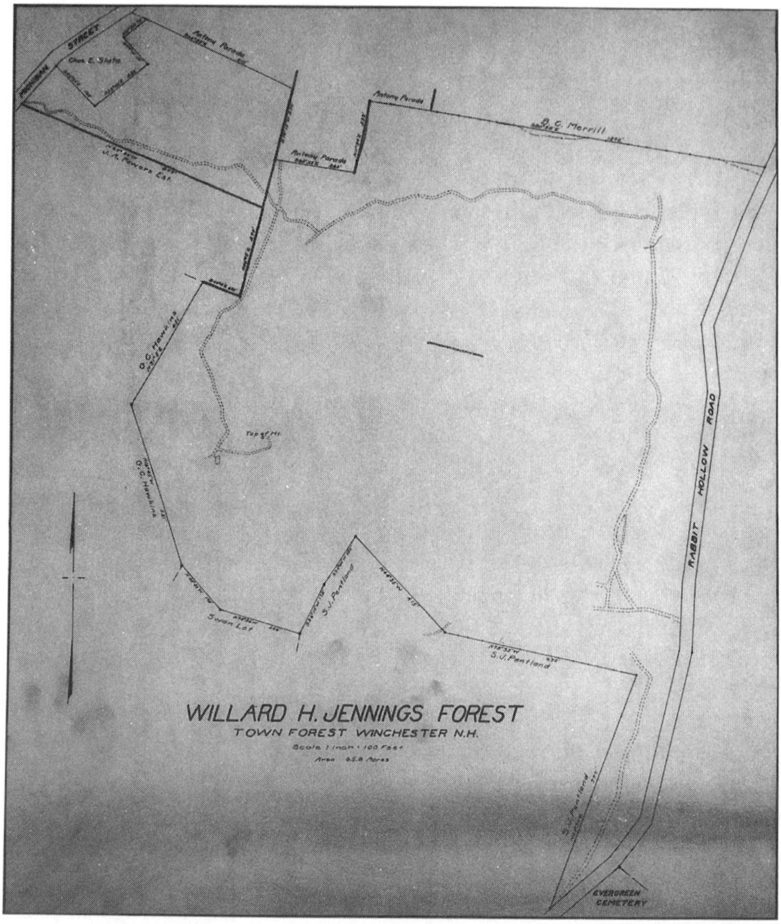

Fig. 4.8. Undated survey of the Willard H. Jennings Forest in Winchester, New Hampshire, mapped by Dwight Jennings, c. 1936. Donated to the town by Lucy Jennings Dickinson, his cousin, and named in memory of her father, a local lumberman, Winchester's town forest is situated on Meetinghouse Hill and is one of a number of community-owned woodlands in New England that provide a verdant backdrop for village centers. Beaman Lumber Company, Winchester, New Hampshire

dedication of Winchester's forest, had begun his own nursery in 1924 and had offered free seedlings to any resident of Richmond interested in reforestation. Richmond's forest, since named the Harry J. Bennett Town Forest, also was mapped by Jennings, who delineated prior ownership of farms converted to woodland; stone wall bounds often remain in such cases (fig. 4.9).[25]

Plagued by cut-over lands left for tax forfeiture, many towns in New Hampshire and Massachusetts opted for forestry management as the only practical means to generate income from otherwise unproductive land. Tax sales occurred often on timberlands, where methods of taxation induced clear-cutting, and towns consistently struggled with the problem. At times, record keeping for remote parcels proved difficult. In southwestern New Hampshire, Lempster officials managed to lose a thirty-acre parcel acquired for nonpayment of taxes in 1901 and located somewhere in the vicinity of Lempster Mountain. Logging on nearby land in 1927 sparked inquiry; the town hired a surveyor to find their missing property, and forestry work was subsequently begun.[26]

Collective concern for visually prominent sites sometimes goaded towns into action. Citizens of Merrimack, situated north of Hollis in Hillsborough County, were faced with this type of predicament when the owner of a twenty-acre tract near the town hall began clearing old stands of pine in 1926. Known as Twin Bridges, the parcel contained a small island formed by a branch of Baboosic Brook before it flowed into the Merrimack River. Access to the island was gained by crossing a pair of footbridges that gave the site its name. Ten individuals rallied to share the property's purchase price and were subsequently reimbursed by the town. Land was replanted with white and red pine supplied by the state, and a survey was prepared by Warren Hale in 1927 (fig. 4.10). Now a forest park, Twin Bridges' proximity to Merrimack's town center has become increasingly important as linear sprawl erodes village form.[27]

Expressions of local sentiment about town forests offer an ideal opportunity to isolate community attitudes toward conservation. One of New Hampshire's exceptional town forests was created in Dunbarton in 1929 when the Winslow family conveyed approximately 225 acres of ancestral timber land to the community (figs. 4.11, 4.12). Their donation was conditioned upon use of the tract as a town forest in perpetuity, with cutting to be supervised by New Hampshire's state forester. More important, all income was to be placed in an account identified as the Town Forest Fund and used only for forest improvement or for buying additional woodland. If conditions were not honored, title to the property would vest in the SPNHF. Not surprisingly, these provisos were received with something less than enthusiasm by a few town members. Correspondence between Arthur Winslow, president of the family corporation, and Frank Garvin, chairman of the town selectmen, clearly identified two points of view. Garvin tactfully wrote to Winslow expressing doubts that town voters would accept an offer that meant potential loss of all income to the community

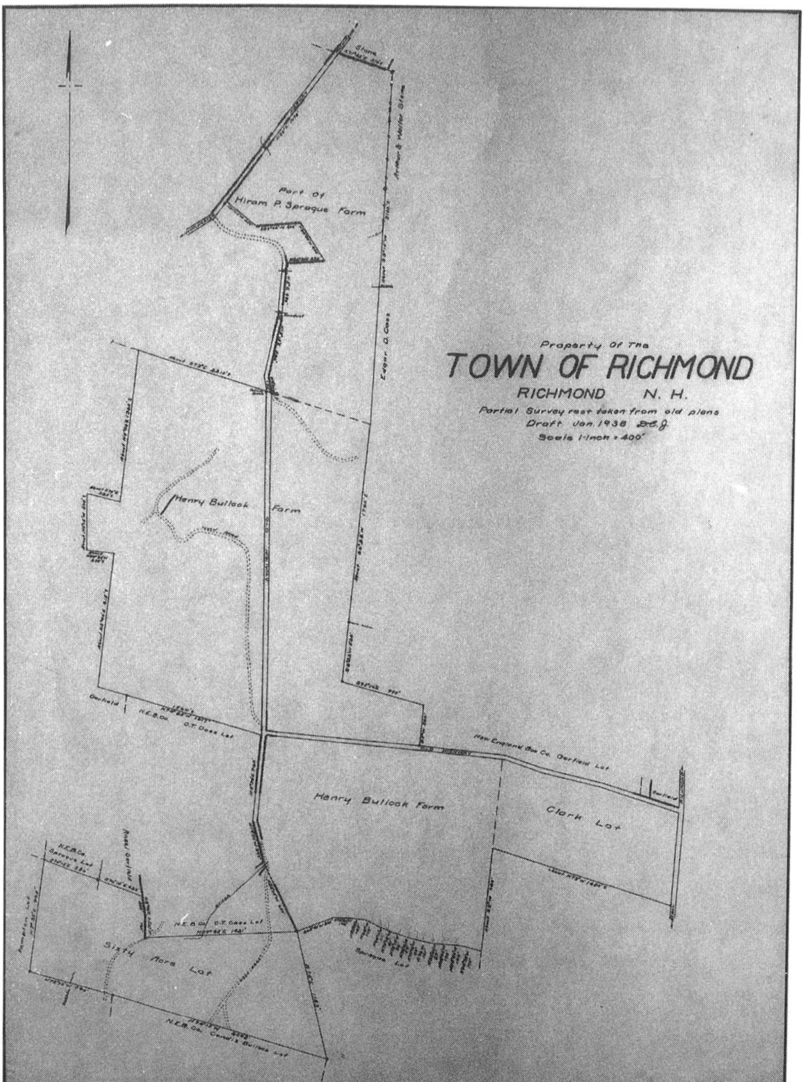

Fig. 4.9. Town forest survey for Richmond, New Hampshire, limned by Dwight Jennings in 1938. Jennings worked as a surveyor and timber cruiser for the New England Box Company founded by Ansel Dickinson, which acquired mills and timberland throughout western Massachusetts and southern New Hampshire. Richmond's woodland was named the Harry J. Bennett Town Forest in 1984. Beaman Lumber Company, Winchester, New Hampshire

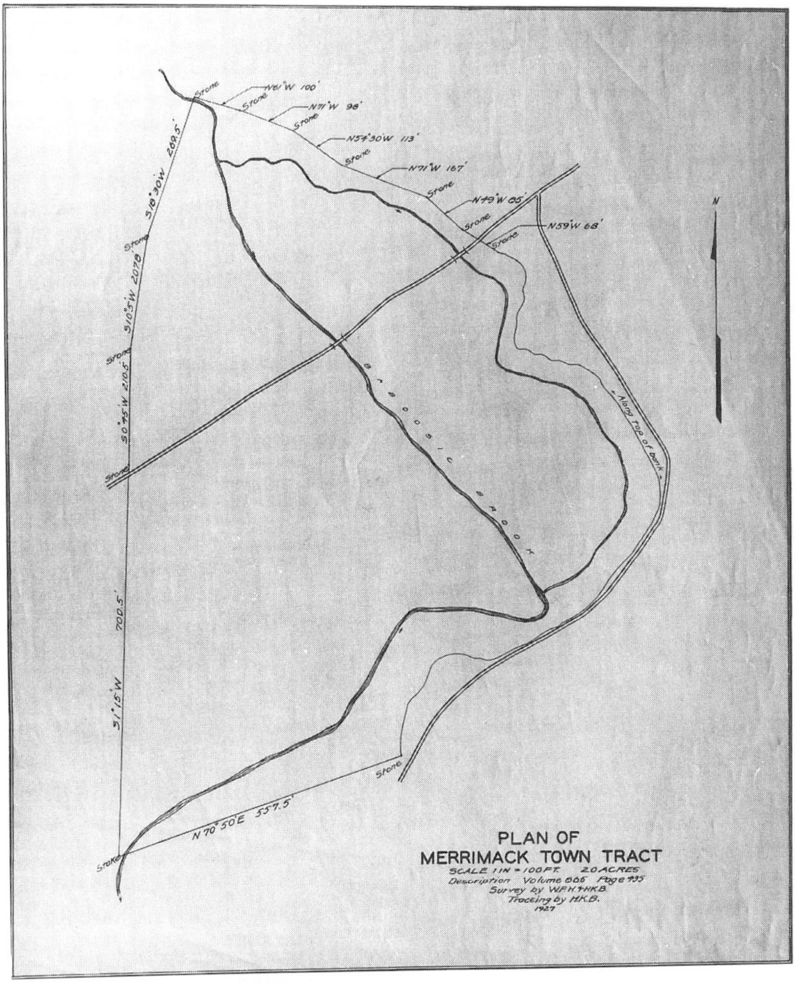

Fig. 4.10. Survey of Twin Bridges, woodland owned by the town of Merrimack, New Hampshire, mapped in 1927 by Warren F. Hale and H. K. Baldwin for the state's forestry department. Located near the village center and currently a forest park, the low-lying area received its name from a pair of foot bridges that cross Baboosic Brook, a tributary of the Merrimack River. New Hampshire Division of Forests and Lands—Concord

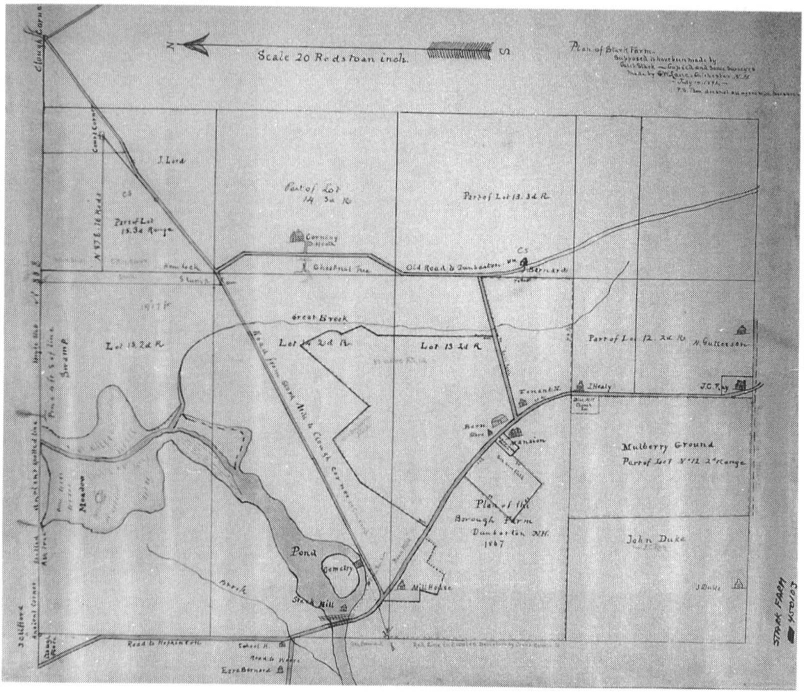

Fig. 4.11. Plan of the Stark Farm in Dunbarton, New Hampshire, attributed to Caleb Stark and copied with revisions by George W. Lane in 1892. In 1930, descendants of the Stark family, whose patriarch, General John Stark, served in the Continental Army, donated their land to the community as a town forest. New Hampshire State Historical Society, Concord

from the property. He suggested, instead, that the village be allowed to use whatever revenue the forest produced. Winslow's reply succinctly captured the essence of community interest in forestry for the period:

> The primary object is to preserve a forest area, undisturbed, indefinitely, which will serve as an animal refuge and place of interest and beauty. Indirectly, there is a further advantage to the town in having such a reserve area as an attractive feature, and also, in the labor it will furnish and the wages which will be distributed among citizens in the incidental care of the forest. The lumber from the land should in time yield quite a good revenue.[28]

Vermont's municipal forest movement eventually coaxed almost as many town forests from its communities as did campaigns in Massachusetts and New Hampshire. Together, these three states formed the nucleus of municipal for-

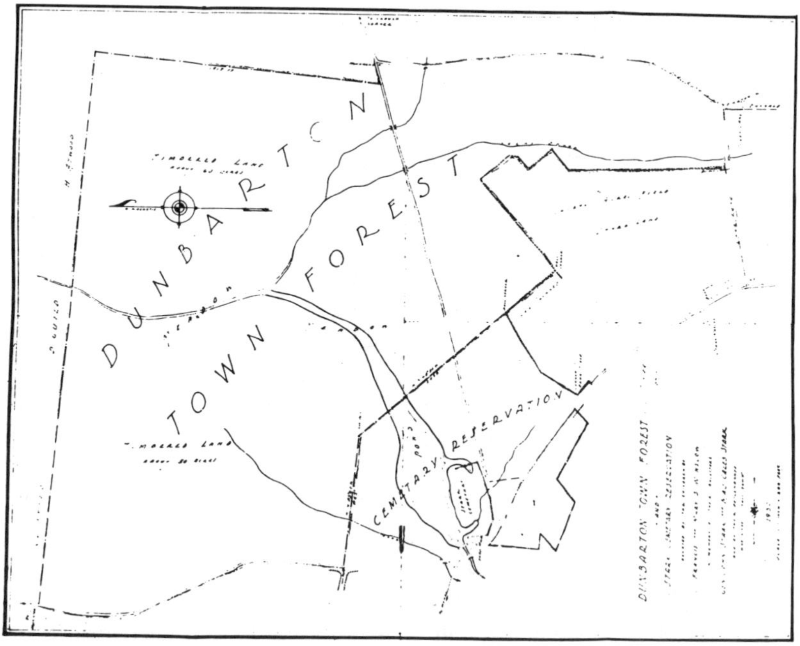

Fig. 4.12. Unsigned survey of the town forest and Stark family cemetery in Dunbarton, New Hampshire, drawn in 1930. More than two hundred acres of woodland were given to the town in perpetuity, subject to the condition that all income be placed in an account available only for timber stand improvement or purchase of additional lands. New Hampshire Division of Forests and Lands—Hillsborough

estry in New England. However, Vermont's period of greatest activity did not occur until after World War II, prior to which the state lagged behind its two neighbors. Reports by Vermont's commissioner of forestry identified thirty-three town forests in 1926 but only forty-two on almost nine thousand acres in 1930, or roughly half the number in the two adjoining states. Part of the explanation may lie in the state's 1915 enabling law. Initially, Vermont's act specifically required that town-owned woodland be designated as school endowment forests, but that provision was eliminated by a 1917 amendment. More significantly, a minimum size of forty acres was established, likely an effort to encourage economic productivity. That standard remained until after 1970. In Massachusetts and New Hampshire, a similar provision would have disqualified a great many town forests. Whether it dampened enthusiasm for town forests in Vermont, however, is unclear. Surprisingly, forestry commission reports in Vermont ignored this condition and listed town forests regardless of their size. As a result, comparisons among the three states are not so skewed as one might otherwise imagine. Also surprising is the fact that plantings in Vermont's forty-two town forests exceeded two million, slightly more than in New Hampshire.[29]

Nevertheless, other factors seem to confirm that municipal forestry was not quite so strong in Vermont during this period. Without an organization singularly committed to town forests, exemplified by the MFA under the leadership of Harris Reynolds, the state lacked a driving force behind its movement and did not benefit from the comprehensive literature available in Massachusetts. New Hampshire had a comparable disadvantage, but its state forestry office enthusiastically promoted town forests in a way that Vermont's did not entirely match. In New Hampshire, Warren Hale and John Foster produced a body of writing that even today conveys a keen sense of appreciation for community forestry. In contrast, biennial reports from Vermont's commissioner of forestry only briefly discussed the topic, did not expound on European community forests, and generally avoided description of forestry work in specific Vermont towns. Nor did Vermont have ancient examples of town forests such as those discovered at Newington and Danville, which generated enthusiasm among foresters and communities alike.[30]

Vermont's enabling law endured without significant amendment until 1945, when the state was required to reimburse towns for one half the purchase price of land acquired for town forests following inspection and approval by the state forester. A separate law, enacted in 1951, required communities to include articles for municipal forests in warnings for annual town meetings. Subsequently, the state was divided into two districts, and a full-time municipal forester was assigned to each. With these changes, Vermont's town forests quickly increased in number, confirming a certain sluggishness before 1945. Vermont's late start was ultimately overcome by its well-managed program organized around the state's two municipal foresters. These differences in the structure of its municipal forest movement distinguish Vermont from other New England states, and evidence of its program is generally more visible. Woodland records and maps are comprehensive, and the location of most town forests has been plotted on town highway maps prepared by the state's transportation agency. Although the forty-acre requirement may have been an impediment at first, it probably deserves at least partial credit for the fact that a substantial portion of Vermont's town forests are still utilized primarily for their timber and wood. In contrast, many smaller tracts in other New England states have, in more recent decades, become important primarily as open space in expanding urban and suburban landscapes.[31]

Despite their initial unhurried pace, town forests in Vermont lacked neither supportive literature nor sincere interest among some state foresters. In 1924, Robert Ross, the state's commissioner of forestry, prepared a short booklet titled *Town Village and City Forests in Vermont*, which briefly identified important examples around the state and quoted liberally from MFA Bulletin 132 in describing ways to create town forests. Writing for *The Vermont Review* in 1927, Kemp R. Flint, past president of the Vermont Forestry Association, outlined the general advantages of municipal forestry and isolated a specific benefit as the perfect answer for one of Vermont's perplexing economic problems, back farms in hill towns. Flint praised residents of Sheffield, located in the state's far northeastern corner, who embarked on a plan that preceded that of Russell, Massachusetts, by several years. Burdened by the expense of plowing a remote strip of mountain road on which schoolchildren were transported, Sheffield voters opted to purchase two farms served by the road. Reforestation with Norway spruce began in 1926, when the community organized a planting bee.[32]

Most writing about Vermont's town forests appeared in *Green Mountain State Forest News*, published first, in 1925, as a typewritten handbill and later, beginning in 1929, as a more polished newsletter from the state's forest service. Articles typically examined acquisition and planting activities or highlighted efforts by specific towns, usually in Vermont but occasionally scattered elsewhere around New England. Methods for encouraging town forests in Vermont more closely matched those employed in New Hampshire than in Massachusetts, due primarily to the role of government foresters dictated by enabling legislation. After considering the suitability of town land for municipal forests, state foresters either advised or controlled most forestry work, including planting, management, and cutting.

At first, yearly distribution of up to 150,000 trees from the state nursery at Essex Junction was authorized but not at a reduced price. A 1927 amendment to Vermont's forest nursery law finally allowed the commissioner of forestry to sell seedlings to towns, schools, and churches at cost, and that was lowered by half two years later. In response to growing unemployment, a special legislative resolution adopted in 1933 offered surplus seedlings to towns at reduced cost. Direct appeals to towns assumed other forms as well. In an open letter to town selectmen in 1926, Governor Franklin Billings urged voters to consider the question of municipal forests at annual town meetings, and similar letters were issued by Billings's successors. At about the same time, the Vermont Forestry Association offered to plant five thousand trees in any town establishing a forest of one hundred acres or more.[33]

Municipal forestry in Vermont began principally on watershed lands protecting reservoirs for the cities of Barre, Bellows Falls, Essex Junction, Montpelier, and Rutland. These represented the state's largest town forests, and in each case planting was started before 1926. Among these examples, the Essex Junction Village Forest is noteworthy as one of the state's most capably managed, often cited for its scrupulous timber stand improvement and productivity. The forest also boasted Vermont's only municipal forest fire tower (fig. 4.13).

Fig. 4.13. Fire tower in the Essex Junction Village Forest, Essex, Vermont, a rare feature in New England's town forests. Improvements such as compartment lines, logging roads, entrance gates and signs, mills, nurseries, footpaths and hiking trails, and recreational facilities assist in the interpretation of cultural landscape history. Artifacts such as stone walls, cellar holes, and even witness trees, together with existing tree species and soil composition, can also help in reconstructing the natural or culturally assisted evolution of forest cover. Private Collection, William Hall

Towers, entrance gates, roads, compartment lines, and many other physical improvements incident to forest management contribute to the visual history of these lands. Acquisition of Essex Junction's forest began before 1900 on property surrounding two springs near the small village of Essex Center. Approximately 750 acres were purchased, and a forestation plan was implemented in 1923. By 1930, more than 400,000 seedlings, mostly scotch, red, and white pine, were thriving in the forest's sandy soil. A painstakingly detailed timber-stand map showing plantation types and ages was crafted by Charles Lockard and Huntley Palmer of Vermont's Forest Service in 1931 (see fig. 4.1). Unfortunately, much of the land was sold for private development in 1978 after the city entered the Champlain Water District. A portion, however, remained open as a condition of the sale.[34]

Occasional gifts, poor-farm woodlots, and outright purchases characterized the origins of other town forests established in Vermont before 1930. Unlike New Hampshire and Massachusetts, very few communities converted tax-forfeited lands, and this too may have contributed to Vermont's slower pace. Despite the state's minimum acreage requirement, residents of Woodstock voted in 1927 to accept a gift of twenty-two acres on the slopes of Mt. Tom. The town's benefactor, Governor Franklin S. Billings, had been appointed chairman of their town forest committee a year earlier. Proctor, a community in west-central Vermont made prosperous by its marble quarries, received 112 acres of neglected farmland from Mortimer R. Proctor and by 1932 had planted more than thirty thousand white, scotch, and red pine seedlings. A management plan and timber-stand map were prepared in 1933 by Wilbur E. Bradder, a district forester (fig. 4.14). Bradder marked composition, age, and history for each compartment and recommended detailed treatments. For instance, compartment 9 represented an area of abandoned pasture dominated by poor quality white pine, with secondary growths of hawthorn and hardwood saplings. Ground cover consisted of grass, berry brush, and shrubs, with spot plantings of white pine from 1926. White ash planted in the compartment's northwest corner during 1927 had failed. Bradder recommended clearing wolf hardwoods for cordwood and girdling the poorest pine. He also encouraged retention of food trees, apples, thornapples, and cherry to support game birds and animals.[35]

By 1930, a number of Vermont towns had converted their poor-farm woodlots to town forests. Many other towns simply purchased old pastures and began reforestation. Calais, a rural community located north of Montpelier, continues to own a one-hundred-acre woodlot once attached to its poor farm. Transplanting began in 1925, and annual reports by the town forester between 1926 and 1930 show regular sales of lumber and bark to an assortment of buyers, among them the United States Clothes Pin Company. When the poor-farm buildings were sold in 1943, an easement was reserved for access to the

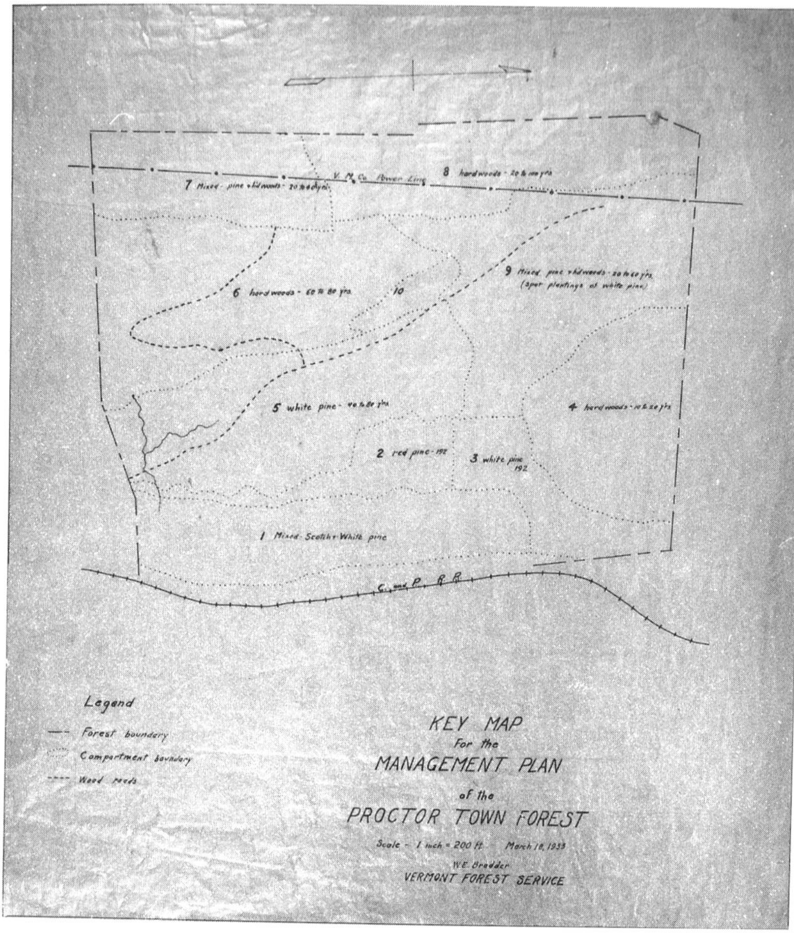

Fig. 4.14. Forest type map of a town forest in Proctor, Vermont, drawn by Wilbur Bradder for the state's forestry department in 1933. Bradder's plans reveal a keen interest in managing woodlands for wildlife conservation, and he took pains to preserve food trees such as hawthorn and bird cherry. Vermont Department of Forest Parks and Recreation—Waterbury

forest. In 1922, the St. Johnsbury's Women's Club organized reforestation of the town's 120-acre poor farm, accessible from Alms Road. Nature trails currently wind their way through plantations of white pine. Poor-farm woodlots in Danville and Thetford also became town forests.[36]

Forestry's relationship with agriculture was not always limited to the simple planting of worn-out pastures but encompassed the cultivated landscape of wildlife conservation as well. Wilbur Bradder spent summers in Tinmouth, Vermont, near the Tinmouth Channel, a sizable wetland region that today supports a multitude of bird and animal habitats. Bradder's fascination with improving neglected farms is visible in his remarkable *Game Food and Cover Improvement Plan of a Typical Abandoned Farm and Woodland Area at Tinmouth, Vt.*, which he prepared in 1934 (fig. 4.15).

Bradder divided the land into separate compartments for hardwoods, evergreens, meadow, pasture, and marsh, and neatly blended woodland management with the introduction of game foods and shelter. Hollow oak trees were left standing in logged-off areas for gray squirrels, and poor soil was planted with hickory and butternut. Red pine and spruce were planted, with grass margins doubling as fire lanes and nesting sites for pheasant, quail, and grouse. Release of cordwood borders, hedgerows, and apple orchards freed grape, thornapple, bittersweet, and cherry and provided food for deer and wildfowl. Swamp hardwoods were girdled to release cedar, grape, viburnum, and nightshade. Old woodlot roads or cartpaths provided gravel for pheasant and grouse or were seeded with clover. Marsh edges, rocky spots, and marginal areas became sites for winter food: corn, cowpeas, and barberry. Brooks were dammed to provide food and winter areas for trout, along with ponds for muskrats and waterfowl. Stream deflectors exposed gravel and food for trout, and discarded tree stumps provided cover. Marsh areas offered ideal conditions for wild rice or cattails and flag for muskrats. Bradder must have relished his work to provide such intricate detail, and his plan is one of the best examples of forestry's expanding influence on landscapes.

New England's three remaining states, Maine, Connecticut, and Rhode Island, had far less ambitious programs and were well behind in numbers of town forests by 1930. Legislative bodies in all three states were very late in passing enabling statutes: Maine in 1927, Rhode Island in 1929, and Connecticut not until 1939. However, the absence of enabling laws should not have posed an impenetrable legal barrier. Instead, tardy passage of laws was more indicative of political apathy rather than inability of communities to at least own forest lands. More pivotal was the fact that enabling laws in each of these states, when finally passed, failed to require management by state forestry organizations, either governmental or nonprofit. Without active and prolonged encouragement from an outside source, town forest activity in these states lacked the crucial ingredient that distinguished vigorous programs in New Hampshire, Massachusetts, and Vermont. To some degree, town forests in Maine and Connecticut followed Vermont's pattern and increased more rapidly, proportionally so, after World War II. Still, the widespread acquisition and planting that characterized New England's town forest movement before 1930, along with the steady flow of supportive literature, never materialized in either state.[37]

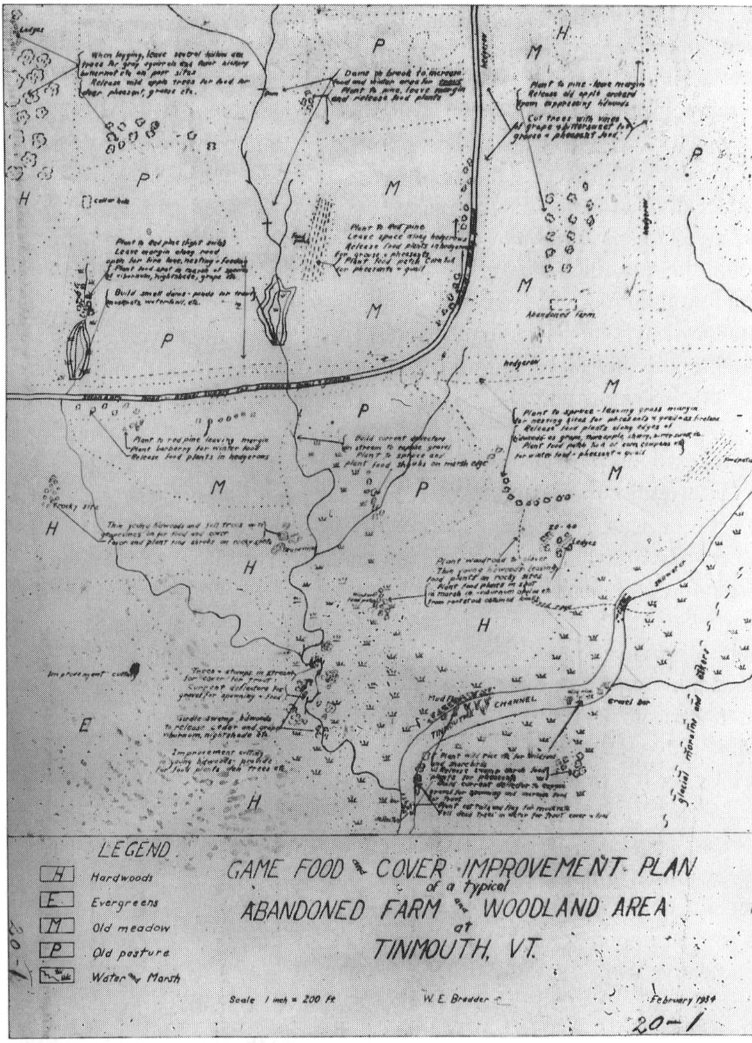

Fig. 4.15. Wilbur Bradder's plan for improving abandoned farmland in Tinmouth, Vermont, prepared for the state's forestry department in 1934. In New England, forestry offered a ready answer to agriculture's problem of abandoned farms, and worn-out pastures were routinely planted with rapid-growing, salable coniferous types, usually white or red pine. Bradder's plan, however, is an unusual example of forestry's contribution to the very detailed planting of game food, such as cowpeas, barberry, and wild rice. The area is now the Tinmouth Channel Wildlife Management Area and is owned by the state. Vermont Department of Public Records—Middlesex

Maine's 1927 town forest law authorized communities to acquire land for the culture of forest trees or preservation of water supplies and required the state forest commissioner to furnish seedlings at cost. Management was simply left to the discretion of towns, which were encouraged to appoint a trained forester. Forest commissioners' reports had begun auspiciously in 1891 with George Talbot's paper calling for a municipal forest program. Ironically, Talbot proposed enabling legislation requiring supervision of planting and cutting by state foresters. However, thirty years passed before the subject of town forests was again broached in a few short paragraphs by Samuel T. Dana, forest commissioner in 1922. Later reports referred to town forests only intermittently and without description of individual examples. For a brief period, Maine's forest service printed a newsletter, *Town Forest News*.[38]

In the biennial report for 1931–32, town forests were identified in eight towns: Bangor, Brunswick, Fryeburg, Kennebunk, Old Town, Skowhegan, Waterford, and Cumberland Center. Austin Cary's plantation on Brunswick Commons had started to produce results by this time, and a forest-type map was prepared in 1926. In 1928, Fryeburg's eighty-acre forest provided a sufficient amount of lumber from stands of white pine to pay for an addition to the town's school. Old Town converted more than two hundred acres of its poor farm to forest and christened it the Theodore Roosevelt City Forest.[39]

Connecticut's 1939 law required towns to create three-member forest commissions to manage town forests and did not encourage involvement by government foresters except to supply planting stock. Reports by the state's park and forest commissioner contained almost no information about town forests other than accounts of trees sold. The Connecticut Forestry Association's quarterly journal, *Connecticut Woodlands*, carried only an occasional article. Prior to enactment of the state's town forest law, a number of towns purchased property under statutes that specifically allowed land acquisition for parks or protection of water supplies. By 1930, roughly thirty town forests had been established, but only a small number were not on watershed lands. Farmington residents voted to create a memorial town forest in 1926 and purchased 180 acres of land from a local philanthropist; management responsibilities were given to a board of trustees. In 1905, Salisbury received a parcel of mountainous woodland that became Barrack-Matiff Town Forest, named after the peak it bestrides; the Appalachian Trail now courses over that summit. In 1924, the town also set aside 140 acres from its poor farm and began managing it profitably with assistance from state foresters. Newtown received a five-acre tract of hardwoods in 1923, and management was supervised by the Newtown Forester's Association. Ellington obtained slightly more than fifty acres of mixed hard- and softwoods in 1927, subject to a condition that management be conducted according to the advice of state foresters for a period of fifty years.[40]

Recreational aspects of town forests seemed dominant in many Connecticut

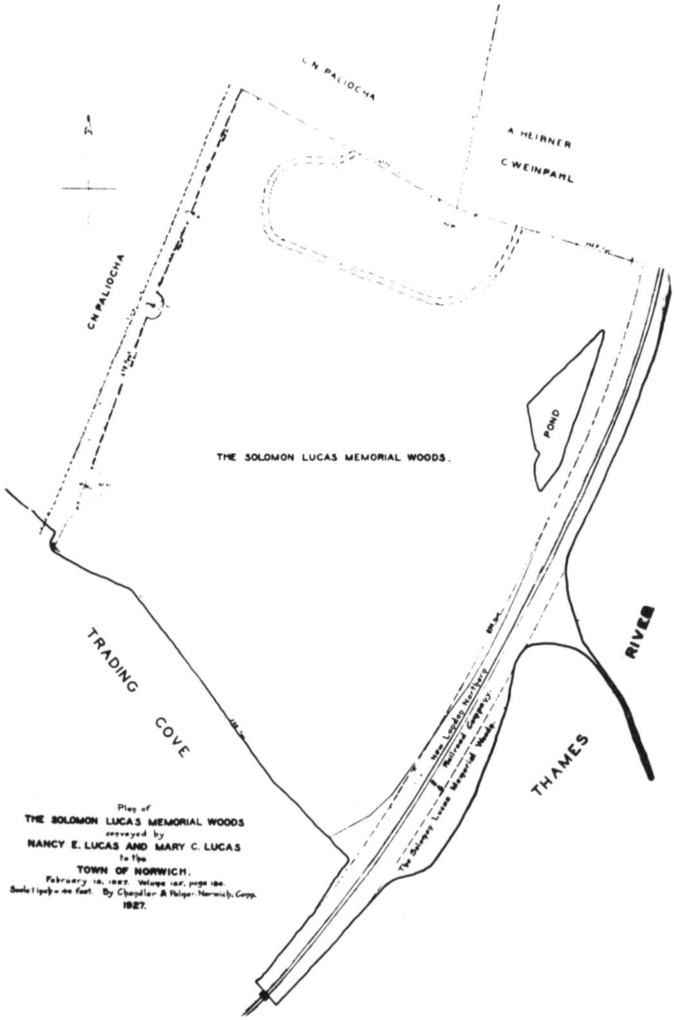

Fig. 4.16. Survey of the Solomon Lucas Memorial Woods, a town forest in Norwich, Connecticut, drawn by the firm of Chandler and Palmer in 1927. Connecticut's town forests generally lacked the emphasis on commercial productivity that characterized the movement in Massachusetts, New Hampshire, and Vermont, where the timber industry was much stronger. The state's most important examples were established as watershed reserves or forest parks. Office of the Norwich Town Clerk

communities. In Norwich, a memorial forest was established at a site bordering the Thames River in 1927 (fig. 4.16). The deed of conveyance to the town set explicit conditions that only hint at a commercial function but expressly oblige recreational use.

> Grantee shall forever hold and maintain the conveyed land in its wooded state as a town forest in accordance with the rules of modern forestry, . . . [and] that said land shall always be open to the public of Norwich for recreation purposes and that owners of dogs shall at all times be allowed to take their dogs and let them run free from leash.[41]

Rhode Island's legislation followed Maine's by only two years, but it provided no mechanism for management of municipal forests until 1954, when town councils were authorized to establish local forest and park commissions. Scituate was among the few towns to establish a municipal forest, but work on its lands was not reported by state foresters until 1933. Public forests in Rhode Island were controlled primarily by the Metropolitan Park Commission in Providence, created in 1904, which acquired a number of sites scattered around that tiny state. The commission was abolished in 1935 and its lands transferred to the Division of Forests, Parks and Parkways, housed in the Department of Agriculture and Conservation.[42]

Municipal Forests in Other Regions

Any measure of success for this phase of New England's town forest movement should take into account, at least briefly, the broader context of municipal forestry in other regions during this same period. The movement was a popular one, advocated by a significant number of prominent foresters and accompanied by a substantial body of periodical literature. Forestry organizations in many eastern and midwestern states, governmental as well as nonprofit, touted the benefits of municipal forests and doggedly encouraged towns to acquire and plant abandoned farms or cut-over woodland. In this respect, the movement followed patterns similar to those that occurred in New England.

The extent to which New England's initial success with town forests swayed other states is difficult to calculate. Influence seems clear if one considers the number of articles appearing before 1930 in journals with national readership such as *American Forestry*, *Journal of Forestry*, *American City*, *Recreation*, *Nature Magazine*, and *American Lumberman*. A significant percentage of those articles were either written by Harris Reynolds or were written about specific town forests in New England. Reynolds also contributed to several state forestry journals, such as Pennsylvania's *Forest Leaves*, Minnesota's *North Woods*, and *Ohio Forest News*, the last, however, not until 1945. At least one of his articles appeared in a Canadian journal, *The Illustrated Canadian Forest and Outdoors*, in 1920. As president of the American Tree Association, Charles Lathrop Pack's endorsement of Reynolds's 1925 book was unquestionably aimed at a nation of forest users who had wastefully consumed America's woodland resources.[43]

One of New England's important contributions to the broader national movement clearly lies with its emphasis on the role of state foresters or forestry associations. With remarkable accuracy, James W. Toumey predicted the important link between state and town forestry in a paper presented to the Massachusetts Forestry Association that appeared in *American Forestry* in 1922: "It is my judgment that communal forestry must be closely linked up with the state forest administration and laws promulgated that will afford a form of co-operation between the community and the state which will insure to each communal forest a reasonably high order of forest management."[44]

Despite New England's achievements, municipal forest campaigns in some states, particularly New York and Pennsylvania, were influenced predominately by their own professional foresters. Activity in New York alone dwarfed that of all New England states combined. By 1930, New York's conservation department reported 386 community forests with nearly 31,500,000 trees planted. New York's forestry journals consistently carried articles about municipal forests, and writings by several of the state's well-known foresters, Ralph Hosmer, Nelson Brown, James R. Simmons, and others, appeared in key periodicals such as *American City* and *American Forestry*.

Hosmer, who worked for the U.S. Forestry Division under Gifford Pinchot and who later became a forestry professor at Cornell, penned two articles with detailed descriptions of community forests in Germany and Switzerland. Nelson Brown first began writing about municipal forestry in 1914 with an article titled "Possibilities of Municipal Forestry in New York." Simmons, forester and secretary of the New York State Forestry Association, was an active writer on behalf of town forests; and George D. Pratt, New York's conservation commissioner, contributed "Communal Forests for America," which appeared in *American City*'s Town and Country edition in 1920. Herbert Prescott, secretary for the state's conservation commission, prepared a short book, *Municipal or Community Forests*, which was published in 1926. Much of this writing was confined to familiar themes: need for reforestation, wastefulness of idle land, potential economic benefits of town forests, European community forests, and examples of towns that had begun plantations.[45]

• • •

New England's town forest movement successfully fostered community interest in forestry and emerged as an effective means for public education, precisely as Harris Reynolds had hoped. Municipal forests also played a part in progressive policy aimed at reversing deforestation, and the results of that policy can be seen today in New England's expansive forest cover. Policy touched more than just wasteful practices on timberlands; it extended to idle farmland and wildlife habitats. Cultural landscapes of conservation epitomized by William Bradder's "Game Food and Cover Improvement Plan," in Tinmouth, Vermont, were the outcome. The movement also sparked a trend toward acquisition of open, undeveloped land by communities. This course has evolved through most of the twentieth century and has had far-reaching implications concerning diverse use and stewardship of these areas. The manner of use and care reflect collective attitudes toward conservation that, in turn, contribute significantly to community structure and habitability.

During the plantation phase, the acquisition of suitable land and planting activities received the greatest emphasis, but forest management was not neglected. From their woodlands, towns received moderate and, on occasion, substantial economic returns generated by sales of cordwood, pulpwood, or saw timber. However, income depended less on forestry management than on the presence, or lack of it, of mature timber stands. Comparisons to European community forests had raised a high level of economic expectation, and Reynolds and other advocates warned against anticipation of short-term profits. More subtle notes of caution by Reynolds underscored the potential for recreational benefits and may have revealed misgivings about whether New England towns could manage timber yields comparable to those in Swiss and German forests. Closing the plantation phase in 1930 is, admittedly, somewhat arbitrary. Effective management of many town forests had begun earlier, and many plantations were started later. Nevertheless, the rate of increase in the number of new town forests declined after 1930 while emphasis on timber management expanded. Economic productivity remained the overriding issue for future deliberation.

Chapter 5

Salesman's Samples for the National Forests

New England's town forests matured in several ways after 1930 (see fig. 5.1). Stewardship acquired greater definition as emphasis shifted conspicuously from acquisition and planting to woodland management. Many towns had neglected their young plantations, and rapid-growing hardwoods unsuitable for commercial sale—gray birch, alder, bird cherry, and scrub oak—sprouted on cut-over lands or old pastures and quickly overtook coniferous seedlings. Communities were urged to employ good silvicultural practices and to weed, thin, prune and release their forests. Appeals to town forest committees stressed the need to keep careful management records, and foresters began to inspect timber yields more closely. The change in focus from planting to management coincided with the inauguration of conferences for town forest committees in Massachusetts, sponsored by the Massachusetts Forestry Association (MFA) and held yearly from 1928 to 1961.

Although beneficial uses still garnered acclaim, the viability of timber cropping received much closer scrutiny than before. Inquiry exposed an assorted collection of economic, administrative, and political problems, ranging from fickle town governments to competing local interests. Municipal forestry was compelled to strike a tenuous balance between a critical need for technical assistance by trained foresters and fear of control over town land by state or federal experts. External problems such as New England's declining timber industry and dwindling local markets for lumber also contributed difficulties. Although commercial productivity through sustained-yield management continued to define the movement, growing skepticism about commercial productivity opened the way for ever stronger public interest in recreational use.

National developments assumed particular importance, influencing opinion on matters of timber production, multiple use, and, ultimately, the movement itself. Franklin D. Roosevelt emerged as a champion of community forests, and

Fig. 5.1. One of three town forests in Calais, Vermont, rises from a hill above Gospel Hollow, where a town hall was built in 1866. Uniquely descriptive place-names associated with many community-owned woodlands often portray the vernacular qualities of New England's cultural environment. RLM, 1990

the Civilian Conservation Corps and Works Progress Administration (WPA) confirmed that local forests could be used to employ those trapped by economic woes. At Roosevelt's request, the U.S. Forest Service finally became a participant in 1938, when it created a formal program for community forests, broadly defined to encompass properties owned and operated for forestry purposes by all types of political subdivisions and nonprofit groups. The service's program stressed public education, successfully so, and disseminated an ample body of literature touting multiple benefits. By 1942, the Society of American Foresters had formed a national committee on community forestry, chaired by Harris Reynolds, and had adopted a definition equivalent to that used by the Forest Service. The movement briefly gained new favor; municipal forests continued to increase in number, albeit at a slower pace; and state nurseries still generated a seemingly inexhaustible supply of seedlings for transplanting. However, skepticism about timber productivity soon tainted the Forest Service's program, and their support withered, shrinking with it the movement's core purpose and, ultimately, the movement itself.

Better to Weed than to Plant

Energy directed toward persuading towns to acquire land and transplant seedlings during the plantation phase gradually yielded to concerted accent on timber management. In his address to the MFA's annual gathering in 1914, William Colton had taken aim at skeptics who doubted the practicality of such projects by outlining a complete management plan for town forests. Whereas contests of 1915 and 1916 had obviously been engineered to launch a planting campaign, rules were arranged to give highest points to those towns that considered future care and protection of their forests. Later, the MFA's standing offer to provide free seedlings had been conditioned upon towns' preparation of land for planting, removal of wild currant and gooseberry bushes to eradicate blister rust, and the promise of reasonable care and protection of seedlings.

In New Hampshire and Vermont, a loosely structured approach to forestry management had been incorporated into enabling laws and depended on informal cooperation between local committees and state foresters. Town forest committees in Massachusetts had much greater control over their forests, and a system for silvicultural instruction developed around yearly conferences for committee members organized by the MFA. At these meetings, often conducted in towns where solid progress in forestry had occurred, committee members typically received a two-day dose of papers and field demonstrations on subjects of timber management. In 1928, an inaugural Boston conference assembled a wide range of speakers, who left little doubt that forestry management was the day's principal topic. Albert Cline stressed that weeding was the most important silvicultural treatment for improving composition and quality of timber stands. Weed trees and inferior sprout growth proliferated in New England's forests, the consequence of repeated logging, forest fires, and abandoned farms. Dollars were best spent, Cline argued, on cutting back worthless stump sprouts and weed species in favor of single-stemmed saplings of pine or valuable hardwood. With a little effort and minimal expense, future crops of cordwood could be transformed into far more profitable stands of saw timber.

Other speakers expounded similar themes. Claude R. Tillotson, a forest inspector for the U.S. Forest Service stationed in Amherst, Massachusetts, viewed in town forest committees an opportunity to demonstrate the practical value of forest management. He exhorted members to achieve the highest possible productivity from their forests. Robert Parmenter, an extension forester in Massachusetts, outlined a recently published MFA bulletin, "Pruning for Profit as Applied to Eastern White Pine," which furnished guidelines for growing clear lumber to command higher prices. Harold O. Cook, chief forester for the Massachusetts Division of Forestry, summarized the assistance available from the state, which included free seedlings and technical advice upon request. He also urged participants to work with district wardens to avoid fires. Approximately 150 committee members attended the conference and listened to representatives from national organizations as well. Ovid M. Butler, executive secretary of the American Forestry Association, cited the educational campaign for town forests in that association's monthly journal.[1]

Subsequent conferences continued to emphasize management concerns, along with other problems connected to the administration of town forests. Cook, Cline, and Parmenter were often key speakers, along with Reynolds, who customarily reported progress in town forests throughout the state. A year later, the towns of Westfield and Russell hosted committee members for two days in early November. Cook disclosed the state's offer to prepare five-year management plans at no cost to communities. Plans covered essential forestry operations and were designed to help committees estimate annual appropriations needed for proper management. During field excursions to the forests in these two towns, Cline demonstrated the importance of weeding and pruning. Roundtable discussions were initiated in the hopes of resolving the various political and economic problems some committees had encountered, such as conflict with selectmen or general apathy. In 1930, the gathering moved to Fitchburg, where Cook reported that five towns had applied for management plans. Cline expanded his discussion to include insect control, notably treatment of pine weevil. Towns also were advised to keep accurate accounts of expenses and income for scientific study. Field trips to forests in Ayer and Groton were conducted, and Governor Frank Allen's address to the conference was broadcast by radio.[2]

In addition to providing an effective means for disseminating forestry instruction, annual meetings of town forest committees provided extra structure to the town forest movement. At the close of the first conference, a committee was established to prepare the following year's assembly and to present a plan for a permanent organization of town forest committees. William Wharton, Guy Hubbard, and Reynolds were appointed and, after weighing the myriad difficulties associated with the formation of a new conservation organization, instead recommended that each year's conference elect an executive committee. As envisioned, this group would arrange each meeting and prepare a program, all the while working closely with the MFA. An executive committee of five was elected at the second conference, including Reynolds, who became the group's secretary. Gradually, this committee assumed greater autonomy and influence in matters affecting town forests while Reynolds and the MFA skirmished along other fronts such as conservation legislation, Dutch elm disease, and a depression economy. In 1933, the association changed its name to the Massachusetts Forest and Park Association (MFPA) to better reflect its expanding activities.

Collaboration among town forest committees, the MFPA, and the executive committee for town forest conferences became increasingly important as the country's economy worsened. In 1931, the association withdrew its proposal to transplant five thousand trees for any town that acquired a hundred-acre town forest and instead offered to weed five acres of woodland for free. At a time when woodland management was most needed in many town forests, communities were least able to afford the work. With the formation of emergency relief agencies such as the Civil Works Administration and the WPA, however, armies of labor became available. The MFPA encouraged the federal government to commit part of this work force to forest improvement in towns, and the association simultaneously helped to organize jobs. In 1933, for example, Reynolds circulated a letter to town forest committees asking them to estimate work needed on their forests. On the basis of their replies, he identified tasks, including roads, fire lines, trails, water holes, weeding and pruning, gypsy moth and blister rust control, clearing for planting, and camp and picnic sites. All the while, the state forestry department continued to furnish an abundance of seedlings.[3]

Toward the end of 1936, economic constraints had begun to loosen, and the MFPA published Bulletin 159, "The Town Forest in Massachusetts, a Twenty-Two Year Record," which announced renewal of the association's free offer to plant five thousand trees on newly established town forests. That summer a survey of town forests revealed continued need for silvicultural operations, primarily weeding, pruning, and improvement cutting in older stands. Emergency employment in many towns had lacked supervision by trained foresters and as a result had been confined to chores such as road and fire-line construction. Reynolds authored the bulletin and concluded that the town forest movement in Massachusetts was far beyond an experimental stage. At the same time, he cautioned that forestry management was necessary if quality products and profits were expected, and he urged towns to appropriate sufficient funds at their annual meetings. Members from several communities had voiced alarm about state control of town forests, and Reynolds offered reassurance on that point. Bulletin 159 also identified several other publications germane to management that had been prepared by the MFPA in cooperation with Harvard Forest. A year later, the association began publication of their newsletter, *Forest and Park News*, which became the primary voice of the executive committee for town forest conferences.[4]

In Massachusetts, the balance struck between strong leadership by the MFPA and well-organized communication among active local committees contributed to able management of forests in many towns. Groton's forest became self-supporting and relied on WPA labor for cutting, thinning, fire-lane construction, and releasing both natural and planted stands of timber. Further south, Norton's reforestation committee provided exceptionally thorough annual reports and closely followed the advice proffered at yearly conferences. In

1929, the committee invited Cook to inspect their forest and the following year recommended acceptance of his offer to prepare a five-year management plan. In 1929, too, Reynolds examined the forest, identified tracts where pruning and weeding were necessary, and selected an area for the planting of five thousand pine trees. In 1933, the committee urged cooperation with the state's welfare board in obtaining employment for planting and silviculture. Three years later, they reported significant improvements to the forest realized through the Emergency Relief Administration.[5]

Without question, foresters and groups attentive to the town forest movement in other New England states recognized a need to counsel proper management in order to achieve commercial profit. However, evidence of any persistent effort to convey this message to local committees is spotted. In New Hampshire, town committee members from ten communities assembled in Warner on a November afternoon in 1930 to discuss their forests. Warren Hale organized the gathering, and John Foster, Ralph Morgan, and Orrin James participated in a short speaking program followed by an inspection of Warner's forest. Hale's report offers no hint that topics of silviculture were emphasized, but various problems associated with local forestry were aired nevertheless. It is not clear whether any other conferences for town forest committees were held in New Hampshire.[6]

Although Foster and Hale must have underscored the need for weeding and thinning at every opportunity, their efforts depended on local initiative. In 1938, the Society for Protection of New Hampshire Forests (SPNHF) restated MFPA concerns from a decade earlier by appealing for improvements to local forests. Removing poorly formed wolf trees, thinning and pruning neglected plantations, and salvaging weevil-infected pine were key responsibilities. That same year, the state forester circulated a questionnaire among New Hampshire communities to obtain information about town-owned woodland. More than half of the acreage reported had been acquired through tax forfeiture since 1932, and an offer of state assistance in planting and management was renewed.[7]

There is even less evidence of any direct effort to isolate the problem of town-forest management in Vermont, Maine, Connecticut, or Rhode Island at this time. The emerging predicament was not so much an inability to discern the issue as it was a failure to sustain communication with local committees and officials. For their part, even towns that received a clear message often did not translate enthusiasm into steady practice. A short article appearing in a 1930 issue of *Green Mountain State Forest News* confirmed that Vermont's forest service did stress the need for careful long-term planning and would prepare free management plans and perform obligatory cutting and thinning. In a few cases, silvicultural operations on well-established municipal forests, such as Essex Junction and Rutland, were widely publicized and helped to spread the

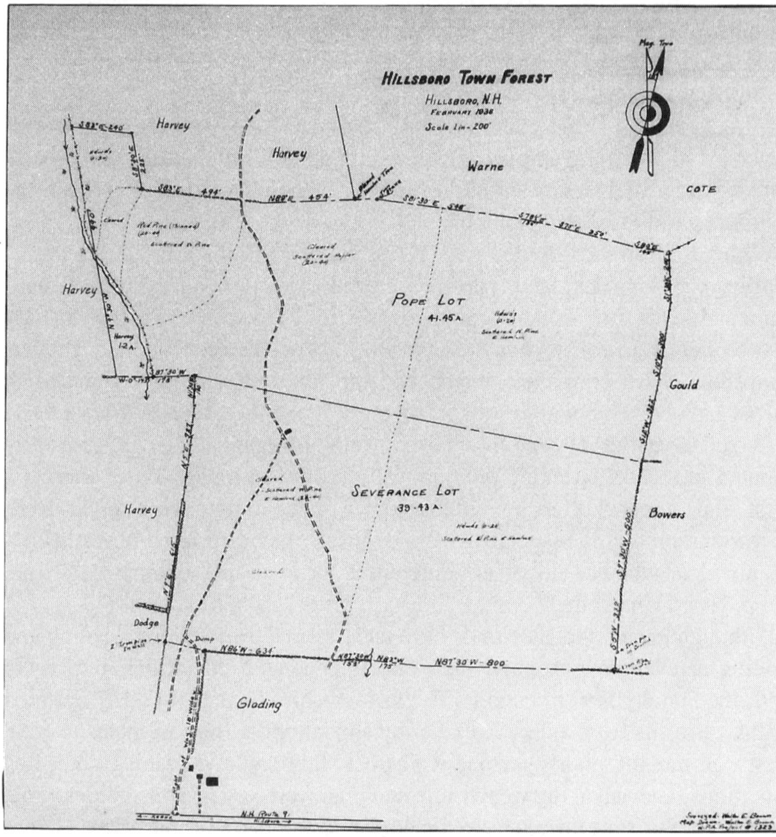

Fig. 5.2. WPA Project No. 1323: Forest type map of a town forest in Hillsboro (Hillsborough), New Hampshire, prepared by Walter E. Brown for the state's forestry department in 1938. Depression-era programs such as the CCC and WPA supplied labor for a number of town forests and arrived at a time when improvements such as weeding and pruning were sorely needed. New Hampshire Division of Forests and Lands—Concord

word. In 1932, the state's forest service circulated articles in local newspapers and pleaded for high quality timber, arguing that low-grade silviculture was as economically unjustifiable as one-crop agriculture. Yet these missives were not aimed squarely at town forest committees. In Maine, communities were permitted to hire their own foresters to supervise management, but statewide coordination appeared to be lacking.[8]

Ironically, the absence of aggressive campaigns promoting silviculture in some states' municipal forests was partially offset as town woodlands became convenient terrain for relief employment. In New Hampshire, state foresters

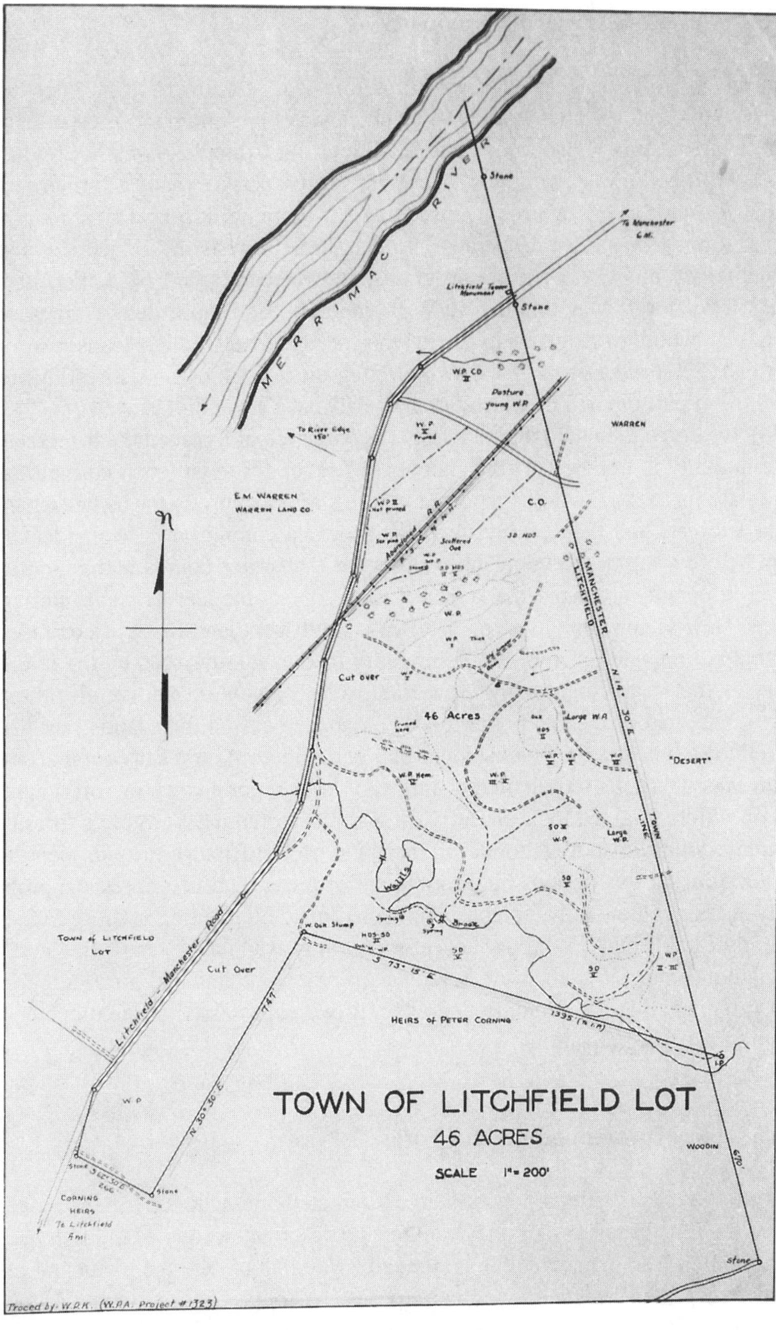

Fig. 5.3. Undated forest type map of land owned by the town of Litchfield, New Hampshire, traced by Walter P. Knoettner for the state's forestry department and part of the same WPA project that involved work for Hillsborough. Franklin Roosevelt championed municipal forestry and prodded the U.S. Forest Service into starting a community forestry program in 1938. New Hampshire Division of Forests and Lands—Concord

and the SPNHF encouraged towns to require work on town forests in return for welfare, and the society offered technical advice. A number of type maps for town forests—Hillsborough and Litchfield, for example—were drafted by WPA workers (figs. 5.2 and 5.3). Vermont's forest service made a similar proposal to mark trees for pruning, weeding, or thinning during winter months when support was most desperately needed. Town forests in Connecticut also benefited from Depression-era programs. Between 1935 and 1938, Farmington's town forest was thinned by WPA laborers, and more than a thousand cords of wood were given to the town's needy. The Civilian Conservation Corps (CCC) released thirty acres of young pine in Ellington's forest during 1937, and participated in reforestation of Salisbury's woodlot as well.[9]

Between 1930 and 1938, New England's municipal forests did not increase substantially in number. At the annual conference for town forest committees held late in 1938, Reynolds reported that 105 Massachusetts towns had established forests and that 4 other towns had formed committees to consider the topic. In his report, Reynolds also observed that water boards had generally been reluctant to change the status of their forests for fear of state interference. New Hampshire's survey of town-owned woodland in 1938 coincided with a nationwide inventory of community forests administered by the Forest Service that year. Regrettably, New Hampshire's study did not identify those towns that had elected to begin forestry management on their lands, verified by either town vote or establishment of a committee. As a result, comparison with Massachusetts is difficult. Overall, 140 towns reported owning forest land, 98 of which obtained their property through tax forfeiture. Vermont's forestry commissioner did not elaborate on the topic of municipal forests in biennial reports during this period, but an estimate of fewer than fifty forests is probably accurate. Generally, emphasis remained fixed on planting in those forests already established. Equally scant information is available for New England's remaining states; if towns were quietly acquiring woodland and practicing forestry management, and some undoubtedly were, it passed unmentioned by state forestry organizations.[10]

Community Forests and the U.S. Forest Service

Prior to 1938, the federal government participated only indirectly in the movement for town forests. Its policy can best be described as one of indifference, and the benefits that accrued to community woodlands were the incidental result of programs designed for other purposes. Although Gifford Pinchot's technical assistance program occasionally found its way to town lands, Circular 21 hardly advanced municipal forestry when its opening sentence flatly identified only three categories of forests: federal, state, and private. To be fair, the small group of men who staffed the U.S. Forest Service during its early years faced enormous political, social and economic challenges in shaping forest policy for the nation. Understandably, ventures in town forests among a few of New England's small states were not likely to command Washington's attention for very long.

Changes in federal policy toward municipal forestry arrived in small pieces. Wasteful practices in privately owned forests presented a conundrum for the Forest Service, and public ownership repeatedly surfaced as one potential solution. When it did, community forests were sometimes included with proposals for state or federal land acquisition. Henry S. Graves, who had succeeded Pinchot as chief forester in 1910, organized a committee within the Society of American Foresters in 1919 to foster public control of timber cutting on private lands. The Committee for the Application of Forestry, as it became known, was chaired by Gifford Pinchot and included Frederick E. Olmsted. In February 1920, it issued a report that recommended legislation establishing regional administration of regulation over private forests. In addition, it urged cooperation with states in acquisition of woodlands by the nation, states, and communities.[11]

During the tenure of Graves's replacement, William Greeley, the Clarke–McNary Act was signed into law by President Calvin Coolidge in 1924. This legislation authorized federal participation in forestry projects lying outside the watersheds of navigable streams and expanded the reach of federal matching funds available to state forestry agencies. Control of forest fires, study of tax laws and their effect on forestry, and cooperation between federal, state, and private woodland owners were all stressed by the law. Reforestation efforts also were encouraged, and part of the cost of seedlings raised in state nurseries and transplanted for "wind breaks, shelter belts, and farm wood lots upon denuded or non-forested lands" was subsidized by the federal government. By its language, the Clarke–McNary Act appeared to exclude costs of seedlings transplanted onto public forests unless used for wind breaks or shelter belts. However, the extent to which states segregated the seedlings they distributed is unclear, and some government-subsidized transplants undoubtedly landed in town forests.[12]

Vermont-born Coolidge, governor of Massachusetts from 1918 to 1920, was well acquainted with New England's municipal forest movement. In addition to endorsing the Clarke–McNary Act, Coolidge was responsible for calling the National Conference on Outdoor Recreation that convened in Washington, D.C., for three days in late May 1924. Declaring the need for a national recreation policy that put out-of-door pleasure within grasp of the rank and file, Coolidge appointed a committee from his cabinet to organize the enormous gathering. Harris Reynolds was one of scores of speakers to address a broad agenda that, among many other goals, sought closer cooperation between federal agencies and municipal governments. In his brief allotment of time, Reynolds outlined the recreational advantages of town forests, remarking that every community should provide accessible domain for tramping or automobile camping. The year 1924 also marked the beginning of the Forest Service's newsletter, *Forest Worker*, published every other month until discontinued in 1933. Its pages often contained articles about New England town forests, a measure of recognition by the service.[13]

For nearly a decade, the federal government's passive role in municipal forestry changed very little. Reluctance to interfere with local interests, best left to state forestry, coupled with a continuing need to shape policy in national forests, were certainly part of the cause. An undercurrent of doubt about timber productivity also may have been a factor. For reasons unknown, state and private forestry groups made negligible efforts to recruit Forest Service assistance. For their part, U.S. foresters were certainly aware of the movement. In his annual reports filed in Washington, Claude Tillotson had observed the progress of the movement. In 1927, for instance, he described Massachusetts's plans for acquisition of more and larger municipal forests. The following year's report contained a summary of his presentation at the first conference for town forest committees. At that gathering, he remarked that a permanent organization of these committees would be a strong supporter of state forest policies and could wield political influence.[14]

Amorphous policy toward community forests began to acquire more tactile form in 1933, when the secretary of agriculture furnished Congress with a document titled *A National Plan for American Forestry*, known as the Copeland Report. The report's underlying philosophy sought a reexamination of forestry and its role in national welfare during a time of social and economic misgiving. Based on an exhaustive, well-conceived survey of American forests, this plan recommended increased public ownership of woodlands as a key for long-term planning and continuity of timber-crop cultivation. Critical of economic pressures that encouraged rapid purchase, harvest, and surrender of forest lands, the report focused on private ownership as the major obstacle to successful management of American forests. The importance of stable land ownership and the inability of private owners to muster sufficient resources for long-term investment made public acquisition of forest lands essential.[15]

Community-owned forests were included in the survey, and questionnaires were sent to experiment stations and state forestry agencies. Each questionnaire divided local woodlands into three categories: Class A represented those with provision for administration and protection; Class B, those without; a final grouping was reserved for watershed forests. Final tallies returned to Washington were generally equivalent to those appearing in biennial reports published by state foresters. However, numbers disclosed by the survey address more pressing questions about progress in the management of municipal forests at the time. New Hampshire, for instance, declared only eleven Class A forests, with seventy-one in Class B. Oddly, that state's tabulation of watershed forests revealed only insignificant acreage, inconsistent with the extensive forestry work around reservoirs for Concord, Manchester, Hanover, Nashua, and Keene. Massachusetts reported seventy-five to eighty in Class A, none in Class B, and approximately 100,000 acres of forested watershed lands. Vermont's count numbered forty in Class A with no data available for the remaining categories, including watershed lands. Maine's eight Class A forests were augmented by fifty-five in Class B, most of which consisted of school lands held in trust. Connecticut disclosed four in Class A; Rhode Island, none. Both states reported substantial acreage for watershed forests.[16]

Unfortunately, the Copeland Report stopped far short of zealously advocating municipal forests and contained no recommendation for federal involvement. Tillotson authored the document's segment on this topic and cast the subject in a cautious light. Typically small, poorly stocked with cut-over and mixed second-growth wood, and lacking technical management, community forests were considered unimportant by Tillotson in a grand plan for timber production. Only the larger ones, he wrote, held the capacity for revenue, a potential that would remain unfulfilled unless drastic changes in conception and administration occurred. Although lukewarm about the commercial value of these woodlands, Tillotson nevertheless recognized their educational and recreational values and acknowledged that some timber, as well as cordwood, could be harvested. With a measure of insight, he predicted that communities would continue to set aside small areas in increasing numbers, with a view mainly to recreational value. These plots, he commented, would be heavily patronized and would be well worth their cost. In the matter of forestry demonstration, the greatest encouragement was reserved for larger watershed forests that received better management.[17]

Roosevelt and Brown

Ironically, just before Congress received the Copeland Report, community forestry enlisted a strong advocate in the person of Franklin Roosevelt, who assumed the presidency earlier that same year. By the time of his election, Roosevelt's philosophies toward conservation and forestry had been carefully molded. As a young member of New York's state senate, he had been appointed to that body's Forest, Fish and Game Committee. Exposed to the magnitude of wasted forest resources in his home state, he began experimenting with timber stand improvement on his family's lands at Hyde Park. Clearing out the "no 'count" trees, as he described them, Roosevelt subsequently managed to realize a profitable return for his efforts. Years later, he promoted "A Grandchildren's Trust," a tree company in which individual shareholders invested in woodlands for the purpose of cultivating trees and left shares to their grandchildren. Unlike the authors of the Copeland Report, he was anything but ambivalent toward municipal forestry. In 1930, while still governor of New York and aware of that state's solid, long-standing support of town forests, Roosevelt had argued vigorously for a widespread municipal forest movement. Referring to European city forests, he suggested that although America had made a start in the right direction, the country ought to be doing a thousand times more.[18]

After becoming president, Roosevelt turned to his friend and forestry advisor, Nelson C. Brown, for assistance in reviewing the Forest Service's policies toward community forests. Brown had served briefly as a U.S. forester in Montana and Idaho before beginning an academic career at Iowa State College in 1911. The following year he became an inaugural faculty member of the New York State College of Forestry at Syracuse. A prolific author of forestry textbooks, he remained at Syracuse until his retirement and directed their program in forest utilization. He also served as acting dean on several occasions. A friendship with Roosevelt developed after Brown was asked to become a forestry consultant at Hyde Park, and in 1933, Brown helped to coordinate Roosevelt's CCC program headquartered on Governor's Island.[19]

As did so many of America's young foresters, Brown traveled abroad to study European management techniques and communal forests. In 1913, on his first of many trips, Brown carried with him a letter of introduction from Henry S. Graves, reciting the purpose of his voyage: "to see the best examples and results of scientific forest management, and to study the methods employed in order that he may be better fitted to carry on his professional work in teaching in this country."[20] Brown's itinerary included stopovers in Berlin, Dresden, Zurich, Munich, Heidelberg, and Frankfurt. Twelve months later, almost to the day, he published an article titled "Possibilities of Municipal Forestry in New York" in his college's quarterly bulletin, remarking that European communities had been managing successful forests for hundreds of years. Brown observed that many of these forests practiced multiple use by supplying wood materials such as fuelwood and lumber, providing aesthetic and recreational benefits, and demonstrating forestry management. Special praise was reserved for the Sihlwald, which piqued Brown's interest in forest utilization.[21]

Frequent jaunts to Europe notwithstanding, the bulk of Brown's labors fell outside community forestry until he was called to Washington. In 1921, he did compose an article, discussing Syracuse's city watershed forest at Glen Haven near the southerly end of Skaneateles Lake and outlining management plans by his college. However, apart from this article which appeared in *American Forestry*, he produced negligible writing on the topic before 1938. Not surprisingly, Roosevelt's invitation spurred Brown to redirect his ample literary talents, and during the next four years he authored more than a dozen articles for periodicals such as *American Forests*, *Outdoor America*, *Forest Leaves*, and

Journal of Forestry. In addition, he collaborated on a monograph for the Forest Service, *Community Forests*, which contained a foreword by Roosevelt. This publication achieved wide circulation, joining the short books by Reynolds and Buttrick as a major work on the subject. It also served as a blueprint for a number of other Forest Service publications that surfaced during the next few years. Clearly, Brown and Roosevelt must have agreed on the need for public education, best achieved through a campaign of published literature. Brown's flurry of articles, together with the Forest Service's new program for community forestry, begun in 1938, opened a new phase in municipal forestry.[22]

Brown's literary contribution was a mixture of old ideas and a few new slants. Just prior to his six-month sabbatical leave in 1938 to assist Roosevelt, Brown once again journeyed to Europe, as a guest of the Carl Schurz Foundation. That organization's journal, *The American-German Review*, also published a number of Brown's articles. He returned, as he had almost twenty-five years before, with the conviction that American towns could benefit from examples set by European community forests. Much of his subsequent writing advanced this conventional theme. Nevertheless, Brown managed to focus more clearly on the patterns of multiple use demonstrated in Europe and, with his academic bent, to articulate the practice more precisely than had others before him. As a result, he avoided the unsteady fluctuation between commercial and recreational use that characterized some of Harris Reynolds's writing.

While stressing the primary importance of timber cropping, Brown reasoned easily that community forests should achieve maximum financial revenues consistent with aesthetic and spiritual benefits. A proper balance between commercial logging and recreation could be maintained, for example, by leaving trees along trails and around lakes in their native wilderness condition. As one might expect, Brown also grasped the significance of efficient utilization through local marketing. He recognized that small sawmills and wood industries connected to European community forests were often the deciding factors in their profitability, and he recommended a similar strategy in America. Brown, too, can be credited for turning a spotlight on Newington's town forest, the history of which he traced in most of his articles.[23]

Notwithstanding the quantity of Brown's writing and the value of his expertise, the substance of his counsel differed only marginally from that of Reynolds and others who had been promoting town forests for more than two decades. True, there may have been little left to say, and Brown was writing to a different, much larger audience that might not have heard all of the arguments before. Roosevelt's aim was to spark a national movement for community forests, and Brown was simply filling the same shoes that Reynolds had worn in Massachusetts decades earlier. Moreover, the focus in New England had shifted slightly from forest acquisition and planting to management. State organizations such as the MFPA were no longer geared for large-scale promotion

Fig. 5.4. Nelson C. Brown on land of the West Virginia Pulp and Paper Company near Summerville, South Carolina, in 1956. Brown inaugurated Roosevelt's community forest program and authored a substantial body of literature on the topic. An advocate for maximum financial return consistent with recreational and aesthetic benefits, Brown joined Harris Reynolds as a key member of the Society of American Foresters Committee on Community Forests. Terence J. Hoverter College Archives, SUNY College of Environmental Science and Forestry, Syracuse, New York

of town forests and were being forced to juggle an ever-increasing variety of conservation issues. Unquestionably, Brown served Roosevelt's purposes well, was a key figure in publicizing community forests around the country, and remained active in the movement until his retirement in 1951 (fig. 5.4).

At the same time, Brown was probably guilty of not taking full advantage of the experience that New England's foresters had developed in working with town forests. Although he urged a combination of state and local supervision, he initially failed to isolate the critical need for consistent technical management, the most pressing issue for New England's town forests at the time. In a 1939 article, "Progress in Community Forests," Brown identified four stages in the movement's development: exploration, public education, enabling legislation, and creation and management. According to his calculations, the last category was just occurring. Presumably, he was thinking about a broad national movement, for his periods are hopelessly inapplicable to New England, New York, or Pennsylvania. Finally, the titles of some of Brown's articles seem to ignore the status of New England's town forests altogether. One, for instance, titled "Community Forests—A New Idea in the American Forestry Program," published in *Southern Lumberman* in 1940, contrasts sharply with *MFPA Bulletin* 163, "The First Quarter Century of the Town Forest in Massachusetts," printed a year earlier. Given the Forest Service's sudden interest in community forests after so long a period of apathy, one wonders whether that bulletin's title was purely coincidental.[24]

Forested Communities

During the closing months of 1937, the Forest Service hastened to complete its plans for a national community forest program. Chief Forester Ferdinand A. Silcox announced its inauguration by radio address and news release, and a short booklet, *Town and Community Forests*, was being distributed by February 1938. By then, the service had already received nearly a thousand inquiries, representing practically every state. Initially housed within the Division of State Cooperation, part of the Branch of State and Private Forestry Divisions, the program became a national clearinghouse, its key objective the accumulation and distribution of information. Toward that end, inventory taking was begun, and more publications followed. Specific data about ownership, use, plantings, management, and revenue were gathered from questionnaires sent to both state foresters and town officials. In 1939, too, the Forest Service circulated a specific definition:

> A community forest consists of lands owned and operated for forestry or allied purposes by a village, city, town, school district, township, county, or other political subdivision, or by other community or group enterprises, such as schools, hospitals, churches, libraries, 4-H Clubs, Boy Scouts, Girl Scouts, and Camp Fire Girls. Locally, a community forest may be known as the town, city, county, school, or municipal watershed forest; village, town or memorial woods; or community forest.

"Community forest" subsequently became an accepted term for a large class of American woodlands. However, the words *municipal forest* and *town forest*, both linked to twentieth-century state enabling laws, are more common in New England. This is particularly true for the latter, which underscores the region's unique pattern of settlement and ancient bond with forest landscapes.[25]

As envisioned, at least by Brown and the Forest Service, chief responsibility for the program would be left to state foresters, who would work closely with local committees or individuals and answer questions about organization and management. No scheme for financial assistance to either towns or states was put into operation, and inquiries were usually referred directly to appropriate state officials. Earl Pierce, chief of the Division of State Cooperation, coordinated initial efforts, but Ernest O. Buhler began supervising the program at some point in 1938, possibly after Brown's return to Syracuse.[26]

While Brown had been fascinated with Newington's ancient forest, Buhler focused attention instead on Danville's church woodlands. By this time, both John Foster and Warren Hale had long been aware of the parsonage committee's meticulous records. A short history of the forest, which appeared in the state forestry commission's biennial report for 1927–28, suggests that its author, probably Hale, had inspected the century-old timber sale records. Buhler managed to obtain copies and late in November 1938 wrote a story about the forest, "Acres of Diamonds." Correspondence with members of the town indicates that Buhler regarded Danville as an opportunity to tabulate exact annual yields from small parcels of woodland over a long duration. Beginning a national program in community forestry by proving that small forests could produce substantial returns seemed providential. Concerned about accuracy, Buhler traveled to Danville to examine the records with the help of an accountant.[27]

Buhler also encouraged the committee to seek technical assistance from John Foster, and town members were asked to begin reforestation. Early in 1938, consulting forester William P. House examined the parcels and urged the use of WPA labor for thinning to avoid any decline in productivity. While enthusiasm for Danville's forest must have been heightened by all of this attention, House's report betrayed his exasperation and revealed a central problem in the management of town forests. Although interested in obtaining income from the woodland, the selectboard's chairman had failed to grasp the long-term importance of appropriating money for timber-stand improvement. Instead, he opted to fund WPA labor for regrading a bank in front of the meetinghouse and cleaning up the town cemetery.[28]

During Buhler's tenure, the forest service assembled a large number of publications about community forests. Most were modest restatements of Brown's work, with instructions added for the benefit of community officials. Cultivation of timber remained the principal objective and was given priority over recreational and other benefits. Many of the pamphlets routinely outlined advantages and then described specific examples, often located in New England. In subtle contrast to Reynolds's tenet that town forests would introduce forestry to the people, the forest service presented them as a way to personalize the conservation movement. In 1939, a lengthy field handbook for foresters also was assembled. This manual contained a series of leaflets describing individual community forests around the country and provided summaries of relevant state laws. A comprehensive bibliography covering both European and American community forests also was prepared that year. Radio broadcasts and filmstrips were subsequently utilized as part of the program's educational campaign. During World War II, Lyle Watts, who became chief forester in 1943, recorded a brief talk with Mrs. LaFell Dickinson, who had become president of the General Federation of Women's Clubs in Washington, a long way from her small hometown of Winchester, New Hampshire. In two five-minute recordings, Watts and Dickinson urged women's groups to sponsor community forests as memorials to soldiers.[29]

It is not certain that Buhler was directly responsible for editing printed material or that other foresters shared the task. Buhler was eventually transferred to the Southwest because of poor health, and in 1941, at Roosevelt's prodding, Secretary of Agriculture Claude R. Wickard promised to assign a full-time forester to oversee the program. Joseph A. Fitzwater and George Duthie subsequently became key figures, but the program's structure was loosely organized. Duthie authored *Community Forests—a Practical Step in Post-War Planning* in September 1943 and later diligently monitored data about the number, type, and acreage of community forests across the country. However, Fitzwater was initially more involved with the program following Buhler's departure. He also became the service's representative to the Society of American Foresters (SAF) Committee on Community Forests, which issued its first report in February 1942.[30]

Fitzwater's assignments were clearly not confined to community forestry. This hazy position aside, his annual plan of work prepared for 1945 sketches a very clear picture of the service's evolving objectives for its young program:

1. Continue policy of selling through the regions community forests to local planning agencies, local governments, welfare organizations and individuals.
2. Seek adoption of programs by national organizations and government agencies such as the Extension Service and the Social Security Board for disseminating information on community forests.
3. Assist the Extension Service with the development of a program for 4-H Club community forest projects.
4. Supply magazines and writers with stories and material for stories about community forests.
5. Collect data for a publication designed to promote more intensive management of the community forests already established.
6. Continue to gather up-to-date census data on community forests.[31]

The Forest Service reshuffled its Branch of State and Private Forestry Divisions at the close of the war. In 1945, the Division of State Cooperation became the State Cooperative Fire Control Division. A year later, the Private Forestry and State Forestry Divisions merged to become the Division of Cooperative Forestry Management. A new category, the Section of State and Community

Forests, was created and placed within that division. Whether community forests had already drifted into the State Forestry Division from the Division of State Cooperation is not clear. In any event, George Duthie emerged as the person in charge and replaced Fitzwater on the SAF Committee at about the same time.

Yearly tabulations from across the country continued to show a slow but steady increase in the number of forests. By 1948, Duthie estimated more than three thousand forests on nearly four and one half million acres in forty-three states. Outside the Northeast, substantial numbers of forests were reported in Arkansas, Illinois, Michigan, Minnesota, North Carolina, Ohio, Oregon, Virginia, and Wisconsin. Michigan, with 846 (627 school forests), led all others, followed by New York (658), Wisconsin (309), New Hampshire (167), Massachusetts (140), and Minnesota (123). Large county forests in Wisconsin on more than two million acres skewed acreage calculations substantially.[32]

The Bankhead Committee Proposal

Despite a genuine commitment to community forestry on the Forest Service's part, there remained an underlying hesitation by the federal government to place full support behind the movement. This was symbolized most conclusively by the absence of any supportive legislation or funding programs. Vague skepticism about the ability of towns and charitable organizations to manage profitable forests, instilled by opinions such as that of Tillotson, surely lay at the core of this ambivalence. Spirited literature notwithstanding, a crucial opportunity for legislation slipped by in 1941 and was never regained.

In June 1938 acting on Roosevelt's recommendation, Congress established a joint committee on forestry to investigate the condition, ownership, and management of American forests. Concern for the nation's unstable forest economy and vast, privately owned woodlands of very limited productivity pushed Roosevelt steadily toward this inquiry. Among his many charges to the Bankhead Committee, as it became known, was a request that it consider the need for extension of federal, state, and community ownership of forest lands and for their planned public management.[33]

In February 1940, at one of many public hearings, the Forest Service offered its proposals to the committee. Attributed largely to Silcox, who died in December 1939, the service's program divided policy into three categories: private ownership, management of public lands, and research. The complex issues surrounding private ownership were addressed by a two-pronged strategy. The first resolved to establish a joint effort at cooperation between public agencies and private owners and included a ticklish plan for public regulation of private forests. The second aspired to expansion of public forests—federal, state, and community—to approximately 150 million acres. States and communities were thought capable of sharing a third of that amount. A specific proposal urged amendment of the Fulmer Act to make it apply to acquisition of land for community as well as state forests. Passed in August 1935, the Fulmer Act had authorized Congress to appropriate $5 million for the purchase of lands by the federal government to be administered as state forests. The service argued for an increase to $10 million and suggested that bonds be issued to pay for lands. By the time bonds became due, mature timber would be available for redemption.[34]

The Bankhead Committee issued its report on March 24, 1941, and, in its findings of fact, noted an opportunity and great need for substantial expansion in both number and area of community forests. The committee followed the Forest Service's advice to the letter concerning expansion of the Fulmer Act and recommended a $10 million appropriation. They also ventured cautiously into the minefield of public regulation on private lands. Silcox's strong program, however, was reworked to turn regulation over to the states, while financial clout and ability to set standards for forestry practice remained at the federal level. Ensuing communications by Roosevelt, Secretary Wickard, and Acting Chief Forester Earle H. Clapp, indicated displeasure with the committee's weak system of regulation and inadequate program of federal acquisition.[35]

Roosevelt, Clapp, and Wickard continued to work for legislation based on the Bankhead Committee's report. Staff members under Clapp and Wickard met with representatives from the Bureau of the Budget and together submitted a revised bill to Roosevelt in May 1942. Acting on their advice, Roosevelt wavered at plans to fund community forests and deleted the provision. Wayne Coy, acting director of the Bureau of the Budget, wrote to the president, reminding him that Congress had been asked twice before to fund the Fulmer Act and had declined. Coy also concluded that the plan to subsidize community forests would not be self-liquidating, as much as stating that they could not generate sufficient revenue. Proposals to regulate private forestry also were dropped, but several of the committee's recommendations were passed into law piecemeal fashion in 1944.[36]

Failure of the legislative initiative on federal control of state and private lands was predictable. Subsidies of state public forests, more particularly the attending restrictions, were unpopular. This was especially true in independent-minded New England. Even Harris Reynolds remarked on occasion that state foresters were capable of formulating appropriate policy and didn't need or want Washington's help. Federal regulation of private woodlands faced even greater hostility, and it is likely that Roosevelt sacrificed these controversial sections to political expediency. Still, Reynolds and others undoubtedly would have encouraged federal involvement with community forests as long as policy issues were left to states or towns. In fact, Reynolds considered local woodlands to be an effective tool for maintaining a healthy balance of power among political entities charged with determining policy. However, Coy's prediction that community forests would not yield enough timber to repay the federal government struck a more ominous note. Coy's opinion was almost certainly rooted in Forest Service terrain at a time when their community forest program was only four years old. Oddly, just six months before, Roosevelt and Wickard had agreed on the assignment of a full-time forester to that program.[37]

Events in Washington augured an eventual decline in the perception of community forests as units capable of viable timber production. At a pivotal time, on a decisive issue, the federal government balked, and the moment was lost. Its decision held important implications for the movement's future. In retrospect, whether more active participation by the Forest Service would have sustained town forests in New England and other regions remains questionable. Inconsistent management continued to be a perplexing barrier, and Roosevelt's verdict intimated that towns were not equal to the task of commercial timber production. Fairly stated, federal dollars probably would not have changed these conditions, and control over local lands by government foresters represented undesirable policy. Yet neither a mood of foreboding on the subject of timber yields nor the wrenching turmoil of war prevented Reynolds, Brown, and Duthie from continuing to work together to promote local forestry, their connections bolstered by the SAF committee.

Society of American Foresters Committee on Community Forests

The national movement for community forests developed along other avenues while the Forest Service continued on its vaguely mapped course. In 1941, members of the SAF garnered sufficient momentum to establish a community forest committee, which issued its first report in February 1942. In addition to Fitzwater, Brown, and Reynolds (the chairman), founding members included Anton Tomasek representing Illinois, Fred Pederson (Virginia), E. Fred Brouse (Pennsylvania), Fred Trenk (Wisconsin), and Paul Schoen (Texas). The opening sentence of its first report offered a new definition but one very similar to that written by the Forest Service:

> Community forests are areas of woodland or of potential woodland, owned and operated by cities, towns, townships, villages, school districts, counties, or other political subdivisions of the state, for the benefit of all the people, through the production of forest products or the conservation of water, the protection of wildlife, the control of floods and soil erosion and the creation of opportunities for recreation.[38]

From the start, committee policies reflected a strong mixture of both Brown's and Reynolds's opinions. This blending marked an important juncture in the evolving see-saw contest over commercial and recreational use of community forests. Their definition clearly leaned more toward woodland conservation than timber cropping. As if to reinforce this policy, a very guarded posture was adopted toward expectation of economic returns. The committee warned: "no greater claims should be made for financial income from these areas than may be expected from any other forest lands in the region having equal qualities and under similar management." Acknowledging that some forests had paid ample dividends, they cautioned that investigation typically revealed only ordinary logging activity, not intensive management. Instead, paramount emphasis was given to conservation education, with an understanding that public interest in conservation was most readily awakened through recreation. In anticipation of postwar economic doldrums, the committee refined this theme even further. The value of community forests lay in two principal areas: employment on self-liquidating work projects and the conservation and development of forest resources for multiple public benefits.[39]

Weaknesses in the movement were identified candidly. Written for a national assembly of foresters, reports tended to omit the promotional decoration sometimes included in other writing. Consistent management remained the critical problem, and states were urged to assume primary responsibility. Strengthening of state forestry programs was viewed as a crucial step in making assistance available. Initially, members warned that a proper balance between assistance and control was necessary. Except for withholding trees until site and species had been approved, regulation was best left to local committees. Later reports, however, looked enviously at German practices and argued for amendment of state enabling laws to place more power and responsibility for management in the hands of state foresters. States also were urged to assign full-time foresters to community forestry. By 1949, on the eve of Duthie's retirement from the Forest Service, the committee set its sights on Washington and suggested that the service should enlarge community forestry by giving it the status of a division or special office along with adequate funds. Promotional agendas were not overlooked, particularly in southern states, where forests ranked high in industrial and agricultural economies.[40]

Unfortunately, the committee was hampered by the same national breadth that supplied its potential for strength. Members were forced to communicate by mail, and Harris Reynolds complained on occasion that the group was impotent to pursue its own recommendations. Regional sections of the SAF were asked to form special committees, and members were exhorted to promote the movement and provide technical assistance to state foresters. At least two sections, Allegheny and New York, complied, but the magnitude of their activities is nebulous. Composition of the national committee fluctuated, and a nucleus of Reynolds, Brown, and Duthie barely managed to sustain momentum. No report was issued for 1947, and Clyde Martin, SAF president, considered dropping the committee. However, Henry Clepper, its executive secretary, intervened and helped to reorganize membership. In his reply to Clepper accepting reappointment in November 1948, Brown acknowledged that the movement had not caught fire in America as it had in Europe. Nonetheless, he confidently predicted that headway would gradually develop, and he reminded Clepper that Europe's success, achieved over centuries, represented evolutionary progress.[41]

At about the same time, the newly reorganized committee asked the American Forestry Association (AFA) to consolidate efforts by state associations and conservation groups and to prepare a special educational bulletin for broad distribution. At a meeting of the AFA board of directors on January 28, 1949, members proposed a special project to assist communities interested in creating forests and recommended an allotment of $50,000. The project was considered part of a long-term plan, and the board anticipated that a portion of the fund would be used for awards to individual towns. Two weeks later, Reynolds wrote the first of a pair of letters to Sherman Frost, AFA's executive director, inquiring about the meeting. For reasons unknown, Frost's reply, which finally arrived in March, mentioned neither the January board meeting nor any strategy for a special project. However, he did disclose plans for a national committee on community forests and for publication of the requested brochure. Frost anticipated that most activity would be channeled to state organizations, and he invited Reynolds to serve on the committee. Curiously, the AFA did not establish its national committee until 1952, and the fate of its plan to offer awards to towns is dubious.[42]

For practical purposes, Duthie's retirement in 1949 ended the Forest Service's community forest program. No major publications had appeared since 1946, and most of Duthie's energies had been directed toward census taking. His departure left a yawning gap in the collection of information across the country and added another reason for the SAF committee to solicit support from the AFA. Initiatives from that group did not soon materialize, however, and Reynolds eventually became discouraged by the state of affairs. His report printed in 1951, the committee's tenth year, carried a tone of irritation. Deploring the absence of national leadership, Reynolds made a firm call for an organized program by a national agency. Obviously pointing in the direction of the AFA, he conceded that any one of several national groups could do the job but argued that forestry rather than conservation should step forward. Although he recognized the financial inability of the SAF to take on such a project, Reynolds did not spare that body from criticism either. Too often, he wrote, committee recommendations were approved unanimously and then quietly deposited in society archives.[43]

To Reynolds, the period for acquisition of national and state forests had passed its heyday. The future for new public forests depended on communities, but few were heeding the call. Whether through frustration or increasing demands on his time, or both, Reynolds had endured enough. In July 1951, he wrote to Clepper requesting appointment of a new committee chairman. He also made a final plea to Washington to revive its collection of data. Shortly before he became Forest Service chief, Richard McArdle had assured Reynolds that community forestry would again become an official activity despite the absence of any appropriation for such work. Reynolds's last report, which appeared in February 1952, matter-of-factly asked the AFA to assume national leadership and to induce the Forest Service and the Association of State Foresters to adopt cooperatively a formal program. He also appealed to the AFA to assume responsibility for census taking. Clepper appointed Duthie chairman in January 1952, and both Reynolds and Brown remained as members.[44]

In March, the AFA voted to establish its committee, and Duthie, as association forester, was asked to supervise. If nothing else, Duthie's dual role offered a unique chance for communication between the two organizations. Unfortunately, the AFA entered the movement at low throttle and stalled soon afterward. Duthie's chairmanship of the SAF committee may have been the single factor that finally goaded the AFA into action; an opportunity too good to miss. No special funds were allotted, however, and Duthie found himself with a vague agenda calling for publicity and cooperation among public and private organizations. Optimistically, he did manage to collect enough data to prepare a census, the results of which were published in *American Forests* in the spring of 1953. The AFA committee produced little else and quietly dissolved.[45]

The SAF committee suffered a similarly inconspicuous demise. In January 1954, Henry Clepper circulated a memo officially dropping the program for lack of a "concrete and specific assignment." Despite its unspectacular history, this committee made several important contributions. The combined voices of Reynolds and Brown pinpointed the evolving status of community forestry very effectively. Reports accurately identified the trend away from timber cropping and toward recreational use under the umbrella of conservation. They also observed that the heyday of acquiring national and state forests had passed and that community acquisition would be the next frontier. More important, the group isolated the movement's problems in a way that only zealous advocates can. Their perspective became even more meaningful after interest by the Forest Service gradually began to fade. As chairman, Reynolds deserves much of the credit, and one can observe the same vision for the nation that he held for his own state. In New England, advice by the SAF committee was observed most directly in Vermont, where municipal forestry matured after 1950. Nevertheless, disappearance of community forest programs from national associations, as well as from the Forest Service, foretold an uncertain future for many town forests elsewhere around the region.[46]

Regional Developments

Between 1938 and 1954, New England's town forest movement was influenced by a great many factors, both national and regional. The effect was not always uniform, often varying from state to state in either significance or period. The Forest Service's program sparked a period of new interest that subsided with the onset of World War II but flickered again briefly after 1945. Although preoccupied by the war, many towns were moved to cull their forests for cordwood and ease the bite of northern winters. Consistent forest management remained an elusive target, and foresters scrutinized potential timber yield as well as problems linked to the administration of town forests. Preserving patches of open land, often wooded, became an increasingly popular means for balancing urban development. The Forest Service's broad definition of community forests caused a minor quandary for New England states, where enabling legislation specifically referred to woodlands acquired by towns or cities. By definition, organizations such as schools, churches or scouting and 4-H groups were excluded from formally classifying their lands as town forests. While woodlands owned by these nonprofit groups were usually not included in tallies, many received the benefit of technical assistance from state foresters and could sometimes take advantage of seedlings supplied by state nurseries. Most state forestry offices dealt with the matter in practical fashion and generally lumped administration for all types of community forests together. A similar problem occurred with watershed forests owned by privately chartered public utility companies. Many were managed by private foresters and consequently were not monitored by state agencies.

Participation by the U.S. Forest Service inevitably led to comparisons of timber productivity among categories of public lands. Harris Reynolds argued that town forests held much greater potential than national forests yet received far less attention from professional foresters and no financial support from the federal government. Compared to national forests in mountainous western states, soil quality on New England's abandoned farms was higher, the growing season longer, and water more plentiful. Transportation costs to large market centers were lower, making shipment of low-grade lumber economically feasible. Also, by-products such as limb wood could be utilized as fuel. With local markets assured, less waste meant more intensive management and greater profit. Reynolds acknowledged that other advantages were less quantifiable but regarded them as important nonetheless. Employment of local labor contributed to community stability, whereas contractors in western regions tended to float from one job to the next. Finally, nearby recreational benefits served more people per unit of land and were readily available throughout the year. Reynolds agreed with Brown that, under reasonable management, town's could expect a yield of about five dollars per acre annually.[47]

Reynolds intended his arguments to provoke debate. Yet the skepticism that daunted municipal forestry never directly responded to his underlying premise that community, not national, forests could produce the highest yields per acre and deserved greater attention. Rather, debate centered on solution-defying administrative and political concerns. One study was conducted by T. Noel Stern, secretary of the Pennsylvania Community Forests Council. This body, created in 1940 following Stern's proposal a year earlier, became a liaison agency linking national, state, and local government agencies; private conservation groups; and the University of Pennsylvania. His study included forests in Danville and Warner, New Hampshire, as well as in Essex Junction and Rutland, Vermont, and identified a detailed list of silvicultural, economic, administrative, and political problems. Apart from normal obstacles facing management of any forest, Stern isolated local fears that undermined consistent policy decisions. Noteworthy were lost tax revenue, controls associated with federal subsidy, competition with private industry, expenditures considered needless, and disturbance of established local interests.[48]

Even under the best of circumstances, sustained-yield management on small parcels of young timber could not be expected to occur overnight. Petersham's ninety-eight-acre forest, one of the most capably managed, illustrates typical problems. Although a twenty-seven-acre stand of mature pine had been cut in 1922, improvement work had produced only nominal returns in lumber and cordwood by 1932. Assuming annual appropriations of $50 per year for silviculture, the committee predicted a thirty- to fifty-year wait for substantial income. Moreover, a treatment plan for the period between 1926 and 1936 showed the difficult choices Petersham's committee was forced to make during the interim. For public forests, the shortest possible period between yields was considered ideal. However, with only mature or very young stands, establishing any yield periods at Petersham was not possible until an even distribution of age classes had been achieved. Although it was desirable to hold some of the mature stands for up to twenty years for the sake of maintaining a semblance of periodic yield, this approach risked a decrease in earning power because of factors such as disease, market decline, and general vigor of the trees. Finally, erection of a portable mill, ultimately a cost-saving step, would not be practical until yield periods were relatively short.[49]

If greater federal or state participation offered the best chance for consistent management and high productivity in towns without a committee of professional foresters, then local resistance to such interference forced a critical impasse. Many proponents of town forests, Reynolds included, were convinced that local attitudes toward forestry were important and helped to check federal and state authority. Water boards, in particular, were reluctant to change the status of forests under their supervision for fear of state interference. Due either to staunch philosophical belief or simple recognition of New England's political facts of life, attempts to alter this balance never gained momentum. Instead, energies were funneled into publicity to keep enthusiasm high and make town officials aware of the need for technical assistance and regular improvements.[50]

Quick to take advantage of increasing national interest, the MFPA and the Conference of Town Forest Committees jointly sponsored a regional conference on town forests, held on October 11 and 12, 1940, in Springfield, Massachusetts. Delegates from fourteen states attended, and speakers included Nelson Brown, Ernest Buhler, William Wharton, Albert Cline, and Shirley Allen from the University of Michigan School of Forestry. Edgar Heermance, secretary of the Connecticut Forest and Park Association, SPNHF forester Larry Rathbun, and Noel Stern offered reports from their respective states. The conference covered a broad range of topics and focused on programs in various states as well as on the need for assistance from professional foresters. Field trips to Westfield, Russell, Northampton, and Petersham helped to temper the lecture docket. Significantly, the conference closed with a resolution addressing the question of local control: "It is the sentiment of this meeting that a community forest should grow out of local interest and meet local needs. We are opposed to any form of federal or state control of such areas. We believe, however, that close cooperation should be maintained with state agencies in drafting and carrying out plans of management."[51]

A second gathering was held in State College, Pennsylvania, a year later. Whether from lack of direction or preoccupation with war, no other regional conferences were convened. Town forest committees in Massachusetts returned to their annual meetings without special emphasis on national trends. These conferences occurred at a time when the movement seemed to be gaining momentum nationally. However, their influence on events in New England was not significant if one considers the number of forests established by towns after that date. Nor did municipal forestry move any closer to unraveling its quandary over consistent management.[52]

Unwilling to rely on government foresters for management assistance on private lands, Reynolds conceived the idea of forming a nonprofit organization to assist woodland owners. The New England Forestry Foundation was established by the MFPA in 1944, with Reynolds as secretary, all the while continuing his myriad duties at the parent association. Geared primarily for small woodlots, the foundation was created to provide complete forestry services at cost, from survey to harvest. In addition, its charter authorized acceptance of woodland grants as memorial forests. Hoping to show owners that modest tracts could produce a profit if managed properly, Reynolds also recognized the foundation's potential for assisting town forest committees. It quickly developed a network of consulting foresters stationed at management centers in four New England States, and many prepared working plans for town forests. In Massachusetts, Greenfield, Groton (fig. 5.5), Needham, North Adams, Pembroke, Pittsfield, Russell, and Walpole were among the many towns that obtained plans from the foundation. In New Hampshire, reports for Dunbarton, Louden, Springfield, Warner, and Weare were prepared under the supervision of Stanley Mesavage, chief forester at the Concord office.[53]

Both the MFPA and the Conference of Town Forest Committees continued vigorous efforts to put forests at the doors of common citizens. Well-established examples around the state were available as models for communities that elected to participate. Bulletins issued in 1939, 1941, and 1949, a number of separate articles penned by Reynolds, and a skillfully edited *Forest and Park News* carried on the promotional campaign. Yet events seemed to conspire against the movement. Town forests suffered along with most of the state's woodlands, which lost an estimated 400 million board feet of white pine to a 1938 hurricane. At least some (Petersham, for example) managed to preserve their stands of young timber. Others moved to replenish lost timber with new plantings. A windfall of a different nature occurred in 1940 when the Massachusetts legislature failed to convene. With his lobbying schedule rewritten, Reynolds had time to visit more than a hundred town forests and helped to start fourteen new ones. With disappointment, he reported pervasive neglect of young plantations. The following year, a law was passed requiring towns to obtain approval of plantation lands before trees could be distributed for planting. That same year, the MFPA outlined a simple, three-point program for town forests: sustaining public support, improving local administration, and procuring management plans.[54]

Unfortunately, programs soon surrendered to the crisis of war, and the topic of cordwood began to dominate town forest activity as oil and coal became scarce. Reynolds, state director of wood-fuel production, regarded the demand for cordwood as a chance to simultaneously improve timber stands through thinning. Wood-fuel production managers were invited to the yearly conference of town forest committees, held in November 1942, in an attempt to introduce one to the other. In response to cries of "no choppers," committees resolved to open town forests to citizens and groups willing to cut their own supplies. Many towns complied, marking or felling selected trees and allowing volunteers to finish the task. Most communities did so without expectation of payment; some charged only the duty of a half-cord for welfare.[55]

Anticipating postwar unemployment, the MFPA once again promoted town forests as ideal opportunities for local labor. Even before the war's end, the association launched a revival of the movement and, borrowing from the New Deal, emphasized planning for self-liquidating work projects. Yearly meetings of town forest committees resumed field inspections, and a forester, David

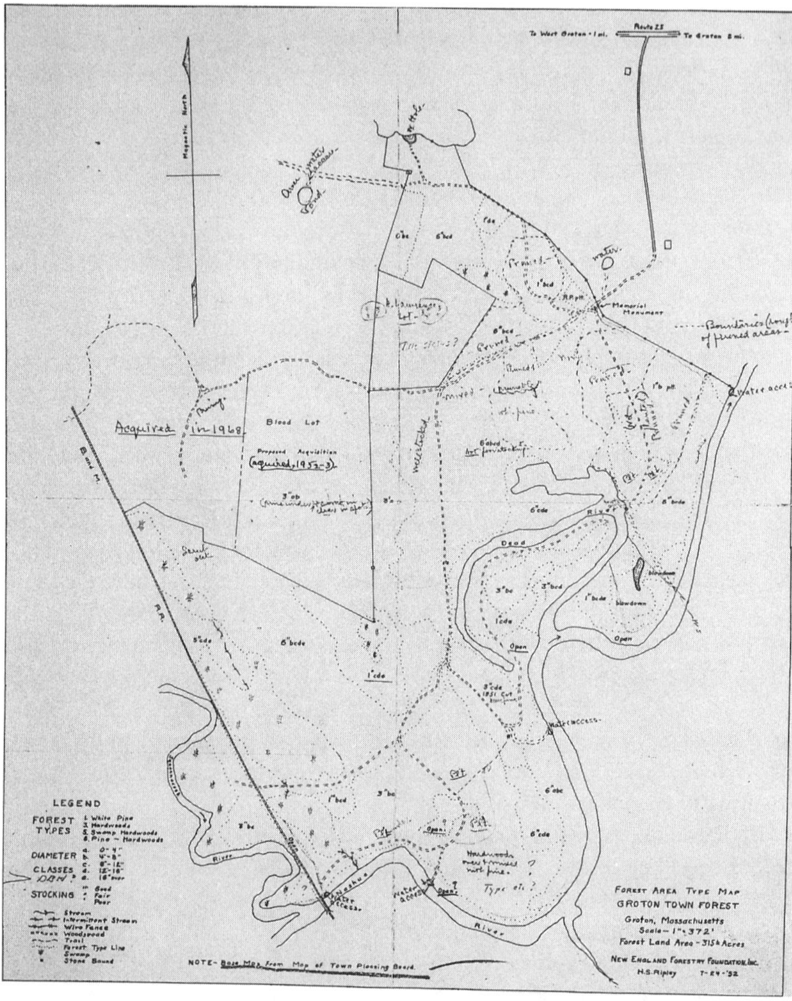

Fig. 5.5. Forest type map of the town forest in Groton, Massachusetts, drawn by H. S. Ripley for the New England Forestry Foundation in 1952. Thirty years earlier, Groton had converted a portion of its poor farm to a forest plantation at the urging of resident William Wharton, MFPA member and later its president. The town subsequently changed the name of its Town Farm Road to Town Forest Road. New England Forestry Foundation, Boston

Miner, was hired to survey town forests throughout the state. The findings, published in 1949, showed that three-fifths of the forests, comprising more than four-fifths of the total acreage had been well managed, surprising even Reynolds. However, practically all forests needed at least some silvicultural treatment. Administrative weaknesses were uncovered in a number of towns that had failed to appoint full committees. Bookkeeping was found to be similarly lax, and committees sometimes failed to submit annual reports. As a result, income and timber yields could not be calculated. Finally, absence of management plans posed the greatest deficiency. At the time of the report's issue, thirty-five years after Fitchburg launched the state's campaign, 127 towns had established forests on 39,839 acres of land and had planted 8,517,978 trees.[56]

New Hampshire's state forester reacted to the Forest Service's program by suggesting reclassification of community forests in 1942, rare parlance for that neck of the woods. Town-owned lands, twenty acres or larger, qualified if acquired by either gift or purchase; tax-forfeited lands, fifty acres or more, were included if held for six years or longer. Using these criteria, calculations were adjusted to show seventy-eight towns holding land for forestry purposes on 32,170 acres, with 2,783,975 trees planted. A second category was reserved for lands of undetermined status in eighty-seven towns on 63,393 acres, with 115,000 trees planted. Special efforts were made to assist towns burdened by extensive tax defaults. Letters were sent to selectmen recommending formation of town forest committees and inspection of lands to determine which, if any, should be retained by towns. Information about cutting and management practices remained elusive, and questionnaires were mailed again in 1950. After that year, however, town forest data was seldom included in biennial reports.[57]

Without a statewide organization, New Hampshire's town forest committees relied on their own initiative for management of local woodlands. Many towns—Louden and Dunbarton, for example—suffered extensive hurricane losses and were forced to redevelop their forests. While technical assistance was available from state or extension foresters or from the SPNHF, spirited promotional campaigns similar to those in Massachusetts did not reemerge. Not surprisingly, management was inconsistent. However, the industry displayed by some committees represented a tribute to forestry traditions in that state. In 1951, Hollis's committee, under the leadership of Henry Hildreth, energetically organized a chopping bee to supply the town library with fuel and began harvesting saw timber the next year. Other communities turned also to the New England Forestry Foundation, and the SPNHF offered to provide a foundation forester, at no charge, for a day's work examining town land. The society's journal, *Forest Notes*, carried most of the state's writing about town forests during this period. School forests also began to receive more attention in New Hampshire, and public high schools in Walpole and Dover embarked on forestry demonstration projects.[58]

Vermont's program advanced bit by bit after legislative amendments in 1945 and 1951 requiring, respectively, partial state subsidy of acquisition costs and propositions for municipal forests in town meeting warnings. As the SAF committee had urged, involvement by state government escalated, much more so than in either Massachusetts or New Hampshire. With both financial and technical assistance mandated by law, Vermont set a progressive standard. Furthermore, classification according to legislative guidelines was more strictly observed. By 1950, sixty-eight towns had established classified forests in excess of sixteen thousand acres, and planting activity accelerated. The Department of Forest and Parks eventually sought to establish a forest in every town, and numbers continued to increase during the next two decades.[59]

There is little direct evidence that municipal forestry in Vermont was influenced by the Forest Service's national program. Yet public institutions such as schools and libraries turned to forestry management during this period, and their lands clearly satisfied the service's broad definition of community forests. Proctor's Free Library acquired more than five hundred acres of well-stocked timberland on East Mountain from the Proctor family. The property was surveyed and later designated as a municipal forest because it retained a quasi-public character. A list of conservation projects was also prepared for Proctor's boy scout troop. In a few communities, enthusiasm for woodland management spawned a distinct class of municipal forests: the fire district reserve. Local fire companies, with charters similar to those of water utilities, were authorized to acquire land, and timber cultivation was viewed as way to subsidize taxpayers' costs for firehouses, equipment, and labor.[60]

Connecticut, Maine, and Rhode Island failed to develop strong state-sponsored community forest programs. Activity, when it occurred, was most often the result of individual town initiative. Generally, towns that had already established forests—Newtown and Suffield, for example—continued intermittent activity. Simsbury began a town forest in 1941, but recreational aspects were given greater emphasis than timber stand improvement. In Troy, Maine, abandoned farms and tax delinquent land spurred town members toward reforestation, and seedlings were transplanted in 1940.[61]

End of an Era

Midday on October 16, 1953, Harris Reynolds suffered a fatal heart attack. His death extinguished the spark that had ignited and reignited the town forest movement in New England for forty years. Reynolds's energy for promoting local woodlands was unmatched, and his life's work did indeed introduce forestry to the people. Yet the task of convincing towns to support consistent management in their forests remained unfinished. Henry Clepper's dissolution of the SAF Committee on Community Forests four months later was symbolic of the movement's defeat in the field of commercial forestry. Although the Forest Service's promotional campaign generally adopted tactics employed by Reynolds and the MFPA decades earlier, it did advance public education and popularize the movement, broadly if briefly so. More important, it contributed an administrative framework that, when disassembled, clearly weakened the movement.

By no means did the decline of municipal forestry occur suddenly. Led by its annual conference of town forest committees, the movement persisted in Massachusetts until conservation commissions gained overwhelming popularity after 1960. *Forest and Park News* continued to chronicle noteworthy events up to the time when yearly meetings lapsed. In some Massachusetts towns (Groton, for instance) local forestry still works for the community. The same is true throughout New England. Harvesting of timber has continued on many forests and does so today under current management plans. In Vermont, certificates for "Good Forestry Practices on Municipal Forests" were given to the towns of Bethel and Vergennes in 1966 by the state forestry department. A year later, award ceremonies were held in conjunction with Forest Festival Week. Today, state foresters occasionally offer public tours through municipal forests in towns such as Starksboro.[62]

In New Hampshire, an encouraging and thoughtful body of literature about town forests has emerged from the state's Cooperative Extension Service at the University of New Hampshire in Durham. Most of this writing has been produced during the past decade or so and emphasizes underutilized benefits available from local woodlands. Surveys of forests and town committees around the state have been conducted, and multiple use has been encouraged. Small programs also exist in Maine and Rhode Island, the latter part of an initiative begun when the state passed its Green Acres Law in 1964. Recreation and conservation are the principal objectives. The Glocester Land Trust, established by Rhode Island's legislature in 1987, manages the Scotstun Town Forest in Chepachet.[63]

At the same time, many town forests have encountered an ignominious fate. Conversions to subdivisions, office parks, or even landfills have occurred more often than one would hope. Memorial woods pay small homage when lumbering slash barricades footpaths. Nor is it unusual to find a town clerk or other local official completely unaware that their community still owns a forest. Origins of some, tax forfeiture in particular, make public access difficult, and only the most determined of users manage to seek out these enclaves. Although the Forest Service has supported a relatively new field, urban and community forestry, its proponents emphasize care for trees and woodland parks that shade city streets or furnish open spaces. Wood-products utilization, if it exists at all, is at best only an incidental part of this program. Town forest advocates in government forestry or nonprofit associations are scarce. Not surprisingly, communities are seldom reminded that their forests can supply lumber and cordwood as well as quiet recreation and nature study or that the former are as much a part of the region's cultural history as the latter.

. . .

In any analysis, one must concede that town forests, as a class of timberland, generated a disappointing record in the field of commercial forestry. Although loudly proclaimed at the movement's outset, benefits such as employment and support of local wood-products trades seldom showed a consistent return. Doubts about the ability of communities to manage their forests efficiently and to produce paying crops of timber, a skepticism that shadowed the town forest movement after 1930, foretold its ultimate demise. As small samples of public timberlands, these forests were thus unsuccessful. Had policymakers gazed into the future with clearer vision, however, they might have recognized town forests to be important demonstration tracts useful in educating the public about timber harvesting in our national forests, currently the scene of heated debate. Nevertheless, these shortcomings were offset in several ways. The movement's driving force, public reclamation of idle lands through forestry, made substantial headway both on the ground and in the collective mind. Many plots furnished dependable supplies of fuel, much of which found its way to welfare. Watershed forestry, the category of communal forests considered next, formed a healthy alliance with the movement. These plantations demonstrated conclusively that commercial timber production on small parcels was, and is, feasible.

Chapter 6

Old Pasture to Young Pine: Municipal Watershed Plantations

Calls for reform in America's forest policy during the late nineteenth and early twentieth centuries often focused on ruinous loss of timber amid predictions that devastating economic consequences were sure to follow. At the same time, the role of forests in an ecological context also received attention. In particular, relationships between forest cover, flow of springs, and resulting water supplies began to receive intense scrutiny throughout the country after the Civil War. Evidence that deforestation directly influenced flooding, protection of groundwater, humidity of air and soil, temperature, precipitation, and erosion was elusive. Disputes about these points were common and sometimes heated. Yet by the close of the nineteenth century, an ample body of written material had been assembled and a convincing link between forests and water resources forged.

Proof that woodlands slowed run-off and controlled stream flow steadily gathered substance, but their influence on rainfall never escaped the realm of speculation. European studies and writing continued to aid development of policy in America by adding credibility to disputed claims that deforestation affected climate and water supplies. Protection of water quality and stream flow on municipal watersheds emerged as an important type of forest use. Stewardship became exacting by necessity, and towns practiced conservation in a very commanding manner. Commercial productivity benefited as well. Reservoirs encircled by deep forests also contributed strong aesthetic elements to communities, particularly so where water bodies are proximate to village centers or well-traveled ways (see fig. 6.1).

Fig. 6.1. View of a young pine plantation at the Mt. Higby Reservoir, Middletown, Connecticut, in 1915. With assistance from Connecticut's state forester, Walter Mulford, a demonstration planting program was begun in 1903. Acorns were seeded in plowed furrows, expanding an area of oak seedlings transplanted from a nursery begun in similar fashion two years earlier. That oak plantation survives today, and watershed lands are currently under forestry management. Connecticut Agricultural Experiment Station, New Haven

Forest Influences

Shifting attitudes toward forests and water resources occurred amid a national migration to midwestern and western territory, increasing industrialization, expanding urban populations, and systematic deforestation. Water resources depended on a panoply of environmental components, few scientific studies had been attempted, and regional differences were critically important. In this broad context, pinpointing the spark that ignited change is difficult, but once again Marsh's *Man and Nature* is illuminating. His work provided a meticulous summation of existing knowledge in 1864 as well as clear recognition of domain governed by incomplete understanding.

On one hand, Marsh recognized that measuring the precise effect of forests on environmental deviation—rise or fall in humidity or temperature, for example—was difficult. He acknowledged this obstacle and noted that historical evidence supporting any general conclusions was slender. Yet he remained convinced that woodland summers would be shorter, cooler, and more humid and the winters longer, milder, and less humid than in those same areas if the forests were destroyed. He approached the relationship of forests to precipitation with similar equivocation, suggesting that wooded regions produced more rainfall than did open country or at a minimum equalized its distribution throughout the seasons. At the same time, he conceded that greater observation was needed to determine whether forest destruction caused a decrease in rainfall.[1]

In contrast, Marsh's discussion of relationships between forests and the flow of springs, streams, and rivers is far more emphatic. Citing several European examples and studies and relying on his own observations in New England, he expressed little doubt that destruction of woodlands diminished the flow of springs by allowing moisture to escape from soil more rapidly. Canopies of leaves and branches protecting the ground disappeared, and gradual absorption of rainfall by undergrowth, mantles of decaying matter, and roots ceased. As a consequence, watercourses dependent on steady percolation flowed unevenly, and the natural reservoirs filled by these springs shriveled. In an eloquent passage, Marsh summarized the general effects of forest devastation:

> The face of the earth is no longer a sponge, but a dust heap, and the floods which the waters of the sky pour over it hurry swiftly along its slopes, carrying in suspension vast quantities of earthy particles which increases its abrading power and mechanical force of the current, and, augmented by the sand and gravel of falling banks, fill the beds of streams, divert them into new channels and obstruct their outlets. The rivulets, wanting their former regularity of supply and deprived of the protecting shade of the woods, are heated, evaporated, and thus reduced in their summer currents, but swollen to raging torrents in autumn and in spring. From these causes, there is a constant degradation of the uplands, and a consequent elevation of the beds of watercourses and of lakes by the deposition of the mineral and vegetable matter carried down by the waters. The channels of great rivers become unnavigable, their estuaries are choked up, and harbors which once sheltered large navies are shoaled by dangerous sandbars.[2]

After *Man in Nature*, much of the nineteenth-century writing that contributed to progress in forest conservation linked the fate of water and forest resources. Topics of weather, stream flow, and water supplies were often jumbled together into rather flimsy conclusions. Frederick Starr's 1865 report to the Department of Agriculture expressed concern for loss of forests primarily in economic terms but also alluded to relationships between forests, equilibrium of temperature, and amounts of moisture in the air. Citing examples in *Man and Nature* and pointing to the world's desolate, deforested regions, Starr specifically suggested that midwestern prairies would benefit greatly from forests. Although offering little to prove his assertions, Starr expressed no doubt that temperature extremes in winter and summer and excessive fluctuations in precipitation would be mollified by forest cover: "shall we leave the lands to neglect and comparative barrenness, when, by adding forests as great modifiers and controllers of temperature and precipitation, they may probably become as desirable as any lands we possess . . . ?"[3]

Studies supporting a direct relationship between forests and climate did not immediately materialize, but the scientific community nevertheless recognized the subject's importance. Franklin Hough's 1873 paper, "On the Duty of Governments in the Preservation of Forests," which he presented to the American Association for the Advancement of Science and which accompanied that association's report to Congress, addressed the topic of forest influences. Emphasizing the effects of forests on rainfall, floods, drought, and water supplies for cities, he remarked that construction and operation of reservoirs for domestic and manufacturing use could not be accomplished productively absent a partnership with forests. Noteworthy, too, was his forceful contention that navigable rivers formed highways of commerce and were owned by the public. Forests at the headwaters of these rivers were in some measure part of the public domain as well.[4]

Although Hough's earlier paper failed to offer any solid evidence to justify his conclusions and frequently repeated sources and information already gathered by Marsh, his 1877 report as Forestry Bureau chief marked a departure from the very general examinations indicative of earlier writing. Hough devoted a substantial portion of the report to an exhaustive review of forest influences and, for the first time, made reference to studies conducted at meteorological stations founded in Bavaria in 1868. Attempts had been made in that country to measure temperature, wind velocity, rainfall, moisture in soil, and flow of springs, rivers, and streams from pairs of observation points erected in close proximity. One station was placed under forest cover while the other remained in open land but near the forest border. Measurements seemed to confirm that forested areas retained greater humidity in air and soil and reduced temperature extremes.[5]

The importance of regional conditions was underscored a year later, when John Warder addressed the American Forestry Association at its meeting in Washington and called for long and continued observations and experiments in America concerning effects of forests on climate and water supplies.[6] Both Hough and Warder can be credited with stressing, in ardent if sometimes overstated terms, the importance of forests to the nation's water reserves. They also linked both to the movement for reform in forestry policy. Convictions notwithstanding, much work remained to expel the mystery surrounding forest influences. A report by Nathaniel Egleston, Hough's successor in the Forestry Bureau, reveals the status of American forestry in 1884, perhaps unwittingly so:

> It is only recently, indeed, that we have learned that the removal of the forests involves anything more than the loss of the forests themselves. Their connection with climate, with the precipitation of moisture, with the flow of streams, with the atmospheric currents, with the temperature of the atmosphere, and consequently with the great interests of civilized life, with agriculture and commerce, was not known—was hardly suspected a little while ago . . . But at length it has been discovered that the forests have meteorological connections of the highest importance.[7]

Egleston was neither a scientist nor a forester, and it may be unfair to criticize his report too harshly. Despite its awkward, ingenuous character, Egleston's writing demonstrated recognition of an issue that offered the government more ammunition in its quest for responsible forestry policies. The Forestry Bureau was burdened by the weighty job of convincing a doubtful public and a willful and contentious lumber industry that America's forests should be protected. Understandably, they relied on whatever arguments were at hand. Notwithstanding lack of substantial evidence linking the two, forest and watershed conservation emerged at roughly the same time and evolved together.

Appointment in 1872 of a state park commission in New York to protect vast watersheds in the Adirondack Mountains signaled one of the earliest efforts at this form of conservation. At the outset, concern for these watersheds was based primarily on economic needs. The Black River Canal, extending northward from Rome to Carthage, was the primary feeder for the Erie Canal and its subsidiary canals, which stretched to every corner of the state. Adequate water levels for New York's canal system depended on a reliable source, and the Black River and its tributaries satisfied this need. A topographical survey of the region followed, under the supervision of Verplanck Colvin, a driven and at times unbalanced man who bent himself, and those who worked for him, to this enormous task.

Colvin relied on his studies of the Adirondack region in a paper titled "Relation of Forests to Water Supply," which he read to the American Forestry Congress in Cincinnati in 1882. Colvin argued that the influence of forests on water supplies in any given drainage area depended directly on evaporation and rainfall, and he offered evidence in the form of monthly precipitation records of the U.S. Signal Service to support his conclusions. From these records, Colvin pinpointed regions where the greatest irregularities in monthly rainfall occurred and noted that these were regions where forest cutting had been excessive. He also attempted to evaluate more exotic relationships between forests and weather and suggested that atmospheric electricity and storm direction were influenced by forest cover. Questionable methodology notwithstanding, Colvin's paper was well received by the congress. Its redemption may lie in his observation that knowledge of regional topography, prevailing wind direction, and average rainfall—all based on consistent study and accurate mapping—were essential if competent decisions concerning forest preservation, replanting, or cutting were expected.[8]

A number of other papers addressing related themes were offered at the Cincinnati congress. On the second day, delegates adjourned to section meetings, where individual papers were assigned for discussion. Section C focused on the topic "Beneficial and Injurious Influences of Forestry." Papers read before that section covered an international range of forest-related topics, including health, climatology, rainfall, floods, droughts, and famine. Another paper, "Forest Reserves at the Head Waters of the Ohio" was assigned to Section A, "Uses of Forests," chaired by Franklin Hough. Subsequent meetings of the American Forestry Congress at Boston in 1885 and Denver in 1886 continued examination of forest influences. At Denver, George W. Minier, a charter member of the American Forestry Association, resumed scrutiny of the relation of forests to water supplies.[9]

Higher Ground

Appointment of Bernhard Fernow as head of a separate Forestry Division within the Department of Agriculture in 1886 elevated the debate over forests and their effect on watersheds to safer terrain. In an 1884 issue of the *Forestry Bulletin*, Fernow had already observed progress being made by the Adirondack commission and had argued in favor of a forest policy extending beyond simple protection of watershed areas. He noted that, with proper forest management, revenue could be produced without any negative consequences to the region's water resources. In the year of his appointment to Washington, Fernow's report to the commissioner of agriculture contained a concise summary (literally in outline form over scarcely a full page) of the effects of deforestation on climate and water supplies. Although brief in content, his report described with precision a sequence of related events, beginning with removal of forest cover and culminating with deterioration in the flow of brooks and rivers.[10]

Fernow's professional training and familiarity with forestry studies being conducted in Germany and other European countries quickly proved valuable amid growing debate over the climatic influence of forests. In his report of 1888, Fernow responded to an article by Henry Gannet, a geographer with the U.S. Geological Survey, who disputed claims that deforestation caused changes in rainfall. In his study, Gannet selected several regions throughout the country that had experienced changes in forest cover and then reviewed measurements of average annual rainfall collected at various stations throughout these regions. Gannett's data seemed to show little or no relationship between extent of forest cover and amount of rainfall. Fernow, however, disagreed with

Gannet's methodology and considered the data unreliable because it was acquired over periods of uneven duration without reference to previous rainfall in a given region and because it depended on measurements taken from rain gauges vulnerable to error.

Irrespective of any flaws in the arguments of either Gannet or Fernow, the latter's report was important for at least two reasons. Initially, it demonstrated Fernow's ability to scrutinize research data and to evaluate the validity of conclusions based on these data. This ability clearly separated the Forestry Bureau under Fernow from the administration of his predecessors, Hough and Egleston. The report also established Fernow's knack for focusing on relevant issues and responding in terms of thoughtful public policy. Significantly, Fernow distinguished the general effect of forests on climatic activity such as rainfall from more important influences on mechanical and hydrologic conditions of soil, which in turn had a very direct effect on water supply and distribution as well as soil quality and health. He insisted that forest interiors and the consequences of deforestation in these zones should be studied separately from effects of forests on their general surroundings. To Fernow, implications of these influences rendered forest conservancy a matter of great public interest, and he pledged that resolution of the debate would be among his first Forestry Bureau duties. Fernow's report thus marks a beginning point for a clearly defined policy toward the influence of forests on watersheds.[11]

Plans for local forest experiment stations under the Forestry Division's supervision had not yet been approved. Fernow admitted in his 1888 report that the government's role was limited, at that moment, to investigation of European studies. He had begun work on a paper summarizing the results of those studies and offered, in that same report, a brief compilation that separated experiments based on data assembled over extensive regions from those applying detailed measurements in confined localities. Fernow suggested that broadly based data was vulnerable to biased manipulation, and he clearly favored the latter. For an appropriate example, he pointed to Prussian experiment stations established between 1874 and 1877 that had begun publishing monthly bulletins and annual reports.[12]

Subsequent annual reports by Fernow and a number of his articles appearing in *Garden and Forest* continued to introduce the results of European research and underscored the need for similar studies in America. Again and again, Fernow emphasized that complex factors—soil condition, topography, and length of water courses, for example—affected hydrological conditions, and he insisted that only careful, scientifically structured observation of local conditions could produce dependable results. His scrupulous concern for scientific documentation bolstered those conclusions he willingly made, particularly that forest cover played a critical role in the mechanical retardation of run-off.[13]

The paper Fernow referred to in 1888 was completed by 1893 and became the country's most comprehensive discussion, to that date, of relationships between forests, water supplies, and climate. Contributing articles by Weather Bureau chief Mark W. Harrington and meteorologist Cleveland Abbe both addressed questions of climate and rainfall, and the composite work was published as a Forestry Division bulletin by the Department of Agriculture. Fernow cautioned that past writers had generalized too loosely about forests. He also reiterated that sketchy summaries of global observations were giving way to a more scientific approach based on systematic, locale-specific experiments. This bulletin considered the influence of forests for three major topics: climate (especially precipitation), disposal and conservation of available water supplies, and run-off.

Harrington's contribution stressed the work of experiment offices established by Germany's meteorological forestry service. These stations were used to monitor soil and air temperatures, humidity, and rainfall over a period of years. Measurements confirmed that in wooded areas, when compared to adjoining open fields, lowest air temperatures were raised, highest temperatures were lowered, and the daily range of temperatures was reduced. In addition, forest litter moderated soil temperature. Although average humidity within forests was increased, studies to determine the effects of forests on rainfall were inconclusive.

Fernow fixed his attention on questions of water supply and retardation of run-off. He noted that forests, at a minimum, helped to conserve groundwater simply by preventing evaporation. Wooded lands conserved air humidity through shade, shelter against wind, and soil cover. Moreover, distribution of water supplies was controlled by complex elements such as topography, elevation, and soil composition, which in turn affected water carrying capacity, permeability, drainage, and surface conditions. Simple influences such as the general unevenness of the forest floor also played a role. Fernow's confident analysis left little room for doubt about ways in which deforestation interfered directly with conditions intricately linked to water resources. Nevertheless, he acknowledged that precise measurement of the effects of forest clearing were difficult because of the complexity of factors involved and variations among different regions.[14]

At about the time of Fernow's bulletin, forest influences began to receive attention from a number of sectors. Protection of water resources was a vital topic in arid western regions, where the effects of watershed cover on irrigation systems were more critical than in water-wealthy New England. In 1896, Secretary of the Interior Hoke Smith asked the National Academy of Sciences to develop a rational forest policy for public domain lands. He posed three questions to the academy, one of which inquired: "How far does the influence of forest upon climate, soil, and water conditions make desirable a policy of forest conservation in regions where the public domain is principally situated?"

Charles Sargent's committee devoted their greatest energies to exploration of western forests. Creation of forest reserves and an administration to manage them were primary goals; protection of water supplies was simply one of many arguments used to achieve these ends. Relationships between forests and water conditions were part of a broader context involving river flow essential to irrigation, and the report underscored connections between forest cover, surface run-off, spring flooding, and low water levels in summer and autumn. The committee relied on a single Austrian study that attempted to measure the flow of major European rivers and to attribute variations to increased cultivation and destruction of forests. Their report offered only cursory assessment of the tenuous relationship between forests and rainfall and in general lacked the substantive analysis found in Fernow's writing. Protection and management of western forests depended on authority to establish rules and regulations, and the committee recommended that the secretary of the interior be given considerable power, including rights to grant irrigation easements and reservoir sites. For their part, irrigation groups readily accepted theories that watershed vegetation directly affected levels of groundwater. For this and other reasons, they actively lobbied for national forest reserves in western states to conserve precious water supplies. These organizations, epitomized by George Maxwell's National Irrigation Association, played an important role in validating developing policy on relationships between forests and water resources.[15]

Writing for Pinchot's Forestry Bureau in 1903, James Toumey prepared a document titled "Relation of Forests to Stream Flow," which appeared in the Department of Agriculture yearbook. Toumey's article summarized observations conducted in California's San Bernardino Mountains, and he approached the subject cautiously. Noting that final conclusions covering the entire field had not yet been established, Toumey reversed direction by suggesting that rainfall, while possibly influenced by the presence of woodland cover, was instead probably the greatest factor in controlling forest density and distribution. Nevertheless, he confirmed that forests provided the most effective agent known for regulating the disposition of precipitation after it reached the ground. The southern California studies demonstrated the enormous effect of forest cover in decreasing surface flow in small catchment basins and measured a material increase in amounts of yearly stream flow in forested areas. Toumey concluded by pointing out that forests also offered a great indirect value by preventing wind and water erosion and by providing an adequate absorbing medium at water system sources.[16]

Relationships between forest cover and water supplies, important to American forest policy in general, assumed very specific dimensions with attempts to create a national forest in New Hampshire's White Mountains. Federal reserves had been established in western states after the Forest Reserve Act empowered the president to set aside public domain lands. However, few such lands existed in eastern states. Shortly after turn of the century, legislation was introduced authorizing acquisition of lands for national forests in the White Mountains and southern Appalachians. Opponents argued that no constitutional authority existed to support federal purchase of forest lands and defeated the measure several times. Led by the Society for the Protection of New Hampshire Forests, an alliance of forest preservationists, hikers, and manufacturers dependent on waterpower responded. With a masterful gambit, they evaded the constitutional question by declaring the law's objective to be forest conservation within river watersheds for the purpose of improving navigation. Couched in these terms, the Weeks Act was a constitutional exercise of Congress's right to regulate interstate commerce, and it was finally enacted in 1911. The law expressly limited authority to purchase lands to those instances in which a direct connection to the protection of navigable streams could be established and also appropriated funds to be used for prevention of forest fires in watershed areas.[17]

Responsibility for proving that individual tracts of land were directly linked to the flow of specific navigable streams fell to the U.S. Geologic Survey, which belonged to the Department of the Interior. Their survey staff, headed by Marshall O. Leighton, lacked an individual with technical training in forestry and was joined by Benton Mackaye, on loan from the Forest Service. Mackaye's assignment sent him back to the White Mountains, the scene of his foot-wearying explorations as a young Harvard undergraduate in 1897. During the summer and fall of 1912, he conducted a survey of drainage basins affecting lands already approved for purchase by the Forest Service. The results of this survey were published in 1913 and paved the way for creation of the White Mountain National Forest.[18]

Percolation in New England's Woods

New England states had already begun to consider similar problems well before struggles to acquire a national forest in the White Mountains reached a conclusion. New Hampshire's first commission on forestry, appointed in 1881, was charged with the task of considering the effects of forest destruction on rainfall and water levels in the state's ponds and streams. Their report, issued in 1885, was based on a crafty piece of homespun research. Circulars were sent to towns throughout the state inquiring whether the flow of streams and rivers had declined within the collective memory of town inhabitants and whether the cutting of nearby woodlands had coincided with this decrease. Predictably, a large number of towns responding had noted a reduction in the flow of water over a period of years and confirmed that timber harvesting had indeed been conducted within the region. Whether mill owners merely took advantage of an opportunity too good to miss or the timber industry was out working in the woods at the time is not entirely clear.[19]

While hardly a model of scientific analysis, the report warrants at least some scrutiny. Water-powered mills had been a backbone for New England's industrial economy through much of the eighteenth and nineteenth centuries, and Yankee millwrights had long demonstrated ingenuity in the development of techniques for mastering water flow. Dams, reservoirs, spillways, canals, flumes, and races were common features across New England's landscape. Even minor changes in water volume were keenly observed. For example, residents of the town of Henniker, located in central New Hampshire along the Contoocook River, estimated that the river's water volume had decreased by one-third during a twenty-year period. Several towns in southern New Hampshire—Richmond, Fitzwilliam, and Rindge—all reported streams that once furnished power for saw mills but that had become desiccated following timber cutting in the region.[20]

Relationships between forests and watersheds created common ground for conservation groups, public water companies, civic improvement organizations, and manufacturing interests. An article that appeared in 1901 in *The New England Farmer*, generally a reliable barometer of trends in that region, observed that matters of forestry were of vital interest for economic and scenic reasons and advised states to establish forestry commissions. Significantly, its author attributed the primary economic importance of forests to their effect on water supplies in streams. Deforestation caused loss of humus available to absorb surface water, increased dangers of flooding following freshets, accelerated rates of evaporation and melting of winter snows, and reduced the flow of springs in summer. All of these factors contributed to losses by manufacturing interests dependent on a constant flow of water.[21]

Introduction of forestry as a means for watershed protection had a very direct effect on cities and towns as well. A great many of New England's municipal reservoirs were constructed during a period that began shortly after the Civil War and closed prior to the onset of World War I. Rapidly expanding urban populations generated by nineteenth-century industrialization needed larger sources of water for domestic, manufacturing and sanitary use as well as for fire protection. Cities began to purchase extensive watershed lands to guard against pollution of newly created reservoirs by industrial or agricultural activity. Smaller towns and villages simply desired water works large enough to supply adequate pressure year-round. Lands acquired were often farms, which typically remained open. Initially, towns often removed deciduous types to prevent leaves from drifting into reservoirs and staining water. Clearing was sometimes accomplished by yearly burning, a practice that lowered soil quality and unfortunately introduced greater amounts of silt from excessive runoff. Hayfields, gravel pits, and woodlots regularly produced a modest yearly income to offset public ownership of land no longer on tax lists.

The founding in 1882 of the New England Water Works Association, a professional body for the region's municipal engineers, coincided with the development of more sophisticated water systems. Scholarly papers on a broad range of related topics were presented at association meetings and subsequently published in its quarterly journal. Beginning in 1887 with an article by civil engineering professor George F. Swain of the Massachusetts Institute of Technology, journal pages became an orderly chronicle of emerging interest in forests as a component of water supply management. Swain's paper painstakingly analyzed swirling opinion regarding the influence of forests on rainfall. He safely concluded that proof of any relationship did not exist and furthermore would be difficult to obtain. He was equally cautious about giving forests too much credit for increasing the overall extent of water resources, noting that many independent factors were at work. Nonetheless, he agreeably recognized the important role of woodlands in retarding run-off and in equalizing the flow of streams. In doing so, Swain focused on precisely the issue that Fernow would soon stress.[22]

Not until 1908, however, did the association turn its attention directly to the matter of forest management. That year, Royal S. Kellogg, an assistant U.S. forester, addressed the group with a paper titled "Forest Conservation." Kellogg recounted the waste of America's timber legacy and called for protection of remaining lands through establishment of state and national forests in the region. Arguments favoring watershed protection in the White Mountains, he observed, applied equally well to water supplies for cities and towns. Presentations emphasizing very detailed methods of forestry practice followed over the years. Edward Bryant, a consulting forester from Boston, outlined "Practical Forestry for Water Works" in 1911, and Sidney Clapp, an engineer for New York City's Board of Water Supply, proffered "Detail Methods Used in the Forestation of Watersheds" in 1915. Two years later, as director of Yale's forestry school, James Toumey asserted that timber production by water companies would not amount to "small potatoes" and urged association members to take up forestry in a serious way.[23]

By World War I, a large number of New England cities already had begun extensive planting campaigns and were well on their way toward proficient management. Many had been assisted by fledgling state forestry agencies looking for opportunities to promote their cause. In several cases, the Forest Service had provided technical assistance and had prepared at least one management plan. Planting with coniferous types—white, red, or scotch pines; spruce; and hemlock—was recommended for a variety of reasons. Softwoods shaded the ground year-round, reducing evaporation and retarding snow melt in the spring. Dense mats of needles protected the soil from frost, and litter was not likely to find its way to water. Rapid growth and high timber yield per acre from pines and spruce also afforded greater commercial value. Already in public ownership and with everything to gain through forestry management, watershed reserves provided ideal opportunities for a young forestry profession. Many of New England's first successfully managed town forests began as watershed districts, which were gradually planted as the benefits of forestry became more widely known.

Private organizations such as the Massachusetts Forestry Association promoted alliances between town forests and municipal watersheds. That group reprinted an address by Frank W. Rane, state forester of Massachusetts, to a gathering of the New England Water Works Association in March 1911. Rane emphasized that the time was ripe for municipal forestry in New England and, declining to lecture on the relationships of woodlands to stream flow, instead focused on economic advantages of watershed forests. He argued, sensibly, that lands already owned by water commissions and standing idle could be converted to profitable use. To support this claim, he provided income estimates based on stumpage yield for several Massachusetts towns. Rane also included a list of fifty-six towns in the state having watershed lands capable of being forested. In addition, he pledged assistance from the state for these communities.[24]

During the decade following war's end, large-scale planting and management efforts were conducted on municipal watersheds throughout New England. Calls for more research on relationships between forests and water supplies periodically surfaced. By this time, however, general consensus on the topic of run-off had been reached, and questions about forest influences were fielded matter-of-factly. In reply to a letter of inquiry about rainfall and run-off from New England Water Works Association president, Theodore L. Bristol, James Toumey summarized the best reasons for water companies to invest in forestry in 1926:

> . . . the drainage area which feeds a reservoir should be kept in forest, preferably coniferous, *not for the purpose of increasing the total water crop* from the drainage area, but, first, for the purpose of distributing the water crop over a longer period following precipitation; and, second, for the purpose of permitting the larger part of the precipitation to be taken up by the soil and the smaller part to flow immediately over the surface, thus decreasing the pollution and sediment which reaches the reservoir.[25]

A large percentage of New England cities obtained their water supplies from surface drainage, and foresters were able to argue convincingly that the initial expense of planting could be recovered by woodland management.

Potable Crops of Water

Evaluation of the origins of local watershed forestry has, until this point, benefited from a relatively logical progression of literature and events that shaped policy. Consideration of specific examples profits from no such orderly sequence. Municipal watershed forests in New England seem to defy any rigid or precise categorization by size, time of acquisition, or date of planting efforts. Clearly, these factors were influenced by concentration of urban population growth, need for larger reservoirs, extent of watershed territory available, or other locale-specific circumstances. Towns of modest size did not require large water reserves, and waterworks were fairly simple. In other cases, widely varying factors, such as newly established industries, proximity to educational institutions where forestry was taught, and even the foresight of a particular individual, were sometimes important.[26]

Ownership of watershed lands was often vested in quasi-public utility companies that received municipal charters to operate. In many cases, these companies were forced to obtain lands located in other towns and frequently paid property taxes as a result. Elsewhere, cities or towns simply organized separate administrative departments and purchased their own land, again subject to taxes if situated in a different town. Apart from watershed lands, the location of which was dictated by geographic features, intentional acquisition of town forests in neighboring towns with lower tax rates was not uncommon. Distinctions between municipal and town administration can often be confusing in New England, particularly where municipal corporations have the same name or one that is similar to towns in which they are located. For example, Essex Junction, Vermont, is a separate municipal corporation located in the town of Essex. Essex Junction's watershed forest was planted on land near the unincorporated village of Essex Center, also in the town of Essex. Taxes on the forest were paid to the town.

In more than a few instances, watershed properties also became forest parks, and landscapes architects were sometimes retained to help with design. Near Boston, Middlesex Fells Reservation was acquired by the Metropolitan Park Commission after local water boards had already set aside substantial areas. Engineering responsibilities were assigned to the Metropolitan Water Works, and they in turn hired Frederick Law Olmsted, Jr., who prepared plans for Spot Pond Reservoir. Lynn Woods, also near Boston, became a city park not long after a devastating fire prompted the community to purchase Breed's Pond in 1870 and create a public water board. Plans for both reservations were presented at meetings of the New England Water Works Association, Lynn in 1895 and Spot Pond in 1900. Boston landscape architect Arthur Shurtleff collaborated with James Toumey on forestry plans for Hartford's ambitious watershed project in 1912. Not too many years later, Toumey opined that water companies should treat forest aesthetics as secondary to timber growth and observed that productive forests usually doubled as pleasant places to visit.[27]

Comparative study of watershed forests among New England states becomes complicated for a number of reasons. Unlike the Massachusetts Forestry Association, which thrust its state into the forefront of the town forest movement, the New England Water Works Association represented an entire region. Moreover, forestry was only a single aspect of the water-supply business and a relatively small one at that. Understandably, association records provide no inventories of those utilities engaged in forestry and only limited information about specific lands. Furthermore, although state foresters were directly involved with a large number of important watersheds, as their reports showed, many companies chose to pursue their own prospects. Consequently, local records hold the largest share of information about these forests. A review of important examples from one state to the next, however, does provide a modicum of structure and suggests at least some points for comparison.[28]

Connecticut

Connecticut played a more central role in the practice of forestry on watershed lands than it did in the movement for town forests. Generous credit belongs to the Eli Whitney Forest, which was cultivated to protect New Haven's water supply. The New Haven Water Company received its charter in 1849 but did not begin any major acquisition of land until the 1880s. Forestry work began in 1901, when Yale's School of Forestry prepared a management plan for a three-hundred-acre tract around Maltby Lakes in the city's western environs. Maltby Division, as it became known, was easily accessible to school foresters and developed into a demonstration tract for intensive management methods (fig. 6.2). Mixed hardwood stands prevailed, with occasional patches of hemlock; open or partially stocked lands were replanted with conifers. Brush-covered areas were permitted to grow until large enough to support a cut for fuel, which then paid for replanting.

In 1907, faculty member Ralph C. Hawley was appointed company forester and assumed responsibility for all woodlands, which by then had grown to eight thousand acres. Holdings in the Saltonstall Division spilled easterly into the nearby towns of East Haven and Branford and enveloped a narrow, finger-shaped lake that gave the sector its name. Over the next decade, sale of wood products (principally lumber, railroad ties, and pilings) netted the company roughly $37,000, three quarters of which was plowed back into timber-stand improvement. During this period, yearly plantings averaged 250,000. Scrupulous management continued, and the forest became a successful and widely cited example for other water companies and for municipal forestry in general.

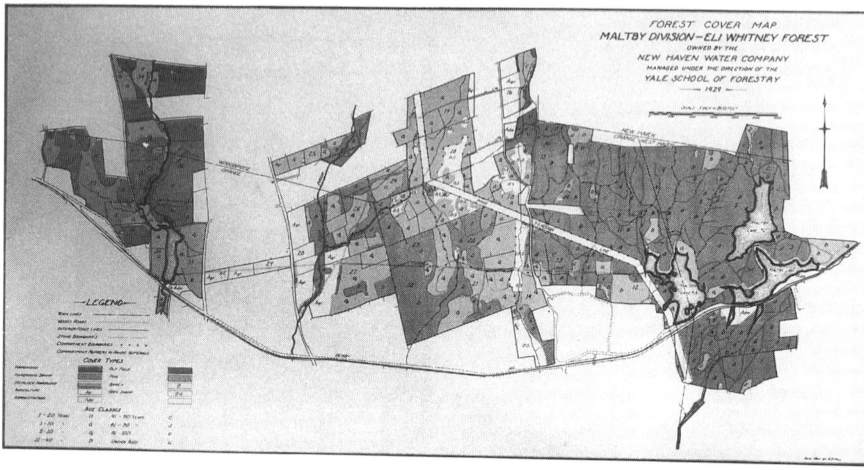

Fig. 6.2. Forest cover map of the Eli Whitney Forest, Maltby Division, drawn by Yale School of Forestry in 1929. In 1901, Yale's forestry program prepared a management plan for watershed lands in the Maltby Lakes region west of New Haven, owned by the New Haven Water Company, and faculty member Ralph Hawley was appointed company forester six years later. Forestry expanded to include company lands surrounding Lake Saltonstall and eventually encompassed more than twenty thousand acres. Yale School of Forestry, New Haven, Connecticut

By 1940, lands had increased in size to 22,000 acres. Forestry management is carried on today under the supervision of the South Central Regional Water Authority, which employs its own forester.[29]

Although it occupies a prominent place in the history of watershed forestry, Eli Whitney Forest was unique neither in date of origin nor in period of planting and management. In Middletown, a short journey north from New Haven, the jutting ridge of Mt. Higby is an easy landmark for those who ply the corridor of Interstate 91. Commencing in 1893, lands along the mountain's eastern slopes were purchased by the town's water company, and construction of a reservoir was undertaken in 1897. This water supply augmented Laurel Brook Reservoir, located southeast of Mt. Higby and pressed into service in 1866, a year after Middletown Water Works was formed. Urged by Robert N. Jackson, a local bank president keen on forestry, the board of water commissioners wielded initiative and began reforestation of farm acreage surrounding Mt. Higby Reservoir (fig. 6.1). In August 1900, an invitation to examine company lands was extended to Henry Graves at Yale School of Forestry, but whether Graves ever inspected Higby's terrain that year is doubtful. Undaunted, and with what little information they were able to glean, the commissioners energetically began experimental seeding and transplanting during the spring and fall of 1901. Superintendent John Broach and caretaker Edwin Bywater tilled nursery beds where walnuts, hickory nuts, and acorns sown in furrowed rows produced a crop of seedlings for transplanting two years later. That first season an assortment of hardwood saplings, along with white pine, hemlock, cedar, and spruce, was transplanted, primarily for ornamental purposes.[30]

Early in 1902, Connecticut's newly appointed state forester, Walter Mulford, viewed the site and agreed to furnish a plan for planting bare fields. Mulford envisioned a demonstration forest and volunteered his services, requiring only reimbursement for travel and board. In April, Mulford, the board of water commissioners, and the Connecticut Agricultural Experiment Station formalized an agreement, and a program was underway by spring of 1903. Hardwoods, consisting of chestnut, walnut, catalpa, and oak, dominated planting activity that year. Tracts bedded with red and black oak seedlings were enlarged by more acorns sown again in plowed furrows, and stands survive today. White pines raised from seed were transferred to a nursery garden for later transplanting. Mulford also directed thinning of older growth at higher elevations. Planting continued over the next decade, but hardwoods were eventually eclipsed by red and white pine. Currently, forest cover encompasses 1,330 acres, 200 of which are planted, and management is supervised by a forestry consulting firm. The hardwood plantation is one of the most important of its kind in New England.[31]

Hartford's board of water commissioners at least pondered reforestation of its water system lands as early as 1901. The suggestion may have come from Allen Hazen, a New York engineer who had been engaged to study means for increasing city water supplies and who later designed a sand filtration plant. In any event, Henry Graves surveyed the site and drafted a management plan in 1902; planting started a year later, mostly with white pine and white ash. In 1910, water department forester L. W. Goodrich described Hartford's plantation to a gathering of New England Water Works Association members, offering technical assistance and recommending forestry as a good business proposition. Progress continued in 1912, when James Toumey issued a report recommending planting of coniferous types on open fields, at that time producing crops of hay. He also counseled the purchase of young stock rather than nursery seeding. When Toumey advised clearing blighted chestnut stands, Hartford's board employed Arthur Shurtleff to devise a strategy for cutting to preserve a parklike setting. At a meeting in April 1912, commissioners resolved "that the whole subject of forestry be studied." In so doing, they revealed a cautious interest that was probably more typical of water companies during the period than the enthusiasm displayed at Middletown.[32]

Elsewhere in Connecticut, a number of other cities turned to forestry to protect quality of their water supplies. P. H. Danaher, superintendent of Waterbury's Wigwam Reservoir in Thomaston, requested assistance from the U.S.

Forest Service in 1906 and received a copy of *Circular 21*, together with Bulletin 42, *The Woodlot*. By 1925, the city had established a 125 acre plantation. Similar work also had been undertaken by public companies at Bristol, Groton, Meriden, New Britain, Southington, and Wallingford.[33]

New Hampshire and Massachusetts

An ample number of cities in New Hampshire and Massachusetts had inaugurated forestry work on watershed lands by or shortly after the century's end. Residents of Concord, New Hampshire, voted in 1872 to spend funds for construction of an aqueduct from Long Pond, located west of the town farm (see fig. 2.20). The newly created Concord Water Board promptly began acquiring farms and woodlots encircling the water basin. In 1895, board superintendent V. C. Hastings planted three thousand white pine and spruce seedlings, but spaced them widely at twenty feet. Philip Ayres assessed the plantation's progress in 1906, and Percy Sanders, successor to Hastings, began extensive reforestation of all open lands a year later. Two thousand additional trees were planted that year, and a nursery was started from seedbeds. Hardwoods were thinned during winter months to release planted areas from cover. In 1913, 31,000 pine seedlings were supplied by the state's nursery.[34]

Private water companies joined in the quest to garner expertise in matters of forestry. Pennichuck Waterworks, situated in Nashua, New Hampshire, transplanted white pine seedlings on its lands in 1896. During the spring of 1906, company officials enlisted Forest Service support. Benton Mackaye was given the assignment and prepared a working plan that summer. His document divided operations into five categories: improvement cuttings to remove inferior species, thinnings to encourage hardy trees, final cuttings, planting, and fire protection. Mackaye discouraged future planting, remarking that successful transfer of wild stock had already been achieved and that seedlings in several plots were too densely placed. His document joins those at New Haven and Middletown among the first formal working plans for watershed forests in New England.[35]

Impetus for much of the forestry work at Pennichuck came from company engineer and superintendent William F. Sullivan. At a meeting of the New England Water Works Association in April 1910, Sullivan tenaciously queried professional foresters, among them Ralph Hawley, about sawmills. The company had inherited a run-down, waterpowered mill located on one of its sites, and Sullivan was determined to find out whether it could be put to profitable use. Wary of the "trimming" done by loggers, who measured stumpage according to a baffling assortment of log rules, Sullivan candidly summarized his opinions and quite possibly those of other company managers as well: "We outside the inner circle are in the pioneer stage, seeking useful knowledge per-

Fig. 6.3. View of a water-powered sawmill operated by the Pennichuck Water Works in Nashua, New Hampshire, in 1909. Pennichuck's engineer and superintendent, William Sullivan, sought to capitalize on the commercial value of company timber by controlling costs of logging and milling. Benton Mackaye, a consulting forester for the U.S. Forest Service, prepared a management plan in 1906, ten years after planting with white pine seedlings had begun. Pennichuck Water Works

taining to forestry and its branches, especially harvesting. We are into it and I don't know how we are to get out of it."[36]

There is little doubt that Sullivan's mind had already been decided on the mill question. Despite recommendations from Hawley that he choose, or at least consider, a portable steam mill, Sullivan issued a report to Pennichuck's president in June 1910, recommending preparation of the waterpowered facility. A series of small ponds not needed for reservoir quotas offered an easy method for transporting logs, and a photograph dated a year earlier appears to show considerable activity already in place (fig. 6.3). At that same meeting, Sullivan challenged the experts on another point. In response to suggestions that companies employ professional foresters, he remarked that his "logging bullies" (fig. 6.4), amounted to natural-born woodsmen, especially those of French-Canadian ancestry. They also had been praised by a government forester, presumably Mackaye. Sullivan steered a true course, and forest management improved steadily. Today, Pennichuck's woodlands encompass more than six hundred acres, primarily in Merrimack and Nashua.[37]

In Massachusetts, Fall River's reservoir commission began comparing water

Fig. 6.4. Crew of logging "bullies" employed by the Pennichuck Water Works in Nashua, New Hampshire, in 1909. In order to combat "trimming" done by private loggers who measured stumpage by a bewildering variety of rules, Pennichuck employed its own woodsmen to cut and move timber to the company's mill. Logs were floated in a succession of ponds and were then retrieved during the slack season when saws were put into motion. Pennichuck Water Works

levels in Watuppa Pond with rates of precipitation during 1899, 1900, and 1901, and estimated that 50 percent of rainfall was lost to evaporation or transpiration. Assistance from the Massachusetts Forest Service was requested in an effort to improve reservoir storage capacity. In response, a working plan with a type map of the five-thousand-acre watershed was drafted by Harold Cook in 1909. Cook began with a review of forest influences and devoted several pages to German practice. He recommended planting on cultivated, pasture, and burned-over lands, thinning suppressed young growth, and clearing roadside underbrush or dead trees to reduce the risk of fire; a fire observation station also was suggested. Cook noted that poor oak and chestnut sprouts covering much of the newly acquired watershed were the progeny of timber rights granted to former owners following city purchase of land, a policy he criticized. He also remarked that coniferous seedlings could be obtained more cheaply and in better condition from Germany than from midwestern nurseries.[38]

Other scattered examples demonstrate comparable work. Keene, New Hampshire, had outgrown its Octagon Reservoir by 1900 and began securing lands around Echo Lake in the neighboring towns of Roxbury and Nelson. A type map, based on a survey by Samuel Wadsworth, was prepared in 1912; it showed primarily coniferous and mixed-hardwood types as well as several tracts of burned land bordering the lake shore, possibly the result of replanting efforts (fig. 6.5). In Massachusetts, State Forester Frank Rane inspected lands belonging to Pittsfield in 1908 and recommended reforestation with white pine. Planting on Westfield's watershed began at about the same time, and a fire tower was later erected.[39]

Timber management on watershed property was not limited to well-populated industrial cities. Hanover Water Works Company, incorporated in 1893 and owned jointly by Dartmouth College and Hanover Village Precinct, acquired more than a thousand acres of open and wooded land by 1904. This included the village's two-hundred-acre town farm with its pastures, mowings, native pine stands, and mixed woods. Robert Fletcher, a mathematics instructor who became director of Dartmouth's School of Civil Engineering in 1870, was appointed president and engineer of the company. He conducted a survey, and his map of 1903, revised in 1906 and again in 1923, portrays an extraordinary visual history of the reservoir district (fig. 6.6). Fletcher experimented with methods of controlling stream flow and preventing erosion caused by floods. During a great rainstorm of May 26–28, 1906, he measured the flow of water over a spillway and observed meadows inundated by overflowing brooks. The lesson learned, he wrote, was to pave main channels in their lower reaches and to keep them free of rubbish. Alders and brush were cleared from banks, bogs drained, and stream banks lined with riprap.

As a successful engineer and surveyor, Fletcher must nevertheless have been frustrated by his efforts at forestry. Clearing of woods near the reservoir had occurred after property acquisition in 1893 and again in 1905 and 1906, the latter clearing suggesting plans for reforestation. After an examination and report by Philip Ayres in 1905, a plantation of five thousand pine seedlings was begun, but more than half were lost to fire and winter-kill. A year later, a thousand spruce seedlings, three to six years old, survived transplanting from College Park. However, that same year a pine nursery, begun from ten pounds of seed planted in an old orchard, failed. Fletcher's map shows only small and scattered plantations through 1926, the year it was last amended. If disillusioned with forestry, the company still managed to resume planting before 1930. This time, they were aided by the state forestry department, which supplied white pine seedlings from its nursery established at Gerrish in 1911.[40]

Pockets of intensive forestry management developed randomly on watersheds scattered throughout New England and did so during periods that varied considerably. Lake Massabesic, which supplies water for Manchester in southeastern New Hampshire, is one such example (fig. 6.7). Manchester Water Works was chartered in 1871, but interest in forestry was slow to develop. In fact, as late as 1903, open fields were still being seeded with grass. An in-

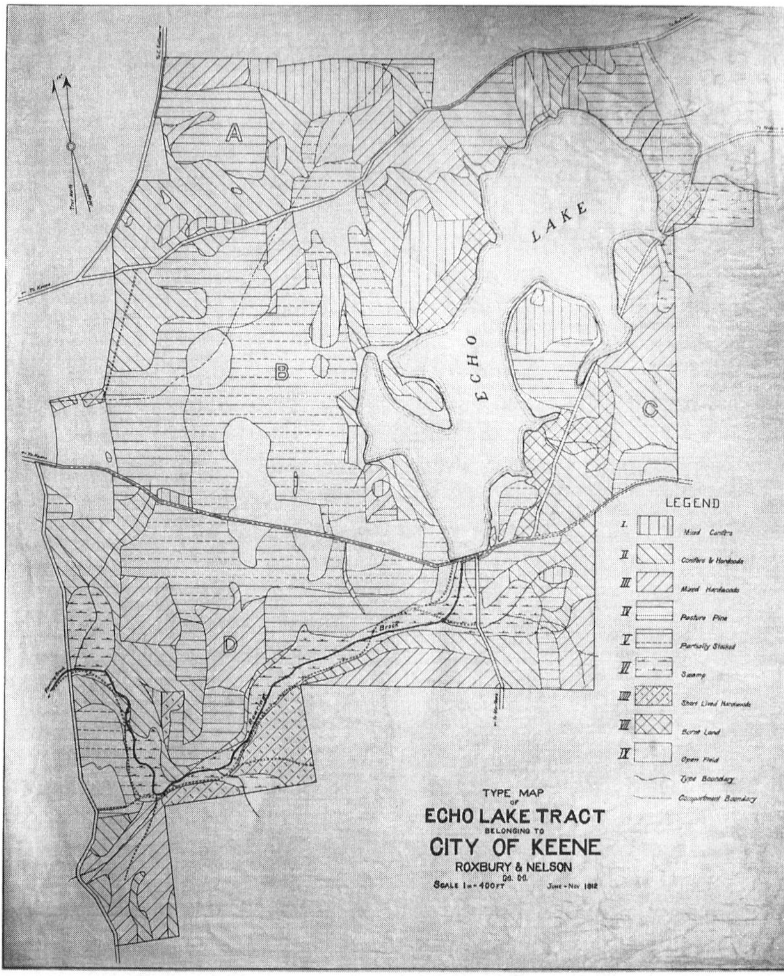

Fig. 6.5. Unsigned forest type map of watershed lands belonging to Keene, New Hampshire, and located in the neighboring towns of Roxbury and Nelson. The map was prepared by the state's forestry department in 1912 but was based on a survey by Samuel Wadsworth. Before 1900, water companies often kept their lands open and controlled wooded cover by burning, especially deciduous types with water-staining leaves. Keene's map depicts burned-over lands bordering Echo Lake, probably a precursor to replanting with coniferous types. New Hampshire Division of Forests and Lands—Concord, and Wadsworth Collection, Cheshire County Registry of Deeds, Keene, New Hampshire

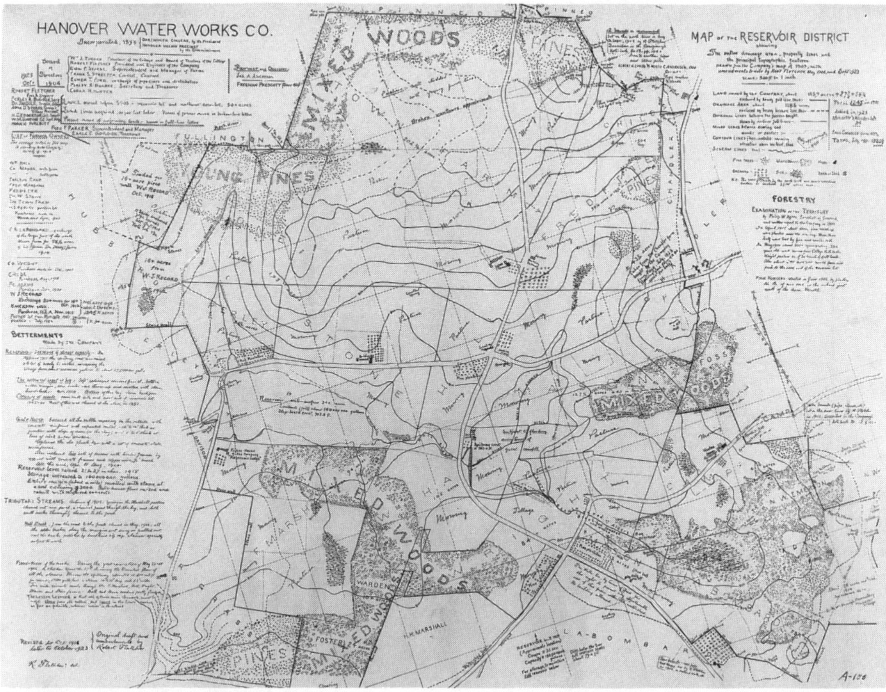

Fig. 6.6. Map of the reservoir district serving Hanover, New Hampshire, and owned by the Hanover Water Works Company. Robert Fletcher, director of Dartmouth's School of Civil Engineering, was appointed company president and engineer, and he prepared the survey in 1903, revising it in 1906 and again in 1923. Fletcher's work is a remarkable piece of artistry and, with extensive notes and observations penned on the map's margins, portrays a landscape history in extraordinarily visual detail. Hanover Water Works Company

spection in 1912 by State Forester Edgar Hirst persuaded company officials to launch a planting campaign, and they quickly made up for lost time. By 1930, roughly 900,000 seedlings, almost exclusively white pine, had been transplanted on nearly 750 acres. More than a third were procured from the Keene Forestry Association. Type maps and management plans were prepared by University of New Hampshire forestry students under direction of Karl Woodward in 1928. During Depression years, Public Works Administration, Works Progress Administration, and Civilian Conservation Corps employees embarked on improvement projects, including draining, weeding, thinning, and road building. A waterpowered sawmill was rebuilt in 1937 and fitted with equipment to churn out both rough and finished lumber. Three years later, a fire

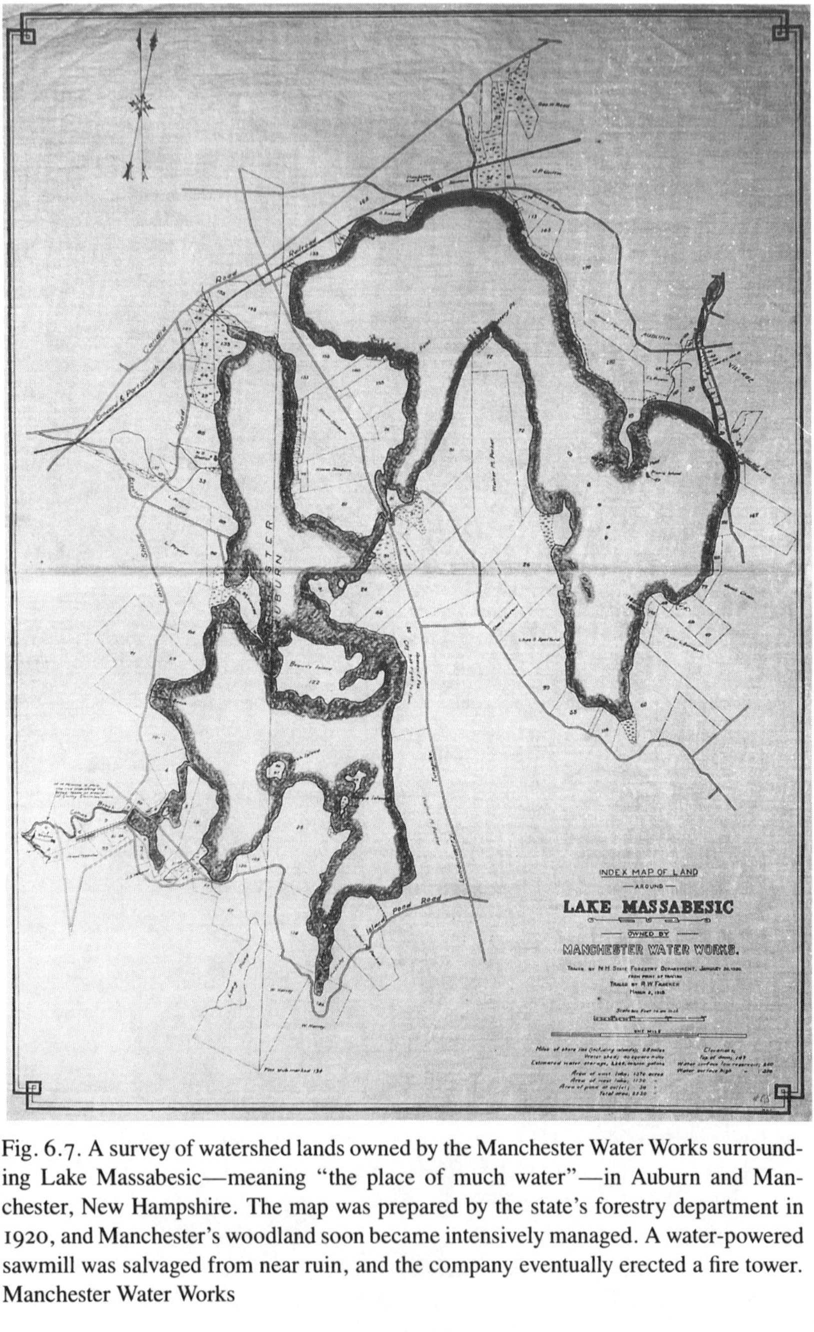

Fig. 6.7. A survey of watershed lands owned by the Manchester Water Works surrounding Lake Massabesic—meaning "the place of much water"—in Auburn and Manchester, New Hampshire. The map was prepared by the state's forestry department in 1920, and Manchester's woodland soon became intensively managed. A water-powered sawmill was salvaged from near ruin, and the company eventually erected a fire tower. Manchester Water Works

tower was erected on Mine Hill in neighboring Auburn from wood materials cut and milled on the site.[41]

Although operations waned during the war years, Nelson Brown and Harris Reynolds examined the plantation in 1944 with company forester Aldus J. Christie and extolled the company's progress. By the war's end, Manchester's forest had increased to 5,300 acres, and intensive management resumed. Area markets for utility poles, pilings, hemlock plank, barrel staves, pulpwood, cordwood, and logs for cabins proved consistently strong. Chips and sawdust were sold to local farmers for litter. By 1960, the forest had grown to eight thousand acres and sustained active lumbering. Although it has since been destroyed by fire, the sawmill continued to operate productively during this period.[42]

Vermont, Maine, and Rhode Island

Similarly isolated examples of diligent watershed forestry exist in the region's remaining states. In Vermont, the cities of Montpelier, Barre, Bellows Falls, Rutland, and Essex Junction and the town of Chester all began construction of reservoirs and acquisition of watershed lands prior to 1900. Active reforestation did not, in most cases, gain momentum until after World War I but was typically assisted by Vermont's forest service. Concluding more than a decade of debate, Montpelier residents finally voted in 1884 to pipe water from nearby Berlin Pond in the town of Berlin. Acquisition of farms and other watershed lands was nearly complete by 1912 and totaled eight hundred acres. The following year, expenses of $63 were reported for the cost of trees. By 1933, lands had spread into neighboring Northfield and had increased to 1,160 acres; boundaries and compartments were identified on a type map prepared that year by the state's forest service (fig. 6.8). Nearby, the city of Barre also began purchasing land in several watershed areas in the towns of Barre and Orange at about the same time.[43]

The municipal charter of Bellows Falls, an industrial center wedged along the Connecticut River in southern Vermont, was revised in 1891, authorizing the village corporation to maintain a water department and to obtain title to lands surrounding Minard's Pond in the town of Rockingham. Forests in Bellows Falls, Rutland, and Essex Junction eventually attained commendation from the U.S. Forest Service during the heyday of the service's community forest program. As early as 1881, the city of Rutland began purchasing woodlots and farms to protect its water supplies. Although an active period of acquisition did not begin until after 1910, several tracts were secured in 1896, including a small parcel of lease land in the town of Mendon. The city's lands gradually spread across the drainage basin of Mendon Brook below the slopes of Mt.

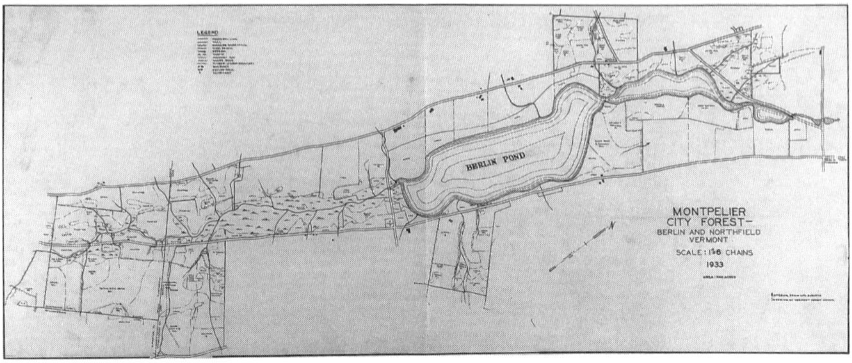

Fig. 6.8. Forest type map of the city forest owned by Montpelier, Vermont, prepared by the state's forestry department in 1933. Located in nearby Berlin and Northfield, Montpelier's reserve was not extensively planted until after World War I, when municipal forestry began to take root in the state. Despite its comparatively late start, Vermont's program eventually became a progressive one and outlasted campaigns in other parts of New England. Vermont Department of Forests, Parks and Recreation—Waterbury

Killington, eventually expanding to several thousand acres situated almost entirely within the town of Mendon (fig. 6.9). On occasion, Rutland's watershed has also served as a demonstration forest where the state's municipal forester, Eugene Keenan, tutored schoolchildren (fig. 6.10).[44]

Purchase of Essex Junction's reservoir lands began before 1900. Roughly equal in size to Montpelier's watershed district, the forest was not extensively planted until after 1923 and then mostly with red and scotch pine. However, it soon emerged as a model that other communities sought to emulate. Unfortunately, its current fate is uncertain. Although a portion of its lands were kept open after sale for office development, abandoned cars and dirt-bike trails now dominate the forest. Smaller towns also undertook water system projects. Chester, located in southeastern Vermont not far from Bellows Falls, purchased a 150-acre farm and several smaller parcels of adjacent land. A plan for the town's reservoir and water main was prepared, but again, forest planting was not initially considered. Lands were subsequently dedicated as a municipal forest. Towns in Maine and Rhode Island also were slow in establishing forestry on watershed lands. In Rhode Island, Pawtucket, Woonsocket, and Providence all reported pine plantations in 1929, although work may have begun somewhat earlier.[45]

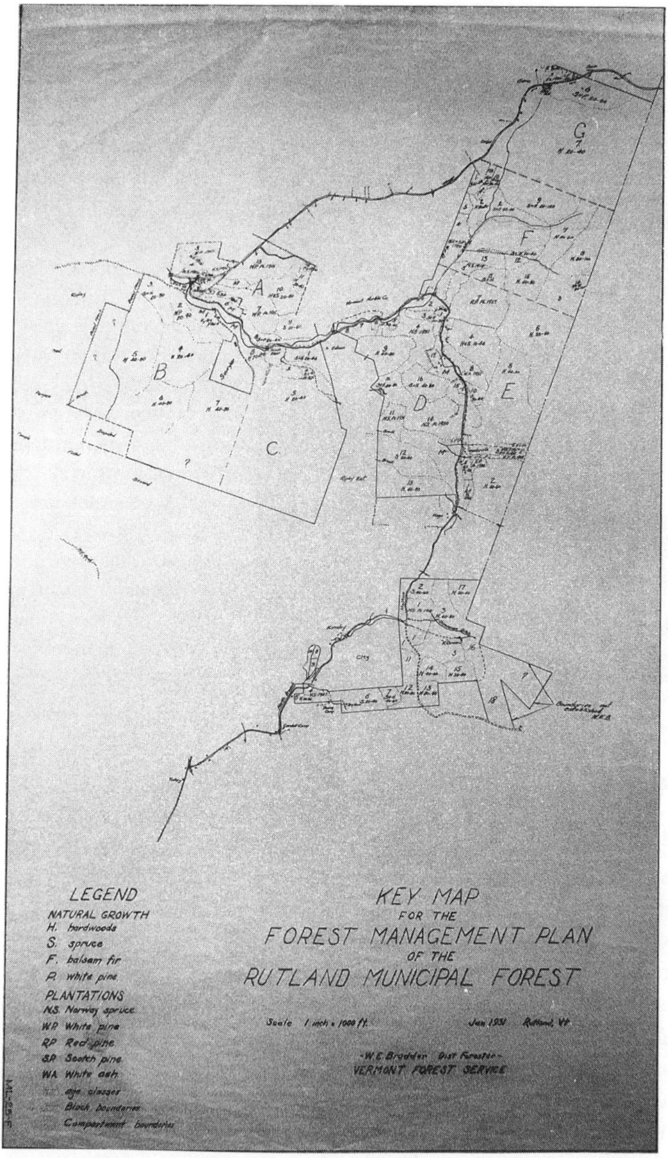

Fig. 6.9. Wilbur Bradder's type map of the Rutland Municipal Forest drawn in 1931 for the state's forestry department. Located primarily in Mendon, Vermont, in the westerly shadow of Mt. Killington, Rutland's watershed reserve had grown to four thousand acres by 1947. Bradder recommended retention of flowering shrubs and fruit or nut-bearing trees in order to improve the land's appearance and provide food for wildlife. Office of the Rutland City Engineer

Fig. 6.10. Vermont municipal forester, Gene Keenan, with sixth-grade students in Rutland's municipal forest in 1970. Vermont was the only New England state to employ full-time municipal foresters as part of its state forestry program. In addition to supplying seedlings from its nursery, the state enacted two important laws: the first subsidized part of land-acquisition costs incurred by towns; the second required that propositions for municipal forests be included in warnings for annual town meetings. Vermont Department of Forests, Parks and Recreation—Pittsford

Woodshore Alliance

Forest plantations eventually became widespread on New England's municipal watersheds during the first half of the twentieth century. More than a few were sufficiently well ensconced by quarter-century to serve as models for town forests while that movement gained ground from state to state. At times the relationship between watershed forestry and the town forest movement has been a curious one. The genesis of the former dawned with recognition that forests protected and improved water supplies, all the while producing a marketable crop from otherwise unusable land. The latter developed from an awakening to America's misspent forest legacy, prompting a crusade to reclaim idle lands.

Although conceived separately, their paths soon crossed. Town forest advocates became ardent suitors of watershed forestry, and the match often worked. Obstacles that routinely shackled productivity on municipal forests were generally absent from waterworks timberlands. For one thing, acquisition of property around reservoirs was a necessity, not an option exercisable by town committees once political consensus had been achieved. Facing unavoidable ownership of land, much of it fallow, water commissions stood to gain from an investment in forestry. Moreover, acreage was frequently extensive, a factor that made silviculture more economically efficient. Just as important, management was conducted under the watchful eye of public or quasi-public corporate entities, a step or two removed from the uneven flow of municipal government. Professional foresters were routinely hired and easily provided tutelage for ground crews already on payrolls. Records were kept in comparatively good order.

Nevertheless, many utilities rejected the advances of municipal forestry. As often as not, a desire to remain independent from control by state foresters was responsible. Yet economic factors were also present. Planting was expensive, return on the dollar a long-term proposition, and the whole business of forestry often something of a puzzle. Choices between full-scale forestry operations from nursery to sawmill, on the one hand, and scaled-down efforts, on the other, were not always easy to make. With the passage of time, many utilities opted for the latter. To help resolve these economic issues, management practices evolved and became more specialized. Speaking to a gathering of the New England Water Works Association in 1947, Ralph Hawley contended that plantations were necessary only along narrow strips bordering reservoirs, not on entire basins. Estimating that such zones averaged only 15 percent of total land area, he suggested that natural forests outside these belts could be cultivated over long periods without the expense of labor-intensive planting. At a similar meeting a few years later, Hawley conceded that dense forests were no longer regarded as ideal watershed cover. Instead, partially open conditions secured by cuttings at intervals of ten to twenty years allowed larger amounts of snow and rain to reach the ground, thus producing maximum water yield. Hawley also recommended that logging operations be conducted by stumpage, noting that waterworks employees generally were more costly than skilled woodsmen. Heavy slash was best left undisturbed on the ground, where it lessened evaporation, decreased surface runoff, and increased soil porosity. Expenses were lowered as a result.[46]

Hawley also was candid about forestry as a business proposition for water commissions. Substantial capital outlays were necessary, and return on investment was far from immediate. He cautioned that plantations should be started only when justified by benefits to water quality, reduced costs of maintenance, or sale of wood products. Harvesting was sanctioned when income exceeded the cost of the undertaking. Ornamental projects were optional, but accounts were to be tabulated on separate ledgers.[47]

Forestry continues today on many municipal watersheds, although advances in technology have produced sophisticated water treatment facilities that make pollution control by woodlands partially obsolete. Federal law also governs water quality, adding new demands on local utilities. Nevertheless, forests can still be managed to increase overall water supplies and to ensure water quality as well. Until very recently, it has been difficult to quantify the effects of forest management practices on ecological changes such as rate of evaporation from tree canopies and transpiration from foliage. Computer models, however, can now accurately monitor stream and overland flow, and management decisions can thus be analyzed more closely. For example, the merit of decisions to minimize coniferous species, which intercept and evaporate more water than deciduous types, can be measured in terms of water-yield savings. The extent to which coniferous types along reservoir borders provide shade and prevent evaporation from winds is also capable of measurement. Moreover, stream flow can be isolated in specific sectors, leading to close management of forest compartments. Properly managed watersheds, too, shield against sources of pollution, thus decreasing need for expensive water purification.[48]

. . .

Watershed forests developed into New England's best managed and consequently most economically profitable communal woodlands. Commercial productivity and ecological conservation have quietly synchronized in these forests and have done so for a good portion of this century. In truth, maximum conservation of surface water demands sedulous management of forest cover. Stewardship has evolved into specialized practice. Opportunities for public education about viability of multiple use abound. Well-kept records by many water departments also provide occasion for future study of timber yields. Finally, watershed lands and related utilities are often a vital, if somewhat overlooked, piece of community structure.

Yet all is not sanguine. The availability of regional water treatment facilities, changing economic conditions, strict regulations governing water quality, dispersed timber markets, and increased land values have become weighty factors for water commissioners to consider. Scenic, secluded reservoirs offer tempting enclaves for development. Not surprisingly, more than a few towns have sold their lands. In contrast, other communities have recognized potential aesthetic or recreational benefits and have committed these woods and reservoirs to public use as forest parks.

Chapter 7

Fountains of Timber in Forest Parks

In the fall of 1859, Henry David Thoreau wandered the slopes of Poplar Hill in Concord. The October air was cold and had already begun to exact its tribute of brilliant color. Ash, yellow birch, and mountain sumac were nearly bare; larch in a close-by swamp were no longer evergreen but a soft, faded ocher instead. Amid these surroundings, he pondered the abrupt change in his town's landscape, a transformation in truth more permanent than seasonal. Thoreau considered New England towns to be institutions worthy of remembrance and wondered why places such as Walden Wood could not be preserved as parks forever. Those about to leave the world without natural heirs would be wise, he believed, to bequeath unspoiled forests to their communities. That day's autumnal musings were faithfully recorded in his journal:

> Each town should have a park, or rather a primitive forest, of five hundred or a thousand acres, where a stick should never be cut for fuel, a common possession forever, for instruction and recreation. We hear of cow-commons and ministerial lots, but we want *men*-commons and lay lots, inalienable forever. Let us keep the New World *new*, preserve all the advantages of living in the country. There is meadow and pasture and wood-lot for the town's poor. Why not a forest and huckleberry-field for the town's rich?[1]

Undeniably, nameless others before him enjoyed roving in local woods, but most left slight evidence of their explorations. Thoreau's advice marks a convenient starting point for tracing the growth of recreational forest use and the category of communal woodlands it inspired, forest parks. In their purest form, these sanctuaries are simple patches of wilderness, just as Thoreau envisioned, adorned by little more than the simplest of footpaths to indulge human use (see fig. 7.1). The substance of this use is recreation, but that term's simplicity is deceiving. Forest adventures such as hiking, camping, picnicking, or nature study oblige unbounded purpose: quiet contemplation, physical exercise, education, spiritual rejuvenation, ascent to panoramic vista, or just rest.

Fig. 7.1. An 1898 view of Pennybrook Bridge in Lynn Woods, Lynn, Massachusetts. Recreational use of forests—exploring, hiking, and nature study—steadily gained popularity during the second half of the nineteenth century and inspired a distinct category of local woodlands: forest parks. Lynn Woods, spared from the clutches of development by its rugged and inaccessible terrain, became a town-owned park and watershed reserve in 1888. Lynn Historical Society

The origins of recreational forest use are as eclectic as the myriad interests fulfilled through woodland treks. Important trends emerged well before mid-nineteenth century; others followed later and continued to develop during the first half of the twentieth century. Recognition of the American wilderness as a fitting antiquity for the new nation helped to shape public opinion about forests. Fascination with the Romantic and the Picturesque, as revealed in literature, art, and architecture, was particularly influential, as were New England's transcendentalists. Dismay over squandered timberlands, concern for ecology, and a newly molded forestry profession synthesized into a conservation movement; forest recreation strengthened public resolve for conservation. Appreciation of wilderness as a unique resource worthy of preservation emerged to counter forestry policies that focused primarily on tree cropping. Well-publicized national parks and forests served as prominent models for community-owned forest parks. Arbor Day became a national event symbolizing the importance of education about the nation's forests.

Urban change was equally dynamic. In the wake of New York's Central Park, designed by Frederick Law Olmsted and Calvert Vaux in 1857, cities began to introduce natural landscape into their overbuilt, congested, and noisy environments. Nineteenth-century pleasure parks featured promenades and tried to duplicate or even heighten nature with pastoral meadows framed by clusters of trees. After Chicago's immensely successful World's Columbian Exposition in 1893, "City Beautiful" became a popular movement, and stately parks appeared as a vital part of energetic drives to improve American cities and towns. City planning developed around this crusade and eventually became an all-encompassing component of urban government. Planners gradually came to view simple forest parks as an alternative to the more formal (and costly) parks of the City Beautiful.

Regional events also were important. A strong network of hiking clubs in New England led to a more conscientious forest ethic. Metropolitan park commissions in several cities, Boston foremost among them, purchased a number of large forest reserves, prompting some nearby towns to acquire similar lands. Especially influential, too, were the methods by which commissions integrated recreational facilities into these reserves. Along another front, the town forest movement stimulated a sizable increase in community-owned woodlands. Local officials soon realized these primitive areas fulfilled a need to keep some land free from development and at the same time provided inexpensive places for recreation and nature study.

New England is full of towns with unpretentious forest parks, many tucked away near an unused marsh or hidden on a secluded hilltop. More elaborate examples borrow, often too heavily, from other types of urban parks and display a surfeit of man-made improvements. When wooded cover or other natural features cease to be a dominant point for recreation, the integrity of forest parks is diminished. Although forestry associations in Massachusetts and Connecticut eventually embraced parks as well as forests, the history of this foliant patchwork lies hidden in local woods; often their intriguing place-names mark a trail-head for searchers. As with watershed forestry, comparisons among the region's states are difficult absent decent inventories. Still, identifiable patterns exist for period of origin, purpose of creation, spectrum of recreational use, character of local stewardship, and contribution to community structure.

Wilderness Antiquity

America's romantic relationship with its forests began in earnest with the work of a great number of authors, poets, philosophers, artists, and architects who viewed in America's untamed wilderness a variety of images uniquely suited to a young democracy. They were inspired in part by emotional appeals of the Picturesque, an aesthetic doctrine favored in England after publication of Edmund Burke's essay, "A Philosophical Inquiry into the Origin of Our Ideas on the Sublime and the Beautiful," in 1756. Recognition that America's landscape was rapidly being transformed also became instrumental. Through visual and literary arts, the nation's forests were interpreted in a freshly creative and energetic manner. Harmony in nature, reciprocal relationships with the environment, humankind's capacity for self-realization through nature, and the value of a wilderness that defined America's cultural heritage were frequent themes.

William Cullen Bryant, a journalist and poet, traveled throughout the Berkshire Mountains and Hudson River Valley between 1820 and 1840. From these journeys, he developed a series of descriptive essays that he hoped would instill an awareness of nature in Americans. His poetry, too, drew from the purity and solitude of wilderness and created imagery that was both descriptive of an environment and suggestive of its importance to a youthful nation. Poems such as "Antiquity of Freedom" expounded an immutable bond between nature and origins of democracy:

> Here are old trees, tall oaks, and gnarled pines,
> That stream with gray-green mosses; here the ground
> Was never trenched by spade, and flowers spring up
> Unsown, and die ungathered. It is sweet
> To linger here, among the flitting birds
> And leaping squirrels, wandering brooks, and winds
> That shake the leaves, and scatter as they pass,
> A fragrance from the cedars, thickly set
> With pale blue berries. In these peaceful shades—
> Peaceful, unpruned, immeasurably old—
> My thoughts go up the long dim path of years,
> Back to the earliest days of liberty.[2]

Other writers—Nathaniel Hawthorne, James Fenimore Cooper, and Washington Irving—revealed their fascination with America's primitive forests by placing characters in a richly depicted landscape that then became a central force in plot development.

Literary and artistic communities converged around wilderness themes during this period. Bryant was a familiar companion of artist Thomas Cole, and the two journeyed together throughout the Adirondack and Catskill Mountains. Cole and other Hudson River School painters, notably Asher B. Durand and Frederick Church, roamed New England's Berkshire and White Mountains as well. In majestic and often brooding paintings, they successfully captured the spirit of an environment untouched by civilization. Cole's *View of the White Mountains*, painted in 1827, and Church's *Mt. Katahdin*, painted in 1853, revealed this untarnished, unbowing superiority of wilderness but at the same time suggested the encroaching influence of human settlement. In *Kindred Spirits*, painted in 1849 by Durand as a memorial to Cole, who had died a year earlier, Bryant and Cole were shown together on a remote overlook in

Fig. 7.2. An "American Cottage" by Alexander Jackson Davis in 1838. Well before mid-nineteenth century, American art, literature, and architecture found creative expression in wilderness, the young nation's true antiquity. Davis's rustic temple portrayed classical themes, recalling Europe's ancient monuments, but stood in a forest primeval. *Rural Residences* (1837; reprint, New York: Da Capo Press, 1980)

the Catskill Mountains. The painting symbolized harmony in creative spirit between literature and art and credited the environment that had inspired this fellowship.

Cole and Bryant were also companions of architect Alexander Jackson Davis, whose romantic interpretations of Gothic architecture achieved a delicate balance between building and nature. Davis relied on elements such as asymmetrical plans, porches reaching outward into the landscape, irregular silhouettes, and ornament in forms borrowed from nature. In his design for an American Cottage, he displayed unusual creativity in an entirely different context (fig. 7.2). There, classical as well as romantic vocabularies were adapted to American forests by employing unhewn trees as columns and planks as exterior cladding in place of marble.[3]

Artistic and literary symbolism expanded into a more abstract philosophy that bound civilization and natural environment together. Ralph Waldo Emerson, in his *Essays on Nature*, published in 1841, envisioned the forest primeval, unsullied by the decay of urbanization, as a key to unraveling the higher destiny of humankind. To Emerson and New England's transcendentalists at Brook Farm, forests were a source of knowledge, and nature represented "the best meaning of the wisest man."[4] Thoreau echoed these themes and regarded woods as a sanctuary, a place of spiritual enlightenment, where one could observe a balance found only in nature. Thoreau lamented the irreversibility of change wrought by civilization and perceptively described this sweeping metamorphosis. In "Chesuncook," based on travels to Maine in 1853, he remarked:

> By his [the civilized man's] mere presence, almost, he changes the nature of the trees as no other creature does. The sun and air, and perhaps fire, have been introduced, and grain raised where it stands. It has lost its wild, damp, and shaggy look, the countless fallen and decaying trees are gone, and consequently that thick coat of moss which lived on them is gone too.[5]

Abstract contemplation of nature soon gave way to more calculated observation. By suggesting that forests were only one part of an extensive ecosystem and that disruption to one segment caused an imbalance in others, Marsh's *Man and Nature* provided a catalyst for translating philosophy, literature, and art into pragmatic policy. To Marsh, and to many others who recognized the magnitude of his work, society's health, fiscal as much as metaphysical, depended on an ecologically sound environment.

Economic and spiritual well-being were fused in the Adirondack Mountains in 1872. That year, primarily to protect watersheds for canal systems, New York's park commission recommended that large areas owned by the state be reserved from sale and that tax-delinquent lands be reacquired. In 1885, legislation from Albany established a sizable forest preserve of 715,000 acres. Seven years later, a new law created the Adirondack State Park and assigned to it more than three million acres. Although controlled logging was initially permitted, an 1894 amendment to the state constitution mandated that the park remain forever wild and prohibited removal or destruction of timber. Significantly, utilitarian motives were eventually overshadowed by arguments that wilderness regions had their own special value.

After 1890, Adirondack State Park became an accessible recreational area for urban populations in eastern states. Verplanck Colvin foresaw the danger of political control over these lands and urged the state to carefully prepare a forestry plan. As the park commission's superintendent, Colvin expressed his concerns persuasively in their annual report for 1899:

> . . . it will behoove the Legislature and the people to use the most watchful care in the protection of these great forest heritages, for every device which cunning can prepare and every argument which avarice can propagate will be employed to induce the people to part with these great forests and to throw into the hands of speculators and schemers an opportunity to enrich themselves from these public lands at the expense of the people. The precious supplies of merchantable timber which will then clothe the

slopes of our mountains will, in the event of a great timber famine or of some international embargo or war, be sought and demanded under every plea and pretext that can be devised.[6]

Similar events were taking place in western states at the same time. California had initiated measures to protect Yosemite Valley as early as 1864, and Yellowstone National Park was created in 1872. By far, the most eloquent pleas for the preservation of western forests were voiced by John Muir, who founded the Sierra Club in 1892 and who urged creation of several national parks, including Grand Canyon and Petrified Forest. Muir celebrated, as did Emerson and Thoreau, the ethereal qualities of the forest primeval and understood relationships between wilderness and human spirit. His descriptive writing achieved wide circulation and contained a vitality and energy that helped to popularize themes of wilderness preservation and forest recreation. Exploring northern California's mountains in 1874, Muir encountered a winter gale and climbed to the top of a Douglas fir to better observe the storm. His enthralling sketch of this adventure, capturing sight, sound, and smell, was published in *Scribner's Monthly* in 1878 and was widely acclaimed.

Now my eye roved over the piny hills and dales as over fields of waving grain, and felt the light running in ripples and broad swelling undulations across the valleys from ridge to ridge, as the shining foliage was stirred by corresponding waves of air.

The sounds of the storm corresponded gloriously with this wild exuberance of light and motion. The profound bass of the naked branches and boles booming like waterfalls; the quick, tense vibrations of the pine-needles, now rising to a shrill, whistling hiss, now falling to a silky murmur; the rustling of laurel groves in the dells, and the keen metallic click of leaf on leaf. . . .

I kept my lofty perch for hours, frequently closing my eyes to enjoy the music by itself, or to feast quietly on the delicious fragrance that was streaming past. The fragrance of the woods was less marked than that produced during warm rain, when so many balsamic buds and leaves are steeped like tea; but, from the chafing of resiny branches against each other, and the incessant attrition of myriads of needles, the gale was spiced to a very tonic degree.[7]

Muir's inspirational prose and dogged advocacy for national parks became increasingly important as forestry policy began to take shape during Theodore Roosevelt's administration. The philosophy of wilderness preservation articulated by Muir countered, and to some extent balanced, a utilitarian strategy advanced by Gifford Pinchot. To Pinchot, trained by European foresters, trees were a crop to be managed, resources a commodity to serve the American public, albeit in prudent fashion. "Highest use" and "multiple use" soon found their way into forest vocabulary. Both Roosevelt and Pinchot regarded themselves as rugged outdoorsmen, bent on engaging and converting an equally vigorous timber industry to more prudent use of America's forests. The success of government plans hinged on the ability of these two men to convince a skeptical western public that long-term economic interests ran parallel with those of conservation. Alliance with groups advocating more restrictive use of federal lands was politically unwise. As a result, those who saw forests as a source of spiritual regeneration or recognized America's wilderness as a cultural monument to a new democratic nation presented obstacles to administration policy.

Despite this unpredictable political climate, concern for a more chaste policy toward majestic western forests developed into forceful opinion. At the same time, realization that America's monuments, natural as well as manmade, were part of a cultural heritage also began to find expression. The American Antiquities Act of 1906 authorized, by executive decree, the proclamation of national monuments of historic or scientific interest on federal lands. Agencies charged with administration of these lands were, for the first time, forced to formally consider recreation as part of their policy.

A Regional Community of Hikers

America's increasing attention to recreational use of forests influenced, and in turn was influenced by, regional patterns and events. From the time of nucleated villages to the writing of Emerson and Thoreau, New England's communities had steadily fashioned an intimate understanding of their woodlands. Whether by necessity or design, the effect of this relationship on the region's literature and art is unmistakable. At times, even village form displays an unusual harmony between natural and built environments. This intimacy is also apparent in attitudes toward recreational forest use. By mid-nineteenth century, at about the same time that alarms for the protection of forests were being sounded, hiking, exploring, and trail building began to appear as activities that would eventually epitomize forest recreation in New England. Weathered mountains in New Hampshire, Vermont, and western Massachusetts persistently beckoned, a backyard woods for city hikers.

New England's hiking community has played an important role in forest protection by expanding the base of political support for responsible, long-range forest policy and by close association with forestry groups. Early hiking clubs—Cyrus Tracy's Exploring Circle in Lynn, Massachusetts (circa 1850), the Alpine Club organized in Williamstown, Massachusetts (1863), and the White Mountain Club of Portland, Maine (1873)—were loosely organized, short-lived associations of men and women, most well educated, who relished any opportunity to explore local mountain peaks or bushwack through rough countryside. The Appalachian Mountain Club, established in Boston in 1876, proved more enduring and from its inception has continuously served as the nucleus of New England's hiking community. Known best by its initials, the AMC was organized by Edward C. Pickering, a physics professor at Massachusetts Institute of Technology. Its first members included Edward F. Morse and Samuel H. Scudder, both of whom had been active with the Portland and Williamstown clubs, respectively, prior to becoming affiliated with Harvard University. Hiking and climbing soon gained popularity during this period and set a pace for recreational forest use for years to come.[8]

From the start, the AMC distinguished itself by commitment to scientific as well as recreational interests, and its early members worked with Charles H. Hitchcock, who conducted New Hampshire's Geological Survey between 1876 and 1898. Activities focused on the White Mountains, where peak elevations were measured, data recorded, and maps prepared. The club also adopted an ambitious agenda of trail building, and a substantial network of footpaths inched forward. Membership steadily increased from ninety-two in 1876 to one thousand in 1898. Many of its members, among them Allen Chamberlain, also were affiliated with either the Massachusetts Forestry Association or the Society for the Protection of New Hampshire Forests. Chamberlain joined the AMC in 1897, became the club's president in 1906, and contributed a steady outpouring of articles to the club's bulletin, *Appalachia*, during his tenure.[9]

As membership grew, the club gradually assembled a collection of land holdings and relied on professional foresters to assist with management. In 1902, the club was given a 250-acre farm in Fitzwilliam, New Hampshire, to protect a ten-acre natural bed of rhododendron from logging operations. Located at the base of Little Monadnock, the Rhododendron Reservation was opened to the public and remaining land held as an endowment. In 1907, Chamberlain, then the club's councillor of forestry, wrote to Gifford Pinchot requesting advice about burning five-year-old logging slash without damaging either the rhododendrons or young pine, spruce, and hemlock seedlings on endowment lands. Replying for Pinchot, forester Overton Price recommended a crew of twenty men for piling and burning slash after a light snowfall and instructed Benton Mackaye to examine the tract. Mackaye, who had completed Harvard's graduate program in forestry in 1905 and had begun working for Pinchot that same year, donated time and energy in dispensing with the troublesome debris. He also conducted a valuation survey and produced a working plan suggesting planting of open areas to achieve a fully stocked forest capable of producing revenue within fifty years. From the time of their meeting in Fitzwilliam, reinforced by a amicable jaunt to the top of Mt. Monadnock, Mackaye and Chamberlain remained good friends.[10]

As hiking paths lengthened, other clubs sprang up around their members' common interest in trekking through the woods. James P. Taylor, headmaster at Vermont Academy for Boys in Saxtons River, Vermont, explored the peaks of the Green Mountains and conceived a plan for a long-distance trail linking

Vermont's southern and northern borders. The Long Trail was begun in 1912, and the Green Mountain Club was organized not long after. The Dartmouth Outing Club, created in 1910 as a ski and snowshoe club, soon expanded its recreational activities to include roving through the hill country surrounding Hanover.[11]

The presence of an articulate, well-educated, and politically influential body of forest users whose recreational pursuits assumed simple patterns—walking, climbing, exploring, and observing nature—served as an important backdrop for the development of forest parks in New England towns. An underlying woodland ethic espoused by this hiking community translated very well into a mood of energy and enthusiasm for local, village-size forest projects. Undeniably, too, recreation's imprint was firmly embedded in the town forest movement from its inception. Yet ties between avid recreational forest use high in New England's mountains and forest parks in less uplifting urban areas also developed in more subtle fashion as trail systems eventually found their way into town and regional planning. Although he was never really part of the region's nucleus of hiking clubs, Benton Mackaye again appears as a key figure. Forest and mountain exploration played a formative role in his later writing about environmental habitability, a subject he described eagerly as "geotechnics."

As a boy of fourteen, Mackaye roamed the woods, hills, and waters surrounding his home in Shirley Center, Massachusetts, during the spring and summer of 1893. Shirley developed, as did so many Massachusetts small hill towns, around a complex of waterpowered mills. Shirley Center, however, occupied a ridge well above the taint of industry, within easy walking distance of Mulpus Brook, the Squannacook River, and Hunting Hill, natural attractions for any young adventurer. Mackaye kept a journal of these expeditions, thirteen altogether, and his narratives represent a fusion of philosophical and geographical observation. Too Thoreau-like not to have been modeled after the author's Concord meanderings, Mackaye's chronicles reveal a humanism that even then dominated his thoughts.

A few years later, in August 1897, Mackaye embarked on a long-distance trek into the White Mountains and was joined by fellow Harvard students Sturgis Pray and Draper Maury. Pray eventually became head of Harvard's graduate program in landscape architecture. The trio crawled through blowdowns and scrambled up a number of summits, including Mount Washington. Following graduation, Mackaye and classmate Horace Hildreth investigated the southern peaks of Vermont's Green Mountains a decade before the Long Trail was begun. In 1902 and again in 1903, Mackaye rejoined Sturgis Pray, who as an AMC member had returned to New Hampshire's Swift River area to lay out trails.[12]

For Mackaye, hiking, exploration, and even professional forestry were no more than contributing players in the larger scheme of habitable human environments. During his incessant search for balance between culture and landscape, Mackaye's idyllic image of New England villages provided a foundation on which ideas for regional and, in his later years, continental planning were supported. Public forests became a means for confining the menacing spread of urban growth, or "metropolitanism"; and, as in his beloved hill towns, woods and trails were just a short leg-stretch away.

Never one to worry over details, Mackaye remained a theorist but one who combined a philosophy of recreational forest use and urban purpose as well as any. Although his most prolific period of writing did not begin until he reached forty (but continued into his eighties), a 1916 essay, "Recreational Possibilities of Public Forests," charted the direction of his early, unbounded thinking. Published in the *Journal of the New York State Forestry Association*, Mackaye's article stressed the measurable benefit of forest recreation to the nation and focused on trail and waterway systems as pathways linking backwoods to big cities. With customary zeal, he pictured these passageways as the skeleton for a national recreation area, a single entity joining ocean to ocean.[13]

Urban Parks

Forest parks did not suddenly materialize in the urban milieu. Instead, they were preceded by a more generalized interest in urban parks that developed during the second half of the nineteenth century. Industrialism fueled the growth of American cities but also forced a recognition that factories created squalor in addition to wealth. Wretchedly dense conditions bred disease and polluted water supplies; fires, once started, burned savagely for days. Parks offered democratic sanctuaries for the nation's confined urban populace and split the mass of endless building blocks with open, natural landscape. By instilling civic pride, park planners also hoped to break the cycle of mobility inherent in America's restless populations.

The success of New York's Central Park prompted other eastern cities to initiate similar plans. Boston's Park Commission, established in 1875, engaged Frederick Law Olmsted, who designed a linear series of interconnecting parks encircling the city and known as the Emerald Necklace. Begun in 1878 and substantially completed by 1890, Olmsted's plan employed open space to unite various city districts. Each part (Arnold Arboretum, for example) contributed unique qualities to the ensemble. Although a few park planners, Olmsted foremost among them, recognized that pure wilderness offered the sharpest contrast to civilization, replication of uncultivated nature in cities was considered impractical and even inappropriate. Pastoral landscape offered a compromise, and trees were used to shape open meadows, frame view corridors, and provide sylvan effects. On a much smaller scale, village improvement societies in New England towns dressed up central commons and planted trees along their streets. Growing interest in civic progress also inspired town benefactors to donate local landmarks, often areas of unique geographic formation or dazzling scenic views, for public use. Undoubtedly, too, publicity heralding the advent of national parks and forest reserves in western states stirred charitable emotions of land owners in New England towns and cities.[14]

Parks became an essential element in urban planning on a grand scale following the Columbian Exposition held in Chicago in 1893. City Beautiful captured the imagination of America's reform-minded population, and planners such as Daniel Burnham, Charles Mulford Robinson, and John Nolen sought order and formality in America's urban centers through classical relationships between buildings and their settings. Parks; wide, tree-lined boulevards; and monumental vistas helped to clear away urban clutter. Well-publicized projects, such as the Senate Park Commission's Plan for Washington, D.C., in 1901 and Burnham's 1909 Chicago Plan, spurred ambitious urban design throughout the country. Planning quickly became more comprehensive and focused on social needs such as housing. Inspired by Ebenezer Howard's 1902 book, *Garden Cities of Tomorrow*, and Letchworth, England's first garden city, planners began placing homes in parklike settings; John Nolen's 1913 plan of Neponset Garden Village in Walpole, Massachusetts, is one such design. By 1917, the American Institute of Architects' Committee on Town Planning had prepared a report on the progress of city planning in America and listed nearly two hundred plans. Not all were carried out, but most included parks and open spaces.[15]

Accents on formality and classical design associated with turn-of-the-century urban planning countered past emphasis on the romantic themes that had dominated nineteenth-century parks. Ironically, forest parks, well founded in transcendentalism and in recognition of America's wilderness antiquity, emerged as a more authentic model for these ideals precisely at a time when the country's enchantment with Picturesque doctrines had begun to fade. As such, they represent an important link in America's history of urban parks and bridge a gap between the natural but organized landscapes of earlier pleasure grounds and the progressive era's reform parks, where organized recreational activities were stressed. More symbolic of nature than man-made thickets and meadows, forest parks offered a specific form of recreation and promoted social reform by encouraging more responsible attitudes toward forest policy.

Once again, European community forests made a contribution. Apart from production of revenue made possible by effective management, so appealing to the nascent forestry movement in America, forests in German, French, and Swiss cities were valued as places of recreation and refuge from urban crowds. These qualities did not go unnoticed by American journals. A report on German communal forests in the grand duchy of Hesse was published in an 1897 issue of *The Forester*, the journal of the New Jersey Forestry Association, edited by John Gifford, forester for New Jersey's geological survey. Forests in the cities of Odenwald and Darmstadt were exemplified as helping to draw tourists and attracting new, well-to-do residents. A few months after this article appeared, *The Forester* became the official journal of the American Forestry Association.[16]

Arbor Day

A melding of groups interested in park planning, recreational use of forests, city beautification, and forestry began to occur annually, with celebrations of Arbor Day, an event with midwestern origins dating from 1872. In that year, the Nebraska State Board of Agriculture adopted a resolution by J. Sterling Morton calling for a day of tree planting throughout the state. Prizes were given to a county agricultural society and to an individual planting the most trees, one hundred dollars to the former and a library of farm books to the latter. The first Arbor Day was held on April 10 with great success as Nebraskans planted more than a million trees; other states soon followed their lead.[17]

Arbor Day was conceived as a measure to improve economic conditions in a region where timber for building, wood for fuel, and shelter from prairie winds often proved scarce. The event became associated with schools after the Cincinnati Forestry Congress in 1882. On its final day, participants joined a procession to Eden Park, where they observed the planting of memorial trees commemorating American presidents and other notable persons. The ceremonial planting was done by women, many of them teachers, assisted by schoolchildren. Throngs of people observed these proceedings, and the event achieved wide publication and popular acclaim. At its meeting in St. Paul in 1883, the American Forestry Congress adopted a resolution favoring proclamation of Arbor Day by schools and appointed a committee to advance that goal. During a span of two years, 1886 and 1887, each of New England's six states inaugurated observance of Arbor Day. Village improvement societies responded with appeals for tree planting around town commons, along city streets, and in newly created parks.[18]

Affiliation with schoolchildren led, quite naturally, to an emphasis on the value of forestry education. Improved techniques in woodland management and dissemination of this information to the general public bred enthusiasm for local forest-planting campaigns. Soon the occasion became a fashionable time to organize town projects. School forests intended primarily for education were a logical outgrowth of links between Arbor Day, schools, and schoolchildren, but other activities benefited too. Large-scale planting programs for watershed lands had begun in a number of New England towns before the close of the nineteenth century and were among the first municipal forests to be planted by students. As professional forestry began to reach communities, both town forests and forest parks became sites for spring tree-planting gatherings.

Boston's Metropolitan Park Commission

To many, forest parks recalled America's primeval antiquity, and unlike the carefully drawn, formal plans for city parks, they were viewed as an opportunity to introduce nature into urban areas without the meticulous and expensive arrangement of landscapes. Woodland reserves needed little of the finicky weeding, trimming, grading, and construction associated with well-groomed parks. Roads and paths were simply a passage through a forest, with borders quietly melting into surrounding foliage. Scenic vantage points offered hikers opportunities for a change in landscape or for rest, and rustic buildings provided occasional shelter.

Forest reserves appeared in a few Massachusetts communities after the state's 1882 enabling law permitted towns to acquire land for the protection and cultivation of timber. Recreational uses were clearly regarded as secondary by this legislation and were mentioned only in a minor paragraph. Ironically, the act was usually referred to as a "public park law." At about the same time, enthusiasm for establishing a metropolitan park system was beginning to gather energy in eastern Massachusetts. Boston, relatively small in geographic area for a major city and closely surrounded by other rapidly growing towns, faced a number of vexing urban problems during the last quarter of the nineteenth century. Solutions for municipal water supplies, sewage removal, and transportation networks were sought on an ever expanding scale, and parks, too, were considered in regional dimensions. Ringed by hills, ocean, and bay and sustained by watersheds of several large rivers, the city's metropolitan environs offered exceptional natural areas for public recreation.[19]

In 1893, Boston's Metropolitan Park Commission was created following legislation introduced by the Trustees of Public Reservations (now the Trustees of Reservations). A quasi-public corporate body, it was established three years earlier by a state legislative act to hold title to scenic and historic public lands. Journalist Sylvester Baxter, landscape architect/engineer Charles Eliot, and various AMC members were all instrumental in its founding. Baxter, Eliot, and Olmsted's firm, joined by Eliot as a partner, were equally principal in planning metropolitan parks. Boston's commission quickly set about identifying sites and selected several large forest preserves, including Blue Hills and Middlesex Fells, substantial portions of which were acquired by 1894. These regional parks received broader public attention than the smattering of local forest parks

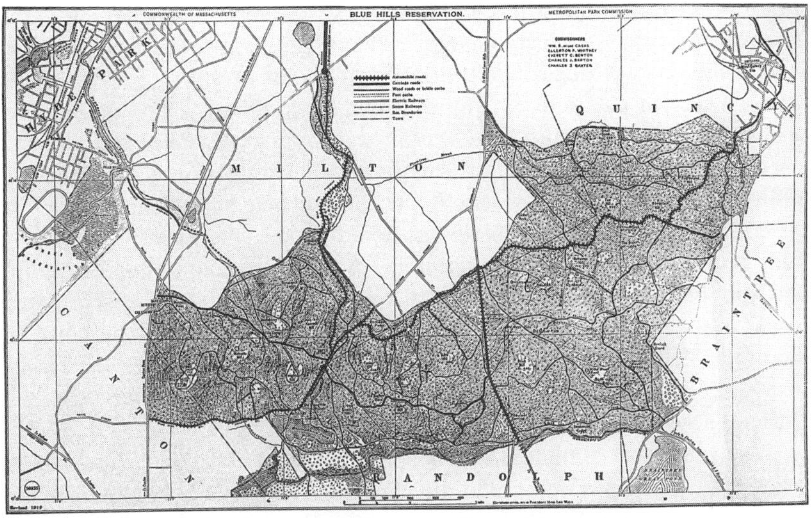

Fig. 7.3. Plan of Blue Hills Reservation in Quincy, Milton, and Canton, Massachusetts, drafted by the Metropolitan Park Commission in 1919. Boston's regional commission acquired a number of forest reservations during the late 1800s, prompting similar undertakings in several communities, notably Lynn, Waltham, and Springfield. Massachusetts State Archives

created a few years earlier under the 1882 law. In addition, commission plans for metropolitan forests influenced, to a great extent, development of other woodland parks that appeared both before and after regional reserves.[20]

Blue Hills Reservation, located primarily in Quincy and Milton, came to be the largest of these woodland parks (fig. 7.3). Area topography was dominated by a series of twenty hilltops, the tallest of which, Great Blue Hill, ascended to 635 feet and straddled a boundary between Milton and Canton. The derivation of its ancient name, often mentioned in seventeenth-century town records, was keenly phrased in the park commission's report for 1894: "it is the vast blue distance which tends to engross the attention,—a distance here of ocean and there of forest, and there again marked by the remote Wachusett and Monadnock."[21] That year, nearly 4,000 acres were obtained, and periodic acquisitions gradually expanded the forest park to its present size of roughly 5,800 acres.[22]

Middlesex Fells, true to its name, occupies an area of broken rock hills and cliffs situated north of Boston. Historian Francis Parkman, who lived for a time in Medford, roamed the Fells, and a group of prominent citizens eventually formed the Middlesex Fells Association, electing Elizur Wright president. As early as 1850, Wright argued that the terrain should be preserved as a public forest for recreation, education, and protection of water supplies. In 1883, a separate group of trustees was organized to hold funds needed to purchase land under Massachusetts's year-old enabling law. As a result, substantial areas were already in public ownership by 1894, when almost 1,600 acres came under park commission control. Boston shipping moguls Peter and Shepard Brooks, developers of Chicago's famed Monadnock Building (designed in 1889 by Daniel Burnham and John Root), donated both land and money for the preserve. While the Blue Hills Reservation offered summit vistas, Middlesex Fells provided a variety of landscapes, including Virginia Wood, Bear Hill, and Spot Pond. The Fells, too, gradually increased in size to its present 2,060 acres. In 1894, the commission secured a third but smaller forest park not far from Blue Hills, Stony Brook Reservation, which encompassed an area known as Muddy Pond Woods. Another wooded reserve, Breakheart, in Saugus and Wakefield, was acquired in 1934.[23]

Over time, park commissioners and their landscape architects have recognized the significance of these unspoiled forests and have generally managed to preserve a semblance of wilderness qualities in large areas. Use by people, a necessary element if the woodland's public character is to survive, has been accommodated by a variety of methods. Electric railways, carriage lanes, wood roads, bridle paths, trails, and an elaborate network of automobile parkways have provided access to every thicket, crag, or pinnacle. In Blue Hills, AMC members constructed a camp at Ponkapoag Pond in 1921 and helped to build a skyline trail three years later. Plans for forestry management were quickly implemented. In the Fells, woodlands were divided into compartments and treated as either "close woods," in which leaf canopy was unbroken, or "open woods," with glades and sunny gaps in cover; early in 1900 extensive cutting was begun to combat gypsy moth infestation. At Blue Hills a nursery was created to provide trees for roadside planting, and extensive pine plantations also were started. Civil Works Administration, Works Progress Administration, and Civilian Conservation Corps workers undertook a variety of forestry chores, including clearing dead and diseased trees, removing brush, controlling white pine blister rust and gypsy moth, firefighting, and constructing lookout towers.[24]

Although the wooded character of these reserves has been emphasized with some degree of consistency, a vast array of recreational facilities also have been introduced. A zoo and amphitheater at Middlesex Fells, a golf course, ski run, and skating rink at Blue Hills, as well as scattered playing fields, tennis courts, swimming facilities, and sundry buildings confirm that recreational activities extend far beyond simple hiking and nature study. The park commission's regional plans were widely publicized and became quite influential during a time of prevailing civic improvement. Very similar recreational improvements were made in several community-owned forest parks, notably Lynn Woods and Springfield's Forest Park.[25]

Their influence also extended beyond the borders of Massachusetts. Comparable regional commissions were subsequently developed in Providence and Hartford. Although Rhode Island's commission acquired a number of wooded reserves, long since converted to state forests, few towns in that small state made similar acquisitions. In fact, the presence of a few large preserves at scattered locations, most relatively accessible from at least three corners of the state, may have deterred some towns from undertaking similar projects. Hartford's commission, headed by Francis Goodwin, became independent from city government in 1895, a year after it acquired almost six hundred acres of pasture and woodland, which became Keney Park.[26]

Communal Forest Parks

In cases where local interest in forest parks began to develop before emergence of Boston's large metropolitan reserves, credit often belongs to key individuals who possessed unusual foresight. This was certainly true in Middlesex Fells, Elizur Wright's kingdom primeval. The same can be said for Lynn Woods, currently a two-thousand-acre sanctuary of forest and reservoir amid an urban wilderness just north of Boston. There, at about the same time that Wright was tramping through the Fells, Cyrus Tracy organized an exploring club and set about the task of studying terrain and flora. From these modest beginnings, New England's premiere example of a communal forest park emerged.

Lynn Woods

Seventeenth-century settlement of Lynn followed typical patterns, with large areas of open and wooded lands held in common by proprietors for farming and for supplies of timber and wood. Uplands that are now Lynn Woods formed an imposing terrain of rock outcrops, ravines, glacial debris, and swamps. Cart paths allowed for little more than harvesting of wood from swamps, among them Meetinghouse Swamp. Wood resources had grown scarce by the close of the seventeenth century, however, and town ordinances prohibited "cutting of common or green wood on the town common for firewood except swamp wood." As early as 1698 proprietors considered but rejected division of wood on the town common. Finally, in 1706, town voters divided common lands among proprietors and those inhabitants who owned property.[27]

Lands remained in private ownership until after the Civil War, but rugged topography continued to discourage settlement. Apparently, people accepted

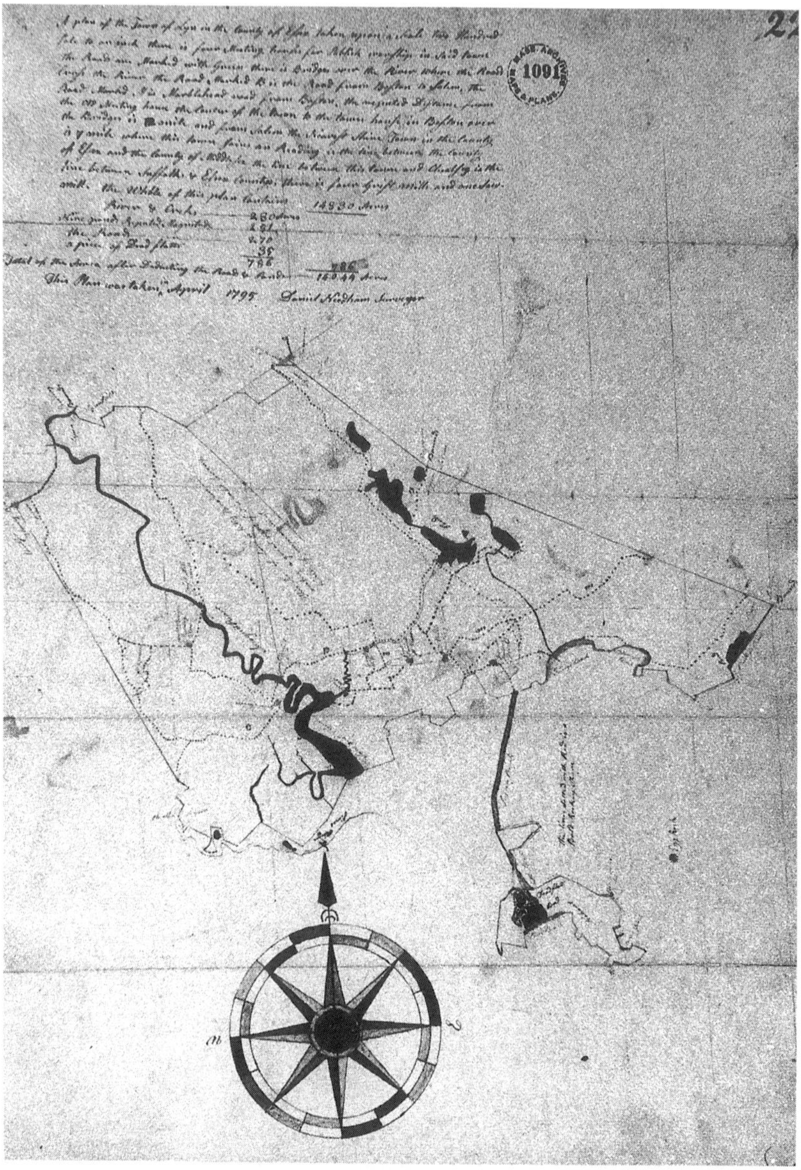

Fig. 7.4. Survey of Lynn, Massachusetts, drawn by Daniel Needham in 1795. Needham pinpointed the region that eventually became Lynn Woods, noting (in script now barely legible) that the land was rocky and unfit for improvement. When viewed with the McClintock survey of 1892 (fig. 7.6), Needham's plan reveals the considerable changes in topography wrought over a century by engineering of millponds and reservoirs. Massachusetts State Archives

the advice of surveyor Daniel Needham, who prepared a plan of Lynn in 1795 (fig. 7.4) and proffered little favor in his topographical description: "This piece of land is hilly, broken, and rocky. A great part of it is unfit for any kind of improvement." Nevertheless, the region's unspoiled natural features managed to excite curiosity in some residents, who joined Tracy's hiking and nature club, which he titled The Exploring Circle. Members began documenting botanical specimens and peculiar geological features such as the Rocking Stone located in an old ox pasture, investigated in 1854 (fig. 7.5). Tracy also compiled an anthology of plants growing in Lynn and adjoining towns that was printed in 1858 under the title *Studies of the Essex Flora*.[28]

Following a calamitous fire in 1869, Lynn began a search for improved water supplies and a year later purchased and enlarged a small mill basin located at the forest base and known as Breed's Pond. A public water board was established, and more artificial ponds were formed. Threats of increased development concerned Tracy, who organized excursions to the woods during "Camp Days" in an effort to build public support for his cause. An association of interested citizens, the Trustees of Free Public Forest of Lynn, was formed in 1881 and, led by Tracy as its first president, began negotiations for the purchase of private lands. In 1888, a public park was established and a board of park commissioners appointed. Philip Chase, a wealthy resident who had donated funds for purchase of lands by the Trustees, was elected chairman.[29]

Lynn's park commission immediately hired a surveyor, Isaac Harris, to map boundaries and clarify ownership. They also solicited advice from Olmsted about a park plan. The latter wisely discouraged introduction of the incongruous and expensive decorative features utilized in formal parks and instead recommended preservation of the area's wild and rugged character. Olmsted's correspondence indicates that he recognized the significance of a true urban forest. In a letter to Sylvester Baxter, by then secretary of the Metropolitan Park Commission, he encouraged a link between Lynn Woods and Boston's expanding regional parks:

> Its value as a roving ground not for Lynn and the northern suburbs only but for the people of Boston, in the future, can not well be overestimated. At least I rate it higher than the commissioners and they are not unappreciative. It shd [*sic*] be to Boston something like Fontainbleau to Paris and Richmond & Winsor [*sic*] to London. The townspeople of Lynn do not appreciate it, I judge. Probably want a park or public garden. It is, what is so much better, a real forest.[30]

Wary of increased use by the region's growing population, Lynn's commission declined proposals by Baxter to connect Lynn Woods with other metropolitan reserves. However, they did observe Olmsted's recommendation to keep the park in its wild state and limited improvements to gradual thinning,

Fig. 7.5. An 1854 sketch of the Rocking Stone in Lynn Woods, Lynn, Massachusetts. At midcentury, naturalist Cyrus Tracy organized the Exploring Circle and with his confederates began hiking the terrain and documenting its flora. Many of the reserve's distinctive features—Ox Pasture, Dungeon Rock, Meetinghouse Swamp, Rocking Stone, and Great Frog Boulder—now entice exploration of its cultural as well as natural history. Lynn Historical Society

introduction of new forms of vegetation, and replacement of unhealthy plants growing in certain swamps. In August 1890, a small handbook of park rules was issued with a copy of the Harris map. Ordinances were restrictive and, among other things, prohibited destruction of any plants or trees. The commission continued park expansion, acquired the ox pasture, and retained John N. McClintock to prepare a second survey in 1892 (fig. 7.6). By 1900, new roads had been built and fire observation towers erected.[31]

Lynn Woods also captured the attention of professional foresters. Bernhard Fernow praised the community in his annual report as chief forester in 1889, and he mentioned the forest a year later in an editorial letter to Charles Sprague Sargent. The latter must have been intrigued with Lynn's park because he included several articles about it in *Garden and Forest*. An 1891 editorial acknowledged its three periods of utility, as common pasture and timber lands, as a source of fuel when farming yielded to industry, and finally as a resource for recreation and protection of water supplies. Unfortunately, throughout most of the twentieth century, Lynn Woods has battled an unending series of problems, including fires, abuse by vandals, devastation by insects, neglect, and limited budgets. Even well-intentioned efforts to increase park usage have created intrusions on its historic wild character. A golf course, built in 1932 under Works

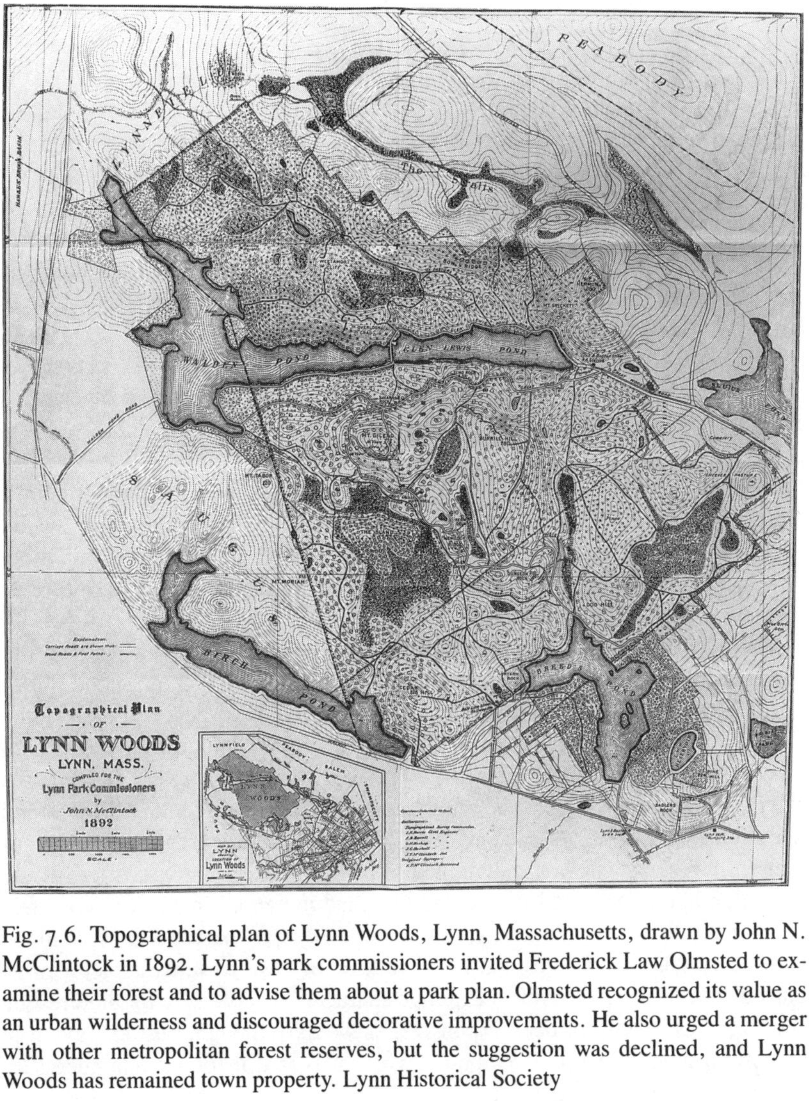

Fig. 7.6. Topographical plan of Lynn Woods, Lynn, Massachusetts, drawn by John N. McClintock in 1892. Lynn's park commissioners invited Frederick Law Olmsted to examine their forest and to advise them about a park plan. Olmsted recognized its value as an urban wilderness and discouraged decorative improvements. He also urged a merger with other metropolitan forest reserves, but the suggestion was declined, and Lynn Woods has remained town property. Lynn Historical Society

Progress Administration auspices, and athletic fields are examples. Reservoirs have gradually been enlarged, also consuming large chunks of woodland. However, much of the park remains as it always has, an unspoiled reserve of woodland, swamp, and precipitous rock ledges. Views such as that of Pennybrook Bridge are as peaceful today as they were in 1898 when the photograph was taken (see fig. 7.1).[32]

Waltham's Prospect Hill Park and Springfield's Forest Park

Lynn Woods was not the only communal forest park to precede regional reserves, nor was it alone in rejecting advances by Boston's metropolitan commission. Great and Little Prospect Hills (and the valley between) in Waltham, a westerly suburb of Boston, also remained outside commission dominion. Community plans for a park took shape in 1893 with purchase of seventy-five acres. Doubts were voiced about authority to clear a summit road until the city's solicitor unearthed town meeting records from 1731 and pointed, with lawyerly flourish, to approval of a town way, unbuilt but never discontinued. In a letter applauding Waltham's commissioners, though with regional as much as local concerns in mind, Olmsted and Eliot encouraged purchase of other lands along the Charles River before spending city dollars on improvements for Prospect Hill. Suggesting that park officials everywhere ought to be regarded as trustees of scenery, the firm's report was persuasive: "playgrounds for youth, concert grounds, public gardens, open air nurseries—all these and the like are good and more or less necessary, and yet the city which, while gaining these things, has permitted its scenic opportunities to be destroyed, may rightly be said to have gained the world but lost its soul." [33]

Forceful argument nothwithstanding, Waltham's government managed to strike an independent course. Intent on acquiring land in the presence of opportunity, they continued to enlarge the boundaries of Prospect Hill and left the Charles River to commission planning. Concurrently, modest park improvements were begun, including a pavilion, well, and rustic seats. Stone walls referred to as "Squadron Lines," representing seventeenth-century proprietor's allotments known as "Great Dividends," were carefully preserved. Management of forest growth was inaugurated, pine trees were planted to offer variety from dominant oak, and by 1904 the park had increased in size to 140 acres. As in many New England towns, prominent geography played a key role in the genesis of Waltham's forest park. Yet practicality was as much a factor in its current wooded and generally unspoiled character as was desire for picturesque scenery: "Care has not been expensive, as we have thought best to leave it in its natural state, and any development which may in time be thought necessary we leave to the future taxpayer." [34]

Elsewhere, similar projects were in the making. Between 1884 and 1898, Springfield, Massachusetts, acquired by gift and purchase more than 450 acres of woodland, meadow, ravine, and bog that someone fittingly christened Forest Park. Located in the city's southern precincts, park boundaries extend easterly from the Connecticut River and currently encompass about 750 acres. An ample donation of land in 1884 by Orick H. Greenleaf, a successful paper manufacturer familiar with parks in European cities, inaugurated the venture,

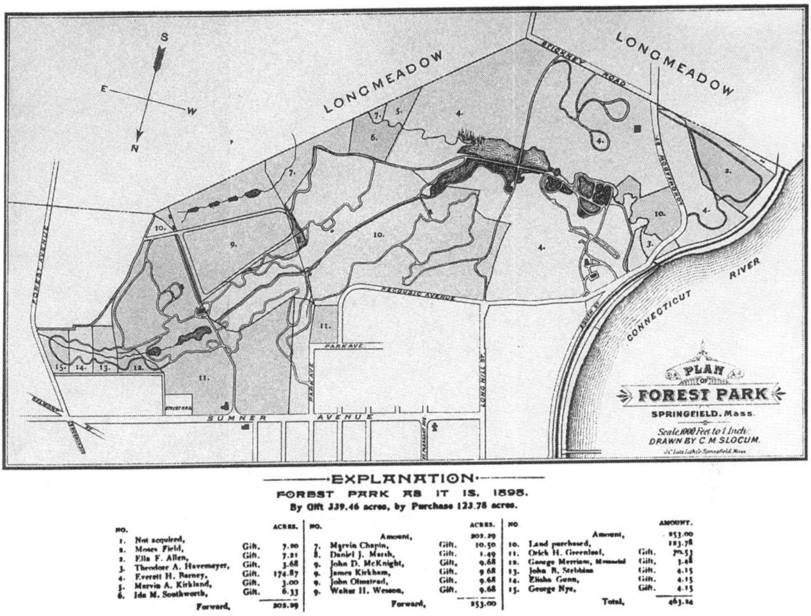

Fig. 7.7. Plan of Forest Park in Springfield, Massachusetts, by C. M. Slocum in 1898, showing land divisions according to former ownership. One of its largest parcels was a gift from Everett Barney, a manufacturer of ice skates, who assisted with the design of a large skating pond. Located well south of the city, the reserve became more widely used after a trolley line leading to the park's entrance was built. Annual Report, Springfield Park Commissioners, 1898

and a park commission was quickly appointed. Other gifts steadily materialized, including an especially large tract presented by Everett H. Barney, who had patented a popular style of ice skates. Barney, appropriately enough, assisted with design of a skating pond on a portion of land he donated. In 1898, boundaries of fifteen contiguous parcels were carefully identified on a plan drawn by city engineer Charles M. Slocum (fig. 7.7).[35]

Forest Park's first commissioners vowed to preserve the area's wild character and retained two local landscaping contractors, George Reynolds and Justin Sackett, who had formed a partnership. Decisions about park plans were apparently reached after collaboration among park commissioners, Greenleaf, Barney, and the landscaping firm. Improvements were initially confined to building roads, bridges, and trails, converting swamps into ponds, and removing debris; trees were cleared here and there to provide scenic overlooks. The park's popularity increased after extension of an electric trolley line to the park's entrance, and ever larger recreational facilities were introduced. By

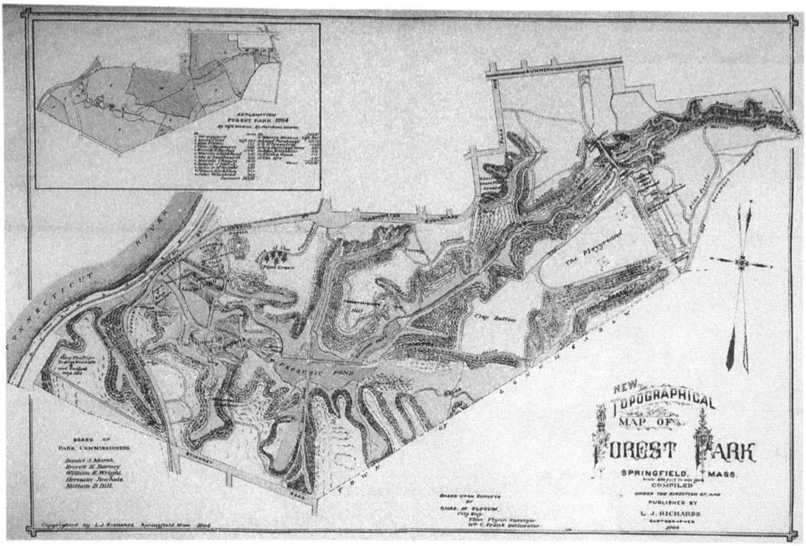

Fig. 7.8. Topographical map of Forest Park in Springfield, Massachusetts, drawn by L. J. Richards in 1904. The primitive wooded character of Springfield's park has succumbed to a variety of recreational improvements and ornamental landscaping. Changes such as these have occurred in many urban forest parks and are explained in part by the desire of park commissioners to encourage substantial public use. The influence of other types of urban parks has also been a factor. Annual Report, Springfield Park Commissioners, 1904

1904, the date of a topographic map by cartographer L. J. Richards (fig. 7.8), the park contained a zoo; a stone grotto; various buildings, including a restaurant and a rustic entrance pavilion; spacious playground areas; fountains; gardens; and numerous man-made ponds. A nursery had been established in 1900, and forestation work occurred periodically. Over time, landscaping has become increasingly ornamental and now dominates all but a few sections, where forest cover is still untamed.[36]

Both Forest Park and Lynn Woods have similar histories of recreational use. Oddly, 1891 ordinances for Springfield's park are nearly identical, word for word, to those used in Lynn a year earlier. Whether Lynn Woods influenced Forest Park or both simply borrowed their material from a common source, possibly Boston's Metropolitan Park Commission, is unclear. Despite their resemblances, Lynn Woods is much larger and has managed to confine its incompatible recreational facilities to more restricted zones. As a result, integrity of its forest has been preserved to a much greater degree. Unfortunately, dense woods are no longer a character-defining feature in many sections of Forest Park.

Town Woods as Parks

Throughout New England, a remarkable number of local forest parks were established during the closing decades of the nineteenth century and early decades of the twentieth century. A number of communities struggled to find a balance, inventive at times, between necessary recreational facilities and conservation of wilderness qualities. For a variety of reasons, many were far less aggressive in their planning than either Springfield or Lynn. Some grasped the value of simple footpaths and woodland scenery; others were just frugal. Smaller cities and towns did not need to encourage or oblige (and protect) multitudes of park-goers, and nature benefited on all scores. In such cases, local residents were often content with a building or two, a picnic area, playfields, and a place to swim or skate.

In Massachusetts, Fitchburg received donations of land and money from Henry F. Coggshall, beginning in 1894. Grants were conditioned on public use of the land, and Coggshall insistently prodded town officials to dispense adequate funds for improvements. By 1913, the park had increased in size to more than two hundred acres and had been named after its patron. Herman Cushing, who labored in Fitchburg's engineering department from 1898 to 1947, when he retired, crafted a splendid plan in 1913 that meticulously identified park topography and natural as well as man-made features (fig. 7.9). Roads and trails, built first, meandered in random patterns and a small dam was constructed to form Mirror Lake. South of this skating pond, paths climbed to Lookout Mountain. Across the park, near its northern boundary, a ball field was cleared. When Cushing limned his fine plan, few buildings had been built. Those erected later, among them a dance pavilion, two generations of band stands, and a stone lodge honoring Coggshall, who died in 1912, were generally confined to perimeter areas or clustered around the lake. Stone pillars were eventually placed at entrances.

Subsequent improvements generally followed guidelines set by Harris Reynolds, who designed a landscape plan in 1913, one of very few by him, shortly before beginning work for the Massachusetts Forestry Association (fig. 7.10). Reynolds suggested enlargement of Mirror Lake and inserted a meadow he identified as "Children's Romp." Formality was added to the park's northern entrance and to the nearby athletic field. The substance of Reynolds's plan, however, represented a compromise between preserving natural beauty and making it publicly accessible through meandering roadways. Park commissioners apparently agreed and remained content with limited improvements.

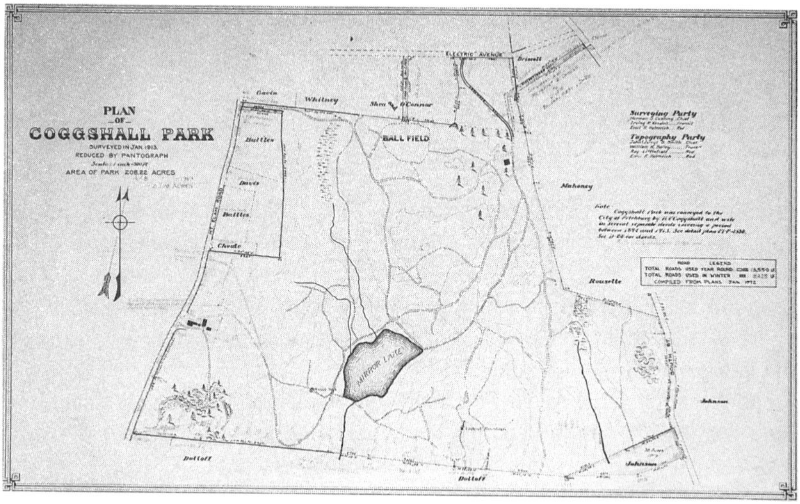

Fig. 7.9. Plan and topographical map of Coggshall Park in Fitchburg, Massachusetts, drawn by Herman S. Cushing in 1913. Cushing's plan, one of an outstanding collection he compiled as a city engineer, depicts an assortment of improvements that typify recreational activities in forest parks: lakes, pavilions, athletic fields, roadways to central locations, hiking trails, towers, and entrance monuments. Typically, these improvements are clustered or confined to strategic locations and the remaining land left to forest cover. Office of the Fitchburg City Engineer

Woodland management was introduced, and blighted stands of chestnut trees were removed after 1920, netting a modest return from lumber and cordwood; cleared areas were reforested with pine. Today, apart from a few scattered clearings, narrowly paved roads that converge at the stone lodge, a small playground, and ornamental landscaping bordering Mirror Lake, the park is mostly thick forest.[37]

Well east of Fitchburg, not far below the northern border of Massachusetts, Andover residents acquired a twenty-three-acre tract of glacial moraine, white pine, and oak, successfully plucking it from imminent timber sale late in 1897. Located near the village center, Indian Ridge was one of a series of kames in that sector of the state. Town members had grown fond of walking through its secluded woods and rallied behind the efforts of Andover's Village Improvement Society, which raised purchase funds. In May of that year, Henry Graves presented a lecture on forestry in his hometown to support their effort, and he agreed to assist with a management plan. A year after acquisition, Indian Ridge Association was incorporated to maintain the reservation, but it later

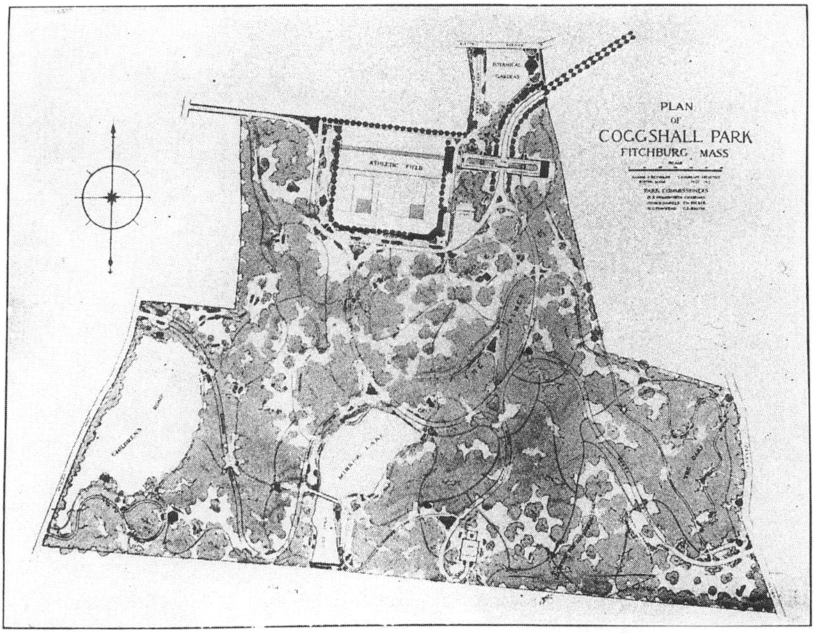

Fig. 7.10. Landscape plan for Coggshall Park in Fitchburg, Massachusetts, drawn by Harris Reynolds in 1913. Reynolds inspected the park not long after becoming secretary of the MFA, and he encouraged public access via meandering roads but advocated preservation of the forest's wild character as much as possible. Town forests, woodlands devoted primarily to cultivation of timber, also generated enthusiasm for recreational forest use and helped to foster development of a specific genre of urban park. Fitchburg Annual Report, 1913

merged with the improvement society. Reforestation work began after 1929, and sections were planted with red and white pine.[38]

Far to the southwest, in the state's opposite corner, Great Barrington made similar strides. In 1894, a gift of seventy acres became East Mountain Park, and reforestation was started in 1919 with assistance by schoolchildren. A deed condition required this acreage to remain in park use forever. Today, white pine, scattered hardwoods, and snarled undergrowth dominate, and the park appears forgotten although it overlooks the city only a short walk from the town center. A second preserve, on Monument Mountain, well known from a William Cullen Bryant poem, was established when 225 acres were donated to the Trustees of Public Reservations.[39]

Forest parks also became popular in New Hampshire towns during this period. In 1884, Jaffrey successfully thwarted private efforts to develop two

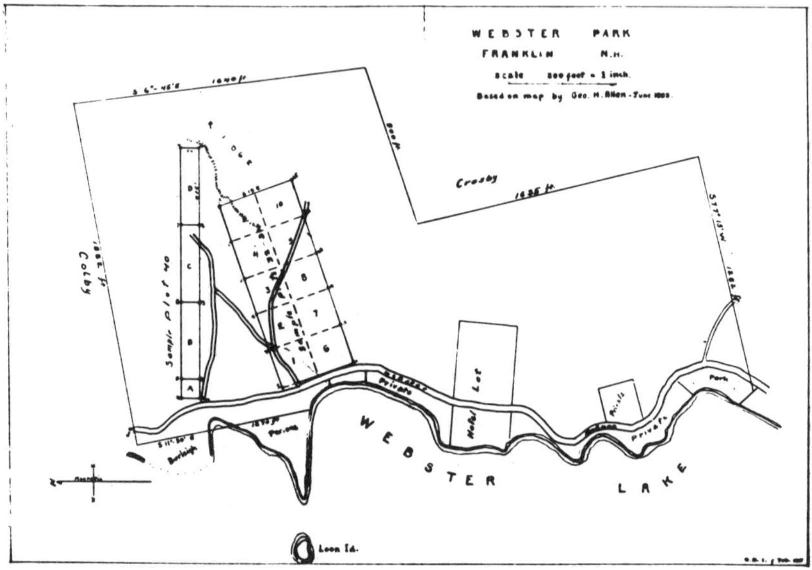

Fig. 7.11. Forest compartment map of Webster Lake Park in Franklin, New Hampshire, prepared by the state's forestry department in 1915, based on an 1893 survey by George H. Allen. Franklin converted a portion of its poor farm to park and requested assistance with timber management and gypsy moth infestation. Compartments probably indicated an area of experimental planting. New Hampshire Division of Forests and Lands—Concord

hundred acres on the steep southern face of Monadnock Mountain, just below its summit, and town land is now part of a large public reserve. Franklin converted 120 acres of woodland bordering Webster Lake from poor farm to park in 1893. Hardwoods, primarily oak, beech, maple and birch, were intermixed with zones of white pine. In 1914, park commissioners asked the state forester to prepare a management plan for park timber. A report issued soon afterward recommended culling large oak trees that overtopped young growth and fostered moth infestation. Stands of young pine were encouraged, with thinning of old trees only, to ease overcrowding. Gradual thinning of remaining forest areas was advised, and white, red, and scotch pine or white ash were suggested for underplanting or open areas, depending on soil conditions. Compartments, probably signifying plantation zones, also were plotted on a map (fig. 7.11). In 1934, the state's forestry department eventually produced a type map that identified compartment types (fig. 7.12).[40]

Similarly diligent stewardship occurred in a number of forest parks owned by New Hampshire communities. In more than a few cases, care of forest trees

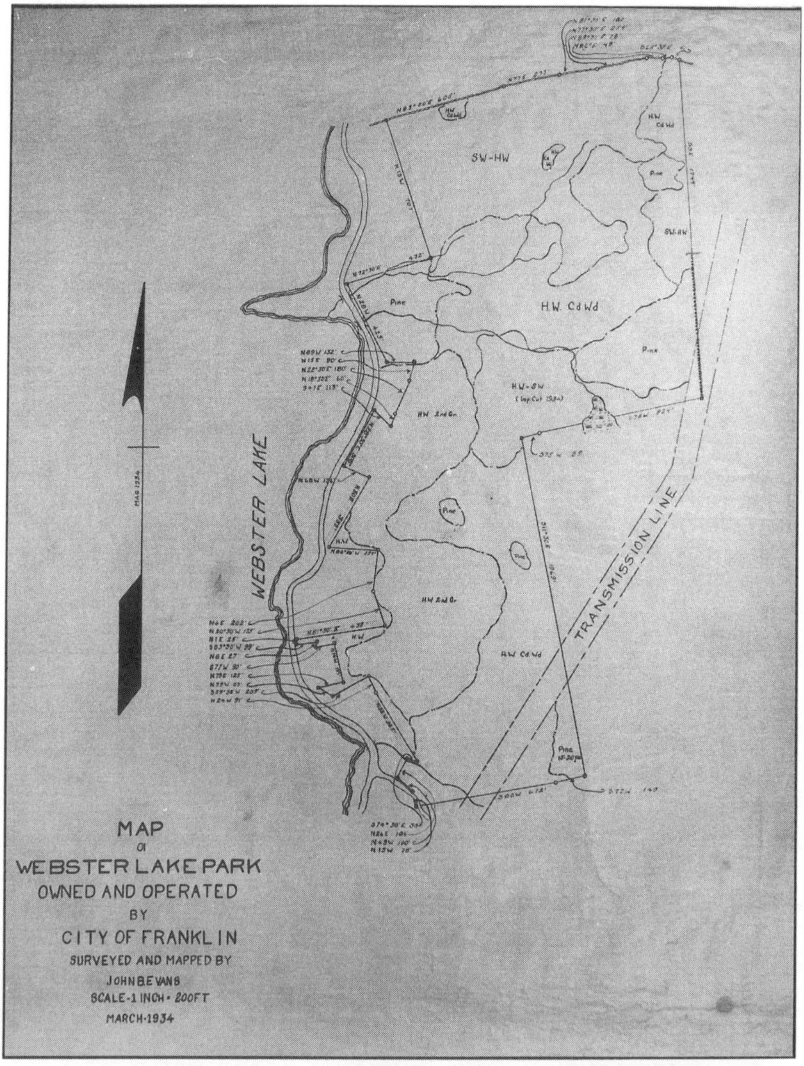

Fig. 7.12. Forest type map of Webster Lake Park in Franklin, New Hampshire, penned by Frank B. Evans for the state's forestry department in 1934. Cooperation between government foresters and local park commissioners occurred with frequency in New Hampshire and was fostered in large part by channels of communication already installed between state offices and town forest committees. Although the two types of local forests generally remained distinct, assistance was typically available for both. New Hampshire Division of Forests and Lands—Concord

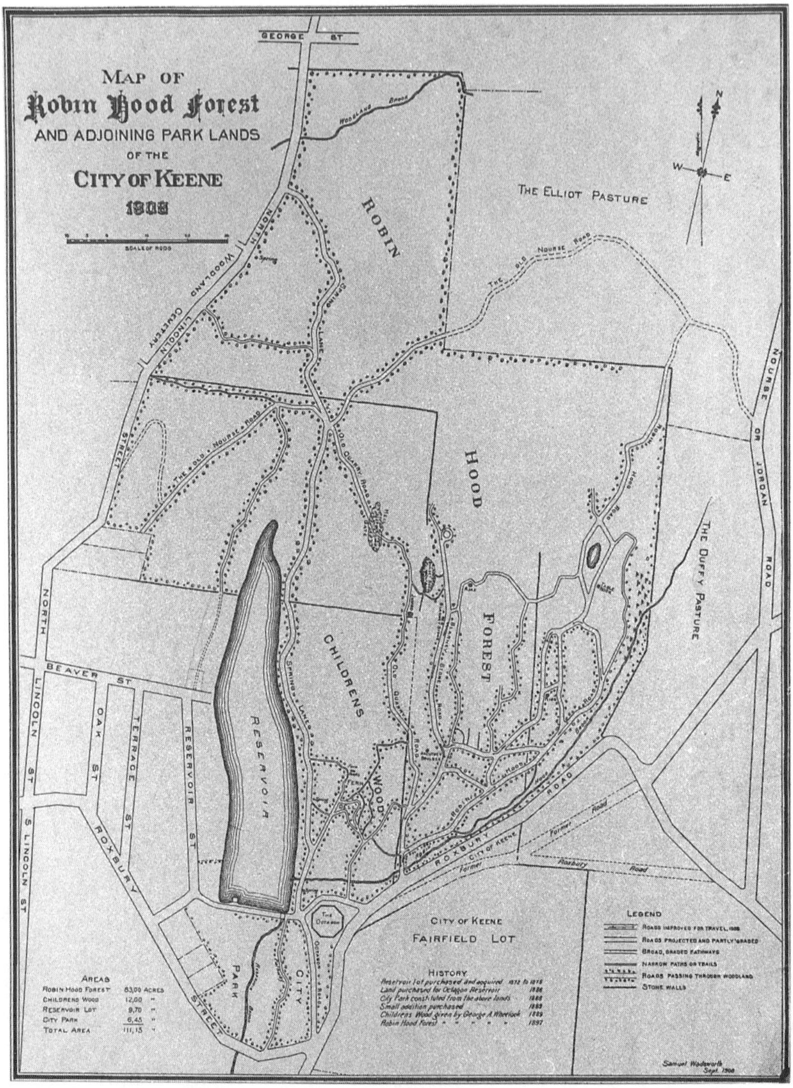

Fig. 7.13. Plan of Robin Hood Forest and Children's Wood in Keene, New Hampshire, designed by Samuel Wadsworth in 1908. Keene outgrew its Octagon Reservoir in the latter decades of the nineteenth century, and surrounding lands, donated by George A. Wheelock, the city's first park commissioner, were converted to a park. Wadsworth planned circuitous, winding paths as well as vistas to distant mountain peaks such as Mt. Ascutney in Vermont. Wadsworth Collection, Cheshire County Registry of Deeds, Keene, New Hampshire

was specified as a condition in deeds for donated land. Keene began acquiring lands surrounding a city reservoir in 1872 and in 1888 dedicated them as a park. Two nearby parcels, Children's Wood and Robin Hood Forest, were donated in 1889 and 1897, respectively, by George A. Wheelock, the city's first park commissioner. A plan of the 111-acre park and reservoir lands was drawn in 1908 by a versatile Samuel Wadsworth (fig. 7.13). A jeweler and clockmaker before becoming city engineer, Wadsworth ambled across much of Cheshire County as a surveyor, naturalist, amateur scientist, and historian. His plan of Robin Hood Forest shows a picturesque network of winding paths with vistas identified by landmarks such as Sunset Rock or Ascutney Boulder. Children's Wood encouraged Keene's youth to combine education with recreation, and Wheelock's grant specified that his lands should remain forested. In two separate gifts, first in 1886 and again in 1889, Wheelock also donated nearly forty acres, a portion of it once a fairground, in the town's western environs; a similar condition was imposed for at least part of the property. The land eventually became Wheelock Park, and a plan by Samuel Wadsworth in 1917 reveals dominant stands of white pine, but twenty-four other species were identified by number (fig. 7.14).[41]

Elsewhere, Stratham Hill, a lone summit overlooking the Piscataqua River in the sandy lowlands of southeastern New Hampshire, was given to the town of Stratham in 1905. The gift was qualified by a mandate for preservation of forest trees "with additions and removals from time to time" and a requirement that the land should benefit people of Stratham, Exeter, Portsmouth, and adjacent towns. The site promptly became a favorite spot for Stratham's "Old Home Days," when residents gathered annually for a picnic and photograph (fig. 7.15). During this period, access to outlying parks often depended on trolley lines. At Stratham Hill, a pavilion built at the park entrance in 1917 served as a convenient trolley stop. In 1931, state funds purchased materials and labor for construction of a fire tower on the hill's 292-foot crest.[42]

In Vermont, Montpelier's Hubbard Park was given to the city by John E. Hubbard in 1901. Although much of the land was neglected pasture, Hubbard stipulated that it be used as a park and requested appointment of a board of park commissioners. Landscape architect Dana F. Dow was hired to prepare a plan for roads, trails, and forestation. Plantations of red, white, and scotch pine as well as Norway spruce were established between 1906 and 1909 with state nursery seedlings. In other sections, pasture sprouted to maple, yellow and black birch, cherry, and hemlock. Park benches were distributed along walks and drives, and a square rubble-stone observation tower, begun in 1916, was eventually completed by 1931. Modest improvements have included a small picnic area and scattered clearings for sledding or athletic games. Today, Hubbard Park remains an unpretentious woodland criss-crossed by footpaths, nature courses, cross-country skiing and fitness trails, and hidden

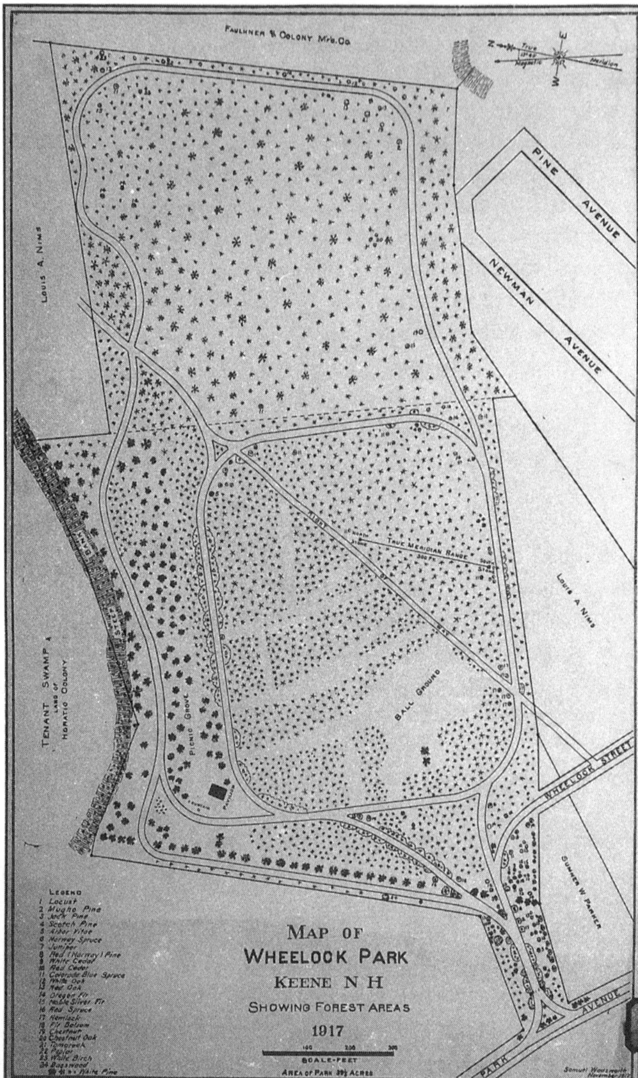

Fig. 7.14. Map of Wheelock Park in Keene, New Hampshire, drawn by Samuel Wadsworth in 1917. George Wheelock also donated lands for this park and, as he had for Robin Hood Forest, specified that the acreage should remain forested. Wadsworth's plan identifies twenty-four tree species in addition to abundant white pine. Wadsworth Collection, Cheshire County Registry of Deeds, Keene, New Hampshire

Fig. 7.15. Old Home Day, Stratham Hill Park, Stratham, New Hampshire, in 1905. Seventy acres of timberland were given to the town earlier that year, with a stipulation that it benefit the people of Stratham, Exeter, Portsmouth, and adjacent towns. A trolley line to the park entrance was built, and improvements, including a pavilion and fire tower, were eventually erected. Stratham Historical Society

moss-covered stone walls. In recent years, logging of mature pine has been accomplished with horses and the public invited to watch.[43]

A number of woodland parks also were created in Connecticut prior to 1900, and stewardship was equally scrupulous. In New Haven, a solitary ridge of outcrop and forest known as East Rock was chartered as a park in 1880 following donations of land and money the previous year. A board of overseers, the East Rock Park Commissioners, engaged landscape designer and author Donald Grant Mitchell to prepare a park plan in 1882. Mitchell left the land in its wild state but introduced stone bridges across water courses and placed rustic benches or arbors along carriage roads and at scenic vistas. A Soldiers' and Sailors' Monument was erected at the summit and dedicated in 1887. In 1889, the East Rock Park Commissioners became the Commission of Public Parks and embarked on preparations for a large-scale park system. A second, similarly picturesque and sentinel-like pinnacle, West Rock, became a park in 1892. Both parks have gradually increased in size and currently sprawl over more than a thousand combined acres. Development and generations of users have steadily eroded the parks' untamed qualities, but each retains substantial forest cover. West Rock, more remote than its companion, is now a state park. Other Connecticut towns followed New Haven's example, and isolated mountain peaks, poor as real estate but rich in scenery, became public sanctuaries. Woodbury's Orenaug Park and Washington's Steep Rock Reservation, the latter now owned and managed by a land trust, are good examples.[44]

Recreational use of town-owned forest parks can extend beyond community borders where trails are linked to regional or statewide hiking networks. In Meriden, just north of New Haven, one of the town's philanthropic residents, Walter Hubbard, began acquiring land in a region known as Hanging Hills and established a public park in 1899. Hubbard shouldered improvement expenses during the park's early years and supervised construction of a crenelated rubble-stone tower, Castle Craig, completed in 1900 on East Peak. By 1901, Hubbard Park stretched to West Peak and covered more than a thousand acres, part of which protected the city's reservoir, Lake Merimere. From the outset, park commissioners emphasized the importance of preserving timber, particularly on watershed slopes. Despite periodic submission to fashionable recreational facilities, among them a greenhouse built in 1940 by Works Progress Administration workers, the park's wild character has remained dominant. Valleys and ridges are connected by a network of hiking trails, including the Metacomet Trail, part of Connecticut's blue-blazed through-trail system, which pushes northward into Massachusetts.[45]

The Metacomet sprang from the earlier Trap Rock Trail, conceived by Meriden's Frederick W. Kilbourne, a librarian, amateur botanist, and associate editor of *Webster's International Dictionary*. Kilbourne and friends began blazing in opposite directions from West Peak in 1916, and by 1919 had completed a circuitous route around nearby urban centers. Kilbourne visualized a connecting trail from Long Island Sound to Massachusetts and won support from Allen Chamberlain at the AMC. Within a decade, the Connecticut Forest and Park Association (CFPA) had established a trails committee. Their group was headed by Edgar Heermance, an 1899 graduate of Yale Divinity School and unlikely pioneer in trailblazing, who concocted a particular shade of blue paint visible in twilight. With blue blazes to guide them, CFPA's committee began the considerable task of organizing long-distance trails through and around the state's densely populated regions.[46]

Clearly, appreciation of simple forest recreation, curiosity about nature, and concern for disappearing woodlands were driving forces behind the establishment of early forest parks in New England communities. Although popularized versions of transcendentalism contributed to an imitation of nature in the plans of many nineteenth-century urban parks, forest parks were seldom designed in any formal sense. Except for footpaths or carriage drives, wooded areas were generally left untouched except for necessary thinning. Once established, however, inconsistent vision for their management sometimes resulted in an undermining of the wilderness character so appropriate for New England towns. Responsibility rested, at least in part, with the difficulty of encouraging use in order to fulfill public function. To some, creating an ever increasing variety of facilities to attract larger segments of the public seemed, logically, a necessary step.

External forces such as reform in general attitudes about urban parks also may have been a factor. Park commissions became administrative bodies in which competing opinions more than occasionally spawned compromise. In such a context, the outcome of debates about whether to introduce organized recreation, gardens, playing fields, or even zoos was predictable. In other cases, growing urban areas swallowed once remote preserves, making control of public use more difficult. In the face of ever-present change and increasing social problems, fading interest in preserving the quality of wilderness patches was understandable. Yet examples such as Lynn Woods, Indian Ridge, and the pair of Hubbard Parks, where cities have clung tenaciously to their wooded preserves, are representative of the strong relationship between forests and New England towns.

Influence of the Town Forest Movement

The appearance of forest parks preceded that of town forests by a decade or two. Progressive-minded citizens in New England responded quickly to themes publicized by America's growing interest in urban parks, the City Beautiful movement, and village improvement societies. Attention to national forest reserves and national parks continued, and local park projects flourished. For well-organized hiking groups adept at shaping public opinion in favor of forest recreation, a leap from park to forest park was an easy one. In contrast, town forests obeyed a pace set by a young forestry profession and were partially handicapped by need for state legislation enabling commercial forestry by local governments. At a time when town forests were first being considered, urban forest parks received national attention at the Second Conservation Congress, held in 1910 in St. Paul, Minnesota. William M. Ellicott, a delegate from the American Institute of Architects, presented a paper titled "Forests as Pleasure Parks," in which he advocated the creation of a national forest reserve adjacent to Washington, D.C.

Ellicott reasoned that western forests were far too remote to be accessible to eastern populations, and he regarded urban forests as an opportunity to demonstrate forest conservation on a greater magnitude. Just as architects and engineers had learned from world's fairs to plan and design on a grand scale, so too could the forestry profession learn from the creation of a great forest outside the nation's capital. Ellicott's ambitious scheme involved assisting nature to achieve a degree of romantic perfection similar to that accomplished in France's public forests. Perhaps as an enticement to foresters, Ellicott suggested the site could become a national headquarters for forestry education. Although his plan never achieved fruition, Ellicott noted that forest parks were an improvement on urban parks and could be introduced to cities throughout the country.[47]

Eleven years later, in an address to the annual convention of American Institute of Park Executives in Detroit, Filibert Roth, dean of forestry at the University of Michigan, commented on the increasing occurrence of recreational forests. Roth titled his speech "Woods as Parks," and in plain terms explained that mergers of the two were escalating because people had begun to realize that "the forest does more and does it better than the park" and for the same or less money. His advice leaned toward the practical, and he saw no reason for separate functions: "Let the people decide, and wherever the mass of visitors go, there fit the woods to the people, while in the rest, let the woods grow timber and serve as haunt of the few, the real lovers of the woods."[48] Focusing on lack of proper management as the issue of greatest neglect, he advocated methodical thinning from one compartment to the next to remove the thickets of coppice that fostered disease and impaired thrifty growth. He also cautioned against overdoing roads in mileage and quality. At the same time, Roth added a measure of the romantic to his recipe for attracting public use:

> By all means preserve the old, the large, the remarkable, the instructive; old oaks, rare hickories, fine clumps of hawthorn, and even blue beach, ironwood and flowering dogwood all should be allowed a little space. Where school children are taken in classes, leave rotten logs, dead stubs, etc., and let them see the wild wood, truly wild. This requires a few acres of land sacrificed to the visitor—it pays big, more than any special growth of timber.[49]

Despite its later start, the movement for town forests influenced development of woodland parks in several important ways. As they matured or became enveloped by urban sprawl, forest parks frequently succumbed to incongruous development. When that occurred, their capacity to provide woodland recreation diminished. Town forests emerged as an alternative that obliged recreational as well as commercial use. Oddly, inconsistent management that plagued the commercial productivity of these forests probably improved their standing as sites best suited for recreation. Furthermore, a tremendous surge in community acquisition of land (if not already wooded, then quickly planted), increased the availability of local forest recreation. In short, town forests emerged as an important category of woodland parks in New England after 1920.

Town forest campaigns also generated forestry programs geared for communities and helped to produce a body of trained foresters, who became proficient at working with local governments. Stewardship benefited as a consequence. Solicitation of management plans for existing parks, such as that prepared for Webster Lake Park in Franklin, New Hampshire, occurred now and again. As towns established wooded reserves, they were able to take advantage of this network of foresters. Finally, forestry and landscape architecture occasionally merged during the planning of new forest parks, sometimes with excellent results. Moody Park in Claremont, New Hampshire, is a superb illustration of joint endeavor by forestry and landscape architecture professionals. Claremont is located in the state's west-central region, almost in the shadow of Vermont's Mt. Ascutney. Land for its 175 acre park, roughly a mile southwest of the town hall, was donated in 1916 by William H. H. Moody, head of an enormous shoe manufacturing firm who had returned to familiar surroundings to recover from poor health. In response to a request by town selectmen, the state's forestry department examined the tract and recommended preparation of a general plan by a landscape architect. Arthur Shurtleff, a former associate of Frederick Law Olmsted, was hired.[50]

Shurtleff, who later changed his name to Shurcliff, earned an engineering degree at Massachusetts Institute of Technology in 1894 and, after graduate study at Harvard, began working in Olmsted's firm in 1896. Three years later, he helped to found Harvard's four-year course in landscape architecture. In 1905, he established his own firm in Boston where he worked closely with the Metropolitan District Commission on a number of park reserves. His varied and fruitful career encompassed town planning as well as park design and included commissions at Colonial Williamsburg and Sturbridge Village. Possibly through his marriage to a woman from Cornish, New Hampshire, just north of Claremont, Shurtleff became acquainted with members of the Cornish Community. This was a loosely confederated group of artists and writers whose numbers included sculptor Agustus St. Gaudens, poet Percy Mackaye (older brother of Benton), and author Winston Churchill, an active member of the Society for the Protection of New Hampshire Forests. The farm belonging to Shurtleff's father-in-law adjoined Churchill's estate, Harlakenden, overlooking the Connecticut River. Whether through Churchill or other contacts, Shurtleff became involved with SPNHF activities.[51]

For his Claremont project, Shurtleff conferred with Alfred B. Hastings, New Hampshire's acting state forester, and prepared a draft report, which he sent to Hastings in February 1918, along with a park plan (fig. 7.16). Stressing a need for long-term strategy, flexible enough to allow progress in stages but specific enough to weather economic and political misfortune, Shurtleff identified the park's distinctive qualities. Foremost was a Great Ravine sculpted along the parcel's northern boundary and fed by a series of smaller ravines.

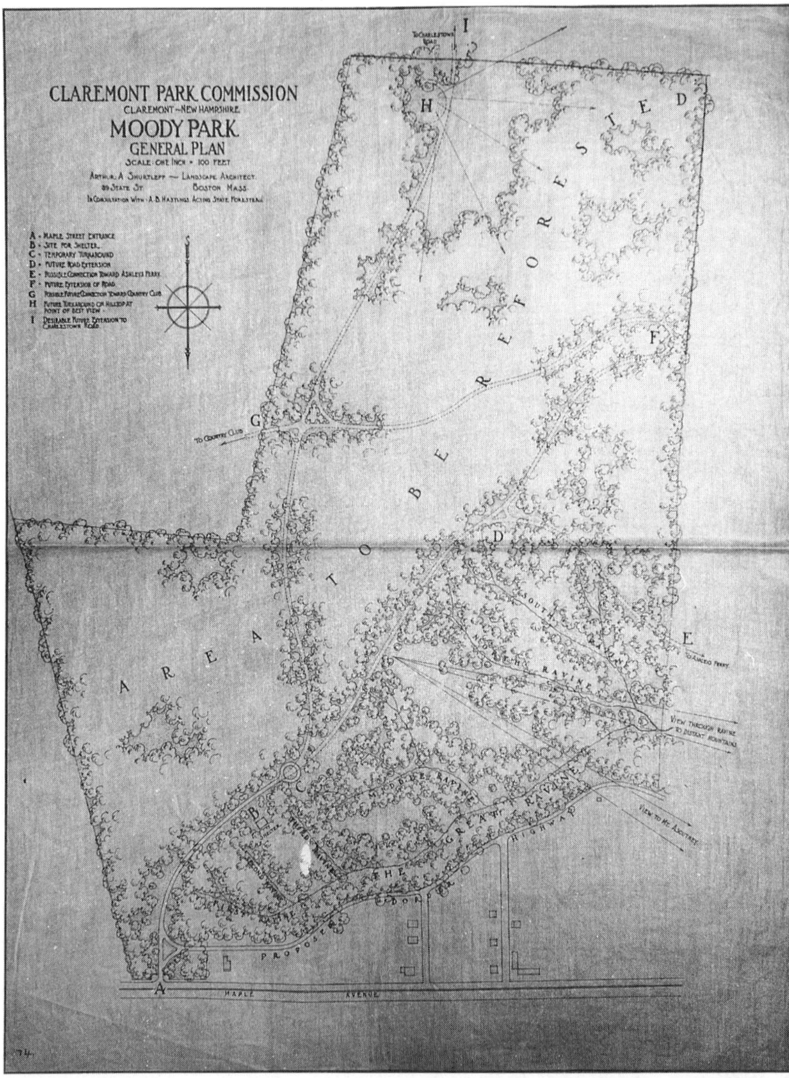

Fig. 7.16. Plan of Moody Park in Claremont, New Hampshire, designed by landscape architect Arthur Shurtleff in 1917. Shurtleff, who later changed his name to Shurcliff, had been associated with Frederick Law Olmsted before establishing his own Boston firm and probably obtained his Claremont commission through contacts with the Cornish Community and members of the SPNHF. He collaborated with Alfred B. Hastings, New Hampshire State Forester and emphasized the park's many ravines and panoramic views; Hastings recommended planting with white and red pine. New Hampshire Division of Forests and Lands—Concord

An elevated region, just below the crest of a hill lying outside the park's southern border, offered panoramic views of surrounding mountains. Protection of vistas, both from hilltop and through ravines, was critical to his plan, and Shurtleff urged arrangement of planting to frame specific rather than sweeping views. He recognized that scenery would become a public attraction and considered a network of connecting roads to be essential. Thinning of ravine woodlands, primarily oak, maple, and hemlock, was advised as a means of promoting natural cover rather than transforming it; coniferous trees planted in open pastures would balance deciduous growth. Level, comparatively open ground between ravines became sites for athletic fields and picnic facilities. In closing, he invited Hastings to cut and paste paragraphs onto the latter's forestry report, which was then submitted jointly to the town.[52]

In its final form, Moody Park belonged more to Shurtleff than to Hastings, but the latter's contributions were important nonetheless. He recommended white and red pine for open areas, planted annually in ten-acre plots, and predicted a substantial yield of lumber in sixty years. Optimistically, he believed that lumbering operations would stimulate interest in the park and serve as an object lesson. Hastings advised against clearing young poplar and birch until they reached a marketable age for pulp or excelsior, thus paying for cost of their removal. He also suggested hiring a local caretaker to undertake thinning under supervision by the state forestry department. Claremont's park was generally developed according to plan and has remained a densely wooded reserve with picnic areas and footpaths. A single paved road leads from its entrance to an open hilltop where a stone pavilion stands. However, as Shurtleff feared, neglected overgrowth has blocked vistas, and scenic ravine areas are not easily accessible. Connecting roads were not built, and elevated land to the south was not acquired in spite of his urging.[53]

A Vernacular Forest Landscape

Wooded parcels of land were set aside and made available for public enjoyment by a great many New England towns during the first half of the twentieth century. For a time, certainly, the town forest movement dominated local interest in woodlands. Nevertheless, many communities placed their forest parks a cautious distance from that campaign. Conservation, woodland recreation, planning, landscape architecture, and sympathy for wilderness preservation all commanded a degree of influence, and all eventually contributed to this trend. Selection of a single prevailing force beyond general appreciation of forests or desire for civic improvement becomes difficult given the independent nature of New England towns. Although distinct patterns to the swelling popularity of forest parks are apt to exist within regions or from town to town, they are well concealed.

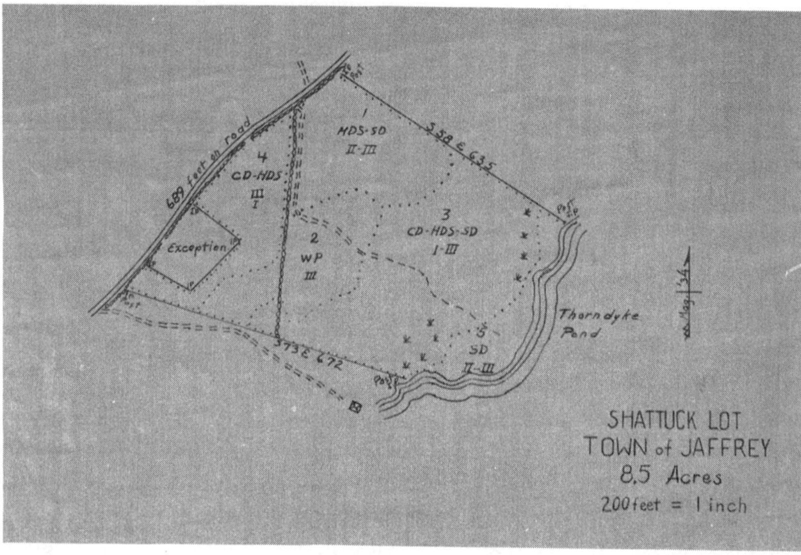

Fig. 7.17. A forest type map of the Henry V. Shattuck Park in Jaffrey, New Hampshire, prepared by the state's forestry department in 1934. Located on the road from Jaffrey to Dublin, the park was bequeathed to the town by Shattuck, who died in 1925. His will specified that if the town refused his gift, intended to be a pleasure resort, the land would go to the state for forestry purposes. New Hampshire Division of Forests and Lands—Concord

Unlike town forests, forest parks lacked the benefit of an organized campaign and were never really monitored in any sustained fashion by government or private forestry groups. Consequently, their history is a collection of scattered pieces not easily arranged into an interpretive mosaic. For instance, although the frequency of land gifts to towns is suggestive of a pattern of origin, it would be unfair to conclude that communities seldom created woodland parks on their own initiative. Moreover, grants or bequests typically carried their former owners' biases in the shape of conditions, and not everyone agreed on methods of forest use. As a result, general conclusions about management of trees or the extent of permissible improvements are equally difficult.

Even within single towns, separate forest parks were created for different reasons. In 1926, voters in Jaffrey, New Hampshire, accepted an eight-and-one-half-acre plot along the shore of Thorndike Pond, together with an endowment for management and improvement of the land as provided by the donor's will. The Henry V. Shattuck Park was bestowed as a public pleasure resort, but the parcel would have gone to the state for forestry purposes had the town chosen not to accept it. Management assistance was provided by

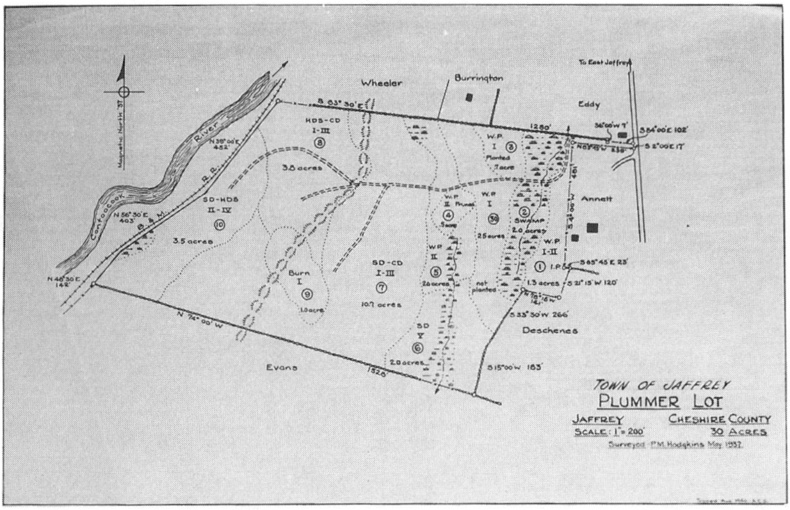

Fig. 7.18. A forest type map of the Plummer Lot in Jaffrey, New Hampshire, by P. M. Hodgkins for the state's forestry department, in 1937. The parcel was given to the town by Albert Annett, for many years a vice president of the Society for the Protection of New Hampshire Forests, whose will required that the land be named "The Children's Wood." Annett mandated that the property be held forever as a forest park or arboretum for cultivation and study of trees, shrubs and flowers. New Hampshire Division of Forests and Lands—Concord

New Hampshire's forestry division, and a type map was prepared in 1934 (fig. 7.17).[54]

Eleven years after Shattuck's gift, a thirty-acre parcel located along the Contoocook River and known as the Plummer Pasture was donated to Jaffrey by Albert Annett, a historian and longtime member of the SPNHF. In his will, Annett named the tract "The Children's Woods" and requested the town's school board to appoint a three-member committee to manage it (fig. 7.18). He specified that the land should be "maintained forever as a forest park or arboretum for the cultivation and study of trees, shrubs and flowers" as an adjunct to the town's schools. He also urged that trees be left to nature's husbandry with only enough pruning to attain ideal growth. His lucid explanation captured the essence of at least part of the thrust behind the creation of forest parks: "While recognizing the practical use of our forests, it is the primary purpose of this bequest to cultivate that love and appreciation of nature that (one) finds in trees when allowed to reach their most perfect development, friends and objects of beauty to be cultivated and enjoyed for their aesthetic, cultural and recreational uses."[55]

With great diversity in size as well as means and period of origin, fre-

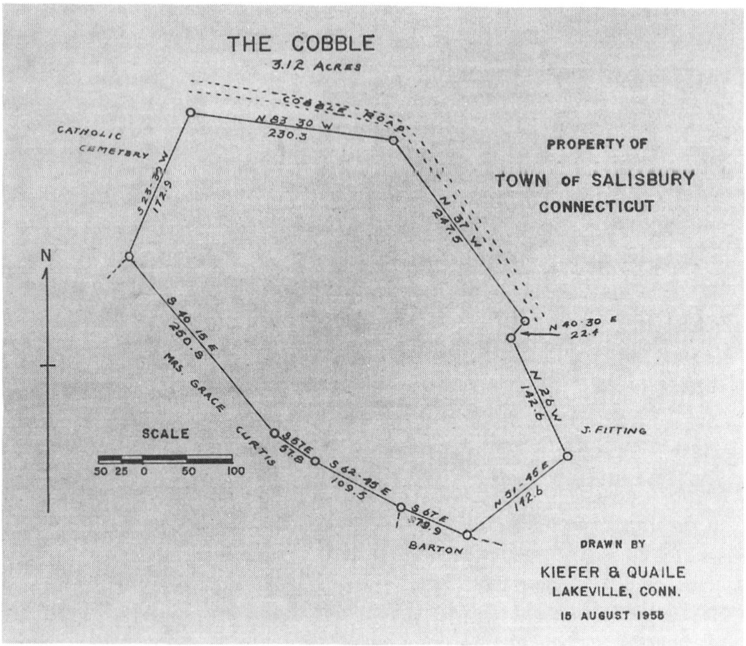

Fig. 7.19. Survey of the Cobble in Salisbury, Connecticut, mapped by the firm of Kiefer and Quaile in 1955. Located near the village center, the park is a small, three-acre reserve of rock and woods, one of several forests owned by the town. Another, Barrack-Matiff Town Forest, blankets part of a mountain bearing that name just north of Salisbury and is traversed by the Appalachian Trail. Office of the Salisbury Town Clerk

quent ties to uniquely scenic locations, and benefactors' often distinct visions for their use and care, communal forest parks represent a vernacular landscape in the word's truest sense. Romantic or associative place names such as Children's Wood, The Cobble (Salisbury, Conn. [fig. 7.19]), Meeting Waters Forest (Springfield, Vt.), Kilburn Crags (Littleton, N. H.), Somerset Woods (Skowhegan, Maine), The Pinnacle (Lee, Mass., and Walpole, N.H.), Kettleholes (Falmouth, Mass.), Whitaker Woods (Conway, N.H.), and Wilbour Woods (Little Compton, R.I.) emphasize the singular qualities of these reserves. As recreation became a dominant activity in communal forests, their character gradually changed from woodlot or sylvan treasury to amenity. A great many bear the marks of consistent use or oblige varied recreational activities. Others exist as little more than hidden patches of woods, known only by those who have taken the trouble to find their simply marked, sometimes overgrown paths. Remnants of stone walls serve as boundaries or as reminders of forgotten pastures in surroundings otherwise isolated from human intrusion.

In New England's panoply of forest parks, the extent or limitation of recreational use seems tied neither to time of creation nor community boundaries. Battell Woods in Middlebury, Vermont, was bequeathed to the town in 1901 by Joseph Battell, whose extensive grants to the state were responsible for creation of Camels Hump State Park. Only a small wooden sign, now overgrown with vines, marks the entry to the town's woods, and paths are its chief form of human improvement. Battell had become distressed "with the evils attending the extensive destruction of the original mountain forests," and he asked that his land "be managed with a view to preserving the trees as far as compatible" with its free use as a public park.[56]

In 1917, the northern New Hamsphire town of Littleton received approximately twenty acres of wild, rocky land atop a ridge separating the Connecticut and Ammonoosuc River valleys. The tract was donated by an heir of Benjamin Kilburn, whose interest in landscape photography led to the founding and success of the Kilburn Stereoscopic View Company. His bequest specified use of the land as a park forever and included an endowment for management. Kilburn Crags, as it became known, has remained untamed throughout the century, with access via an abandoned town road.[57]

Adjoining Long Pond in Lempster, a short distance from the Second New Hampshire Turnpike as it cuts a narrow, winding course through the state's southwestern corner, almost seven hundred acres of land were forfeited to the town for unpaid taxes. Voters agreed, in 1945, to establish a town forest and two years later sold spruce and hardwoods as well as a modest number of Christmas trees. Part of the land was later subdivided and sold for campsites. Remaining acreage is currently a nature preserve, and a single mowed but otherwise unmarked path weaves its way beneath the tracery of maple and birch leaves until it reaches a small marsh. Although lacking the scenic qualities of Kilburn Crags, Lempster's forest park seems equally unaffected by use.[58]

In Connecticut, Norwich's Mohegan Park is conspicuously different from Lempster's secluded, almost concealed woods and represents the opposite end of the forest park spectrum. Initiative for the park belonged to a group of individuals who convinced city officials that residents would benefit from a site for outdoor recreation. The park was established in 1906 with land both donated and purchased, and improvements quickly materialized: pavilions, skating and swimming ponds, athletic fields, playgrounds, a concert arena, a refreshment stand, gardens, and a menagerie. As in Springfield's Forest Park, ornamental landscaping began to take over large sectors. Yet the desire of many to protect the park's wild character was recorded in 1909 as park commissioners battled to find a suitable equilibrium for park use:

> It is not difficult to understand the desire of many, who are greatly interested in the park, that there should be few, if any, structures located there, it being a very general and wholesome wish that natural conditions should be disturbed as little as possible. But

it must be remembered that now the park is visited, and hereafter will continue to be visited, by thousands of people.[59]

While efforts to accommodate public use continued, reforestation was begun after 1920, and a large portion has remained thickly wooded. Where trails lose their way among stands of hardwoods, pine, and hemlock, comparisons to places such as Lempster's reserve are reduced to little more than quality of footpaths or timber management.

Between examples such as Battell Woods and Mohegan Park lie a multitude of forest parks, many with widely varying characteristics. In Vermont, Hartness Park in Springfield and Paradise Park in Windsor have both been classified as municipal forests at one time or another. In northeastern Massachusetts, Methuen's town forest offers well-maintained recreational areas for swimming, picnics, and scouting camps. A single paved road winds through timber stands until it reaches Forest Lake, where buildings and playgrounds are generally confined. Pembroke's J. J. Shepherd Forest has also combined forestry with varied recreational facilities and offers a community center as well as open fields for baseball and carnivals. In Boston's westerly environs, a portion of Natick's hundred-acre town forest was set aside as a natural science park with permanent study plots. A manual explains aquatic and terrestrial animal life, soil, geologic history, tree species, and forest influence on climatic conditions.[60]

In Skowhegan, Maine, the Somerset Woods Trust was created in 1927 to hold forested areas for the benefit of Somerset County towns in which trust lands are located. One of its parcels, a thirty-five-acre woodlot bordering the Kennebec River just below its Great Eddy, frames a verdant entrance corridor for travelers approaching Skowhegan from the east along U.S. Route 2. The tract has become a demonstration forest for northern white pine timber management and offers education in cultural history as well as natural science. Its trails are primarily those left by logging trucks. Walpole, New Hampshire, owns a number of forested lands used for hiking, camping, and nature study, among them Hooper Woodland near the town's reservoir, Mason Forest, the Merriam Lot, the Pinnacle, and a large pine plantation owned by the town's water company and seeded by high school students in 1917.[61]

. . .

Forest parks symbolize the genesis of recreational use as a legitimate and recognized form of public activity in New England's town woods. They also represent a vital link in the transformation of communal forests from seventeenth century woodlots to twentieth-century patches of wilderness managed today by local conservation commissions. The latter comprise a final category of communal forests. One can most closely observe the evolution of public involvement with forests through recreation. As that form of activity has gradually matured, so too has the relationship between forests and community environments; forest parks are often a visible, frequently used, and carefully tended segment of community structure. Now a dominant use of local woods, recreation encompasses all of the activities and interests that began to emerge a century ago. A full spectrum of uses can exist in one forest, ranging from nature study to hiking and to cultural education in the form of logging demonstrations. Currently, use also encompasses the recognition that open spaces provide a critical balance to relentless change in human environments. Mere ability to observe these ecological preserves has become a form of recreation. The context for growth of this class of use is also a context for origins of conservation spaces.

Chapter 8

Working Plans for Habitable Communities

For New England, most of the twentieth century has been an era of consistent and dynamic escalation in numbers of communal forests. Whether for recreation, watershed protection, timber growth, or, ultimately, conservation, wooded lands gradually began to represent an important aspect of civic progress and attracted planning as well as forestry and park professionals. Town forest advocates tried to draw local and state planners into their fold, reasoning sensibly that both planning and forestry were long-term endeavors and should be pursued together. National journals, *City Planning* and *American City*, also carried the banner. On a more conceptual plane, Benton Mackaye (see fig. 8.1) formed alliances with the Regional Planning Association of America and with a coalition of New England planners and conservationists. An assortment of inventive schemes for uniting forest and town were an outcome, and many of his ideas were revealed in articles for journals such as *The Survey*, *The Nation*, and *The New Republic*. Public recognition that forests were an important part of community structure steadfastly increased.

On another front, decline in the momentum of New England's town forest movement, not a uniform event among the region's states, coincided with the appearance of laws enabling communities to establish conservation areas. Local commissions were permitted to acquire and use undeveloped land to conserve natural resources, balance relentless urban sprawl, and monitor society's effects on ecosystems. These included wetlands, watersheds, or simple open land, as well as forests. Although woodlands continued to epitomize a primeval environment, the accent shifted slightly, and a full spectrum of ecological settings were construed to symbolize pure wilderness. Public stewardship of community lands thus grew to encompass concern for the fragile balance of ecology in a world of unyielding change. One can trace the recurring cultivation of environmental stewardship, in all of its patterns, throughout New England's

Fig. 8.1. Benton Mackaye on Hunting Hill in Shirley, Massachusetts, in 1937. Ever the theorist, Mackaye fused forestry, conservation, wilderness advocacy, and planning into a lifelong quest for habitable communities. From his "Sky Parlor," a tiny office in the Shirley Center cottage he called home, Mackaye often mused about boyhood roamings around nearby spots such as Hunting Hill, Mulpus Brook, and the Squannacook River. Dartmouth College Archives

history of communal woodlands. A context for the origins of conservation areas, in many ways evolutionary progeny of the separate types of local forests, thus becomes apparent. Efforts by theorists such as Mackaye to fuse planning and conservation into a broader quest for habitable communities are a vital part of that same context. Whether expressed as wilderness, as Mackaye often did, or simply as ecological preserves, conservation areas symbolize the need for achieving a satisfactory balance between humanity and our environment.

Planning for Communal Forests

With public appeals already aimed at many constituencies, disciples of the town forest movement began somewhat sporadic efforts to enlist the support of planning bodies charged with civic improvement. Any measure of success at wooing these groups is difficult to calculate because local interest in forestry was sufficiently strong to propel the movement by itself. However, unions between forestry and planning did occur and helped to create a foundation on which later coalitions between conservation and planning were forged. In 1917, spurred by success of John Nolen's creative efforts, the planning committee for Walpole, Massachusetts, assembled a comprehensive book, *Town Planning for Small Communities*, and devoted a lengthy chapter to town forests. Much of it repeated material already in print: an examination of European community woodlands, James W. Toumey's summaries of town and school forests, MFA contests, and Walpole's prototype. It also included a substantial and previously unpublished bibliography by Raphael Zon, by then chief of forest investigations for the U.S. Forest Service. More important, it established a bridge between municipal forestry and planning at an early date in the town forest movement and trumpeted forests as an essential part of any complete town plan.[1]

Planning groups represented a union of individuals who, presumably, were motivated by a variety of civic interests. Advantages of addressing them as a single body were obvious, and articles that previously had tried to sway only specific groups were addressed more broadly without changing content. Following the American Tree Association's printing of Harris Reynolds's short book about town forests, Charles Lathrop Pack leveled an appeal at planners in "The Community Forest and the Community Plan," which appeared in *City Planning* in 1926. Pack construed planning's primary goal to be that of directing community growth and development in ways that maximized social, economic, aesthetic, and recreational advantages. With no difficulty at all, he then explained how various benefits from town forests fit neatly into those categories. Apart from their economic value, Pack emphasized woodland recreation and education as opportunities that deserved earnest consideration.[2]

For his part, Harris Reynolds occasionally noted contributions made by City Beautiful in educating municipal bodies about planning as a means for controlling future development. He produced a more focused argument, however, in a 1939 article published in *A Planning Forum*, the journal of the Massachusetts State Planning Board. In addition to his tried-and-true summary of European origins and multiple blessings of town forests, he pointed out that both planners and foresters dared to look many decades into the future and could assist each other by combining their vision. Urging cooperation between both professions, he specifically asked planning boards to help identify town lands suited primarily for production of trees and the by-products of forest management. Less than two years later, the state planning board disclosed its findings from a Park, Parkway, and Recreational Area Study conducted with assistance from the National Park Service. The report stressed recreational projects that did not require large public expense and recommended greater numbers of town forests that produced an economic return.[3]

Both Reynolds and Nelson Brown can, to a limited degree, be credited with pondering fairly imaginative uses of local woodlands within the realm of planning. Reynolds visualized town forest plantations along the Lincoln Highway, America's transcontinental route between New York and San Francisco, begun in 1914 and later designated Route 30. A desire for picturesque effect, appropriately symbolic for Lincoln, who spent much of his boyhood in the woods, dominated Reynolds's conception. Unfortunately, he failed to expand an implicit theme that public borders along highways prevented uncontrolled and unsightly development and helped to confine transportation corridors. Still, his 1916 proposal, published in *American Forestry*, preceded similar but more sophisticated plans for townless highways from Benton Mackaye and Lewis Mumford by more than a decade. Brown, too, hinted that policies of public land acquisition could stabilize community growth and improve village appearance, and he suggested that forests placed at outskirts of towns would help to control haphazard urban development.[4]

Toward a Comprehensive Ordering: The Contribution of Benton Mackaye

Exploration of relationships between forest and community tapped a more philosophical vein after formation of the Regional Planning Association of America (RPAA), established in New York City during the spring of 1923. Among its progressive founding members were Mumford and Mackaye; architects Clarence Stein and Henry Wright; economist Stuart Chase; real estate developer and philanthropist Alexander Bing; Charles Harris Whitaker, editor of the *Journal of the American Institute of Architects*; and Robert Bruere, an editorial staff member of *Survey*. The RPAA was born in counterpoint to England's International Garden Cities and Town Planning Federation, inspired by Ebenezer Howard's book *Garden Cities of Tomorrow*, and experiments for garden cities at Letchworth and Welwyn. In fact, Stein initially chose the name "The Garden City and Regional Planning Association of America," but the group instead settled on an informal affiliation with their English colleagues.

Throughout its existence, the RPAA remained a loosely organized federation, and membership alternately swelled or shrank around a core of fewer than a dozen persons, including Stein, Mumford, and Mackaye. Despite such fluctuations, it sponsored frequent conferences at Hudson Guild Farm near Netcong, New Jersey, where lengthy invitation rosters typically included men and women, European as well as American, who were prominent in their fields (fig. 8.2). At one time or another, Howard, Scottish planner Sir Patrick Geddes, and English planner/architect Sir Raymond Unwin all attended. Housing specialists Gertrude Stein (not related to Clarence Stein) and Catherine Bauer also were frequent participants. Its charter papers reveal the extent to which England's garden cities influenced RPAA members, and a preamble drafted by Clarence Stein explained their mission:

Fig. 8.2. Benton Mackaye (*right*) and Lewis Mumford at the Hudson Guild Farm near Netcong, New Jersey, 1923. With Mumford and architects Clarence Stein and Henry Wright, Mackaye helped to found the Regional Planning Association of America. Modeled after England's International Garden Cities and Town Planning Federation, the RPAA hosted frequent gatherings at Hudson Guild, inviting prominent guests such as Ebenezer Howard, Sir Patrick Geddes, and Sir Raymond Unwin. Upon Mackaye's invitation, Harris Reynolds also became a member. Cornell University Archives

A regional plan calls for new population centers, where natural resources will be preserved for the community, where industry may be conducted efficiently, and where an adequate equipment of houses, gardens, and recreation grounds will ensure a healthy and stimulating environment. The foundation of these communities will not only benefit the immediate population, it will also relieve the depressions and congestion of the present centers.[5]

Stein envisioned comprehensive, enlightened planning to promote common welfare, and he believed that private initiative was incapable of improving conditions in American cities. As had Howard's detailed blueprint for garden cities, RPAA members based their objectives on the principle that sales of land and provision of workers' homes should be removed from the domain of speculative profits. Convinced that proper housing was a public necessity, on a par with education, they argued that government subsidy was required if private capital did not become available at minimum interest rates. Finally, the group

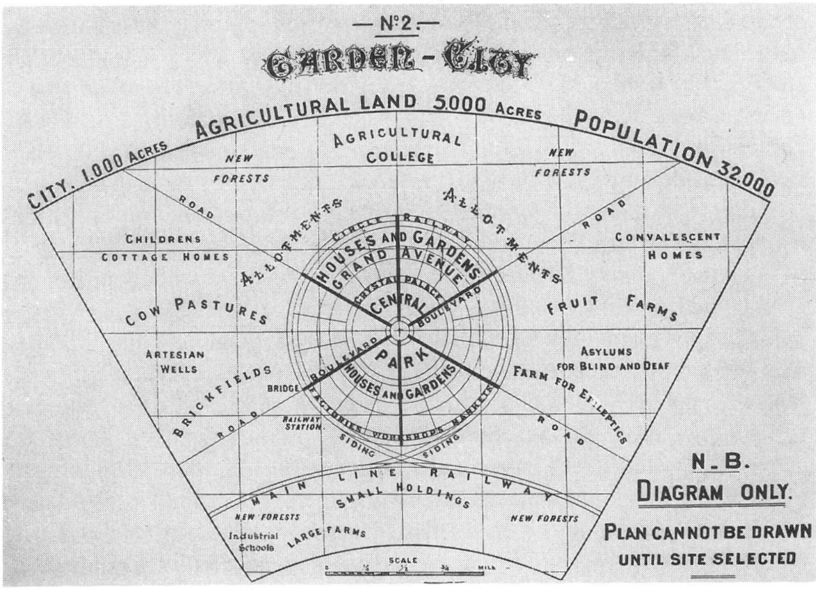

Fig. 8.3. Plan for a Garden City from Ebenezer Howard's *Garden Cities of Tomorrow*, published in 1898 (under a different title) and 1902. Howard's scheme sought planned dispersal of industries and people into communities of limited population. Surrounding fields and forest plantations ensured a rural setting, and the city owned all lands as a cooperative venture, thus eliminating speculation. Mackaye observed that American forests were uniquely suited to fulfill a similar function and developed this theme in many of his regional plans.

vowed to assist in forming corporations designed to finance and control land needed to implement their ideas.[6]

Today, almost a century after Howard's book appeared in print, many of his concepts have been misapplied or borrowed out of context, and terms such as "garden city" and "green belt" have succumbed to an American usage that does not always correspond to Howard's intentions. His ideas are key if one is to understand Mackaye's contribution to urban and regional planning and its connection to New England's communal forests. The essence of Howard's scheme lay in planned dispersal of industries and people into communities of limited population representing various cross-sections of society (fig. 8.3). Houses were placed within open, gardenlike settings, and parks or parkways, comparable to America's well-known green-belts, helped to divide wards or neighborhoods. Community infrastructure—roads, zoning, densities, building design, and landscaping—was preplanned. All land was held in a quasi-public capacity, and residents secured rights to use property through leases. Individual enterprise was permitted, but cooperative ventures also were encouraged. Howard's city was completely surrounded by an expansive area of open land reserved permanently for agriculture, thus securing a market for farmers and a countrylike setting for residents. Certain quadrants of farmland were set aside as forest plantations. Financing was secured through incremental increases in land values, which accrued to the entire community rather than to individual investors.[7]

Although the RPAA never produced its own garden city, individual members applied Howard's concepts in creative ways. It was Mackaye, with his intuitive appreciation for New England's native woodlands and quiet villages, who devised a uniquely American model for urban planning. He theorized that forests, if publicly controlled, could be employed much as Howard had used agricultural land to enclose communities and limit their size. Watersheds and geographic features such as mountain ranges or river bottoms, all part of regional ecosystems, confined urban growth. Recreational trail systems linked valleys to ridges and joined civilization to nature in a carefully balanced symmetry. Transportation corridors from one region to another were bordered by woodlands and were physically separated from towns. Development along their paths was restricted, thus preventing "wayside fungus" or "motor slums," the latter a term Mackaye borrowed from naturalist Walter Prichard Eaton. These elements represented only a small part of the city, regional, and continental plans that he expanded ceaselessly throughout his lifetime. With a writing style almost as unrestrained as his thinking, Mackaye scattered many of his ideas throughout dozens of published articles. However, his 1928 book, *The New Exploration: A Philosophy of Regional Planning*, assembled many under a single cover.[8]

Although Mackaye's involvement with RPAA members and guests helped to shape his ideas, that period in his career was really one of synthesis for themes he had already begun to articulate. For Mackaye, community and forest had never been very far apart, as his earlier writing confirms, and what remained for Stein, Mumford, and Geddes (the latter especially influential and responsible for introducing Mackaye to "geotechnics") was an opportunity to channel his intellectual energy. In 1918, he had switched from the forest service to the labor department, a move precipitated by concern for large-scale economic and human problems connected to migratory clear-cutting by logging companies. Several years before, he had been sent to the barren, cut-over timberlands of northern Minnesota and Wisconsin, where his assignment called for a plan to colonize the region. There he began to explore methods for shaping new communities from empty lands. In 1916, with midwestern lessons in sharp focus, he traveled to Pacific Northwest forests, where he tried to encourage a more satisfactory relationship between the timber industry and the transient shack towns struggling in its wake. Through application of sustained-yield management, attainable with public control of forests, Mackaye sought lumberjack utopias where camps could develop into complete, well-balanced, and self-sustaining communities.[9]

At the Department of Labor, Mackaye began to put his ideas into written form, and a flurry of articles expounding themes of forest use and community architecture followed. Several focused on opportunities for peacetime employment after World War I. In a 1919 article, "Employment and Natural Resources," he outlined an experimental forest village located in the Wild River watershed of the White Mountain National Forest, a plan somewhat akin to Howard's garden city but on a much smaller scale in terms of population and community structure. His proposal anticipated an area of thirty thousand contiguous acres and a permanent, planned village with limited boundaries to accommodate a minimum population of two hundred people. Subsistence homesteads were assigned to nearby agricultural lands, probably outside village perimeters. Employment for at least forty men was treated as a corporate venture, and jobs were divided between forest, sawmill, and a plant for woodenware products. Cordwood supplies were a driving force behind the scheme, but Mackaye observed that other markets existed: spruce and poplar for pulpwood, white birch for bobbins used in textile mills, and hardwoods for other turnery products.[10]

In that same article, Mackaye advanced a program for new agricultural communities in America and borrowed Canadian plans devised by Thomas Adams, then a consultant to Ottawa's Commission of Conservation and former manager of Letchworth. Adams's blueprint (fig. 8.4) offered four designs for planning quarter-sections of townships, land units equivalent to the mile-square divisions that blanket western and midwestern America and date from its public land survey. Conceivably, Mackaye had learned of Howard's plan for a garden city while working on Forest Service assignments in the Midwest. If so, this knowledge clearly unfolded during his stay with the labor department, for he noted that Adams had been markedly influenced by that English movement. Pared to its kernel, the unique Canadian plan was really an attempt to fit Howard's round city into a square grid. For efficiency's sake, Adams placed roads leading diagonally to community centers and aligned farm plots in orderly geometric patterns. Leftover, mostly triangular plots were assigned as forests. Mackaye added the suggestion that these wooded areas be designated as community forests with sustained yield management.[11]

In 1920, not long after resigning from government service, Mackaye returned to his home in Shirley Center, where reflections on primeval forests, villages, trail systems, watersheds, and humanistic planning blossomed into his best-known regional plan, the proposal for an Appalachian Trail. While at the labor department, Mackaye became acquainted with Charles Whitaker,

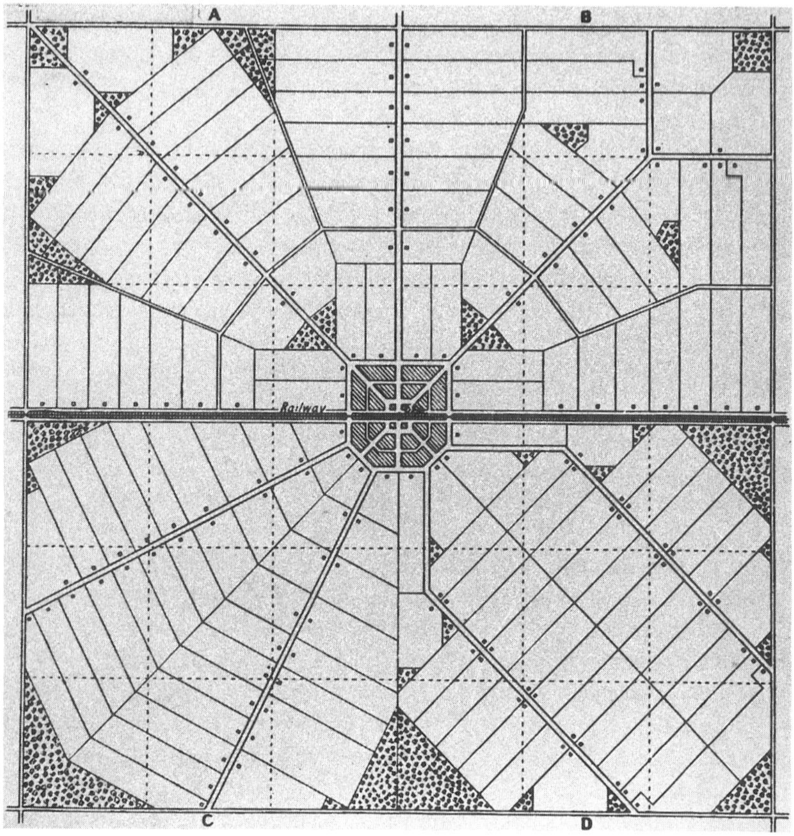

Fig. 8.4. Plans for Canadian townships conceived by Thomas Adams and borrowed by Benton Mackaye in the latter's *Employment and Natural Resources*, published in 1919 by the U.S. Department of Labor. Adams attempted to fit Ebenezer Howard's round city into a square grid, and Mackaye adopted the idea as suitable for new agricultural communities in midwestern states. There, cut-over timberlands had been abandoned by a timber industry pursuing the more lucrative forests of the Pacific Northwest. Although he did not create the plan, Mackaye did suggest that triangular remainder plots become community forests. Dartmouth College Archives

who, after listening to Mackaye's ideas, agreed to publish the plan in his journal; he also introduced Mackaye to Stein, chairman of the American Institute of Architects Committee on Community Planning. Stein furnished a preface to Mackaye's article, which appeared in 1921, and the two began a very close, lifelong friendship.

Apart from the significance it has acquired as a shining piece of Americana, the Appalachian Trail project represented a turning point in Mackaye's life. His vision of a long-distance trail fired the imagination of many supporters, who helped pull him from a somewhat murky period during which a simmering of ideas had, as yet, produced little certainty of direction. Almost spontaneously, the interdependence of wilderness and civilization in creating habitable human environments surfaced to become a theme that would dominate his later writing. Support from RPAA members also was key. Mumford, in his inaugural 1923 report from the association's program committee, regarded the Appalachian Trail proposal as the organization's flagship and considered it an ideal opportunity for public education about regional planning. However, oversight for the project soon fell to the Appalachian Trail Conference and volunteer clubs, and the RPAA had almost no direct involvement.[12]

Mackaye's scheme was far more complex than the wilderness footpath that now links Springer Mountain in Georgia to Katahdin in Maine. He selected the Appalachian region as a belt of underdeveloped lands, large enough to allow redistribution of urban populations and, concurrently, offer opportunities for a productive balance between recreation and employment—in other words, "a new deal." From its skyline, a strategic battle line to be reinforced against encroaching urbanization, one could observe a large part of the nation at work and play. To Mackaye, the chance for more efficient use of spare time represented a solution to the "problem of living," as he put it, and planned communities were a critical step to that end: "The project is one for a series of recreational communities throughout the Appalachian chain of mountains from New England to Georgia, these to be connected by a walking trail. Its purpose is to establish a base for a more extensive and systematic development of outdoor community life. It is a project in housing and community architecture."[13]

Mackaye envisioned groups of rustic, communal settlements, beginning with primitive shelter camps that would expand naturally into community camps. This second category was designed to include separate domiciles where members could focus on every possible type of nonindustrial activity, such as recuperation, foot travel, and scientific study. Food and farm camps, located either adjacent to community camps or in nearby valleys, were proposed as experimental subsistence units modeled after places such as Hudson Guild Farm. Finally, forest camps would eventually supplement farm centers with fuelwood and lumber acquired through communal timber rights sold by the federal government from its national forests.[14]

Although it is true that most of Mackaye's plans were, at the very least, conceived on a regional scale, many were constructed with small, livable communities in which surrounding forests played an essential role. His emphasis on relationships between forests and cities is evident in "A New England Recreation Plan," which he presented at the Third New England Forestry Congress held in Hartford in 1929. By then, both writing and thought had matured to full style, and he contrasted cities, "the end product and cross-section of a mechanical civilization" and "a marvelous product of the evolution of mankind," with forests, "the root of man's society." Convinced that in balanced societies, forest and city must grow side by side, Mackaye argued that wilderness inroads should mark the beginning point for any recreation plan. Thoreau-like observation, he believed, provided the impulse for understanding nature's schemes. Without this stimulus, forests were "a country without a prophet," and the impulse alone, without its primal setting, "a prophet without a country."

> The forest as an entity must be woven within the matrix of our territorial development. Our early settlers first planted civilization by inroads of population through the forest; we today, in order to restore civilization, must develop forest inroads between our population centers.
>
> They should form a series of embankments, or "levees," for controlling the flood of mechanistic civilization which is pouring forth, in the shape of temporal chaotic structures, along the motor ways outward from our metropolitan centers. This "flood" is getting out of hand.[15]

In its substance, imagery, and prose, "A New England Recreation Plan" represents Mackaye at full philosophical stride. No less Mackaye-like, however, is its absence of any specific, tangible method for implementation. Although intensely sincere in devotion to humanistic ideals, Mackaye made contributions that settled more in theory than in practice. Even on projects such as the Appalachian Trail, one that stayed close to him during the remainder of his life, he did little organizing, detailed mapping, or trail building. Similarly, although village and woodland in New England formed the seed that sprouted to creative thought about habitable environments, he never became an active participant in the movement for town forests.

On the one hand, Mackaye was probably unsuited for the day-to-day drudgery associated with encouraging local committees to acquire, organize, plant, and manage their forests. Credit for those tasks, and loads of it, belongs to workhorses such as Harris Reynolds. To be completely fair, however, more subtle factors also may have been at work. In 1920, reforestation was a primary impetus behind town forests, and by that time Mackaye had already moved away from forestry and was considering regional rather than woodlot planning. His chief venture with town forests occurred close to home when he joined a committee in Shirley appointed to study the acquisition of a series of wooded parcels along the Squannacook River. The idea was proposed by Harris Reynolds as a means for extending the Willard Brook Reservation in Ashby, and late in September 1931, Mackaye guided Reynolds and William Wharton on a scenic canoe trip along the river. However, the project never advanced beyond mapping stages.[16]

The RPAA suffered from a corresponding lack of visible achievements to serve as touchstones, and it drifted into dormancy after 1933, with a brief resurrection in 1948. Its last major conference was held at the University of Virginia in 1931. In attendance was Franklin Roosevelt, who, as governor of New York, had been influenced by the association's programs for housing, planning, and employment. In fact, some of the group's ideas found their way into Roosevelt's New Deal. Stein, in a 1948 monograph about the organization, revealed at least a partial explanation for their meager record: "It sought to attract no one to it by making expedient compromises and concessions. On the contrary, it devoted itself to long term programs and far-sighted plans, believing that without such programs and goals even the most "practical" plans would fall short of their very limited objectives." [17]

Absence of collective success did not prevent a sharing of ideas that led to individual achievements by RPAA members and produced results far outside the group's immediate circle. Stein designed a great many innovative communities, including Radburn, New Jersey, and Sunnyside Gardens in New York City, and compiled a book, *Toward New Towns for America*, which he published in 1957. Henry Wright pioneered a statewide plan that eventually was put forward by the New York State Housing and Regional Planning Commission, chaired by Stein. For his part, Mumford generated an incredibly prolific flow of writing about architecture and urban planning.

Mackaye experimented with a number of government positions across the country, including a short stint with the Tennessee Valley Authority. While in Knoxville in 1934, he helped Robert Marshall, Harvey Broome, and Bernard Frank draft a charter for the Wilderness Society and joined those three, along with Aldo Leopold, Robert Sterling Yard, Harold Anderson, and Ernest Oberholtzer, as co-founder. Between 1945 and 1950, he served as the society's president and, after 1950, as its lifetime honorary president. Always, Mackaye returned to "The Empire," his cottage in Shirley Center, where he presided, often in near poverty, over a small office christened the "Sky Parlor." There, nourished by a pervading harmony between village and natural environments, he gave free rein to the inner philosopher.

Conserving Wildland Patches

While RPAA members were investigating the failings of American cities, a cadre of regional planners were making similar forays in Massachusetts. Boston architect William Roger Greeley, planner Edward T. Hartman, and journalist-economist William Leavitt Stoddard charted headings for several organizations, among them the Massachusetts Federation of Planning Boards and the New England Regional Planning Committee. They were joined by ornithologist and natural resource activist Laurence B. Fletcher, Reynolds and

Wharton from the MFPA, and other key figures, such as Allen Chamberlain. Together, these groups and individuals managed to form a long-lasting alliance between planning and conservation in the Bay State. Assisted by Mackaye's introductions, they also forged links with the RPAA, a connection that helped to pull Mackaye toward projects in familiar territory.

As did RPAA members, Greeley and company emphasized orderly regional development and concentrated on the interdependence of broad issues such as commerce, industry, transportation, and housing. Regional focus notwithstanding, they also developed informal coalitions with local planning boards in an effort to build constituencies, much as the MFPA created branch organizations to promote local forestry. As a result, ideas sifted at regional levels more than occasionally drifted into town programs, sometimes in a slightly altered form. In contrast to the RPAA, however, New England's regional planners did not toy with utopian schemes for new cities but instead embraced more pragmatic objectives. In matters of conservation and outdoor recreation, a distinct emphasis on preserving open spaces emerged, a trend influenced considerably by the Trustees of Reservations and their many successful acquisitions.

Open spaces were just that, encompassing marshes, waterways, beaches, and scenic spots as well as parks and forests. A number of plans, particularly those by Mackaye, tied selection of public lands to geographic features or placed them in critical locations where they became an integral part of regional fabric. As conceived, they served multiple capacities, providing sites for recreation as well as influencing direction, extent, density, and, hopefully, the quality of urban growth. In practice, vision yielded to practicality, and wild lands were often captured from looming development rather than planned in anticipation of orderly progress. More often than not, they surfaced as scattered, randomly chosen fragments rather than as parts of a cohesive whole. Both town and regional planning were correspondingly vulnerable to these circumstances, and open spaces evolved as a means for balancing, rather than shaping, irrepressible growth.

The Massachusetts campaign to preserve open spaces developed as an outgrowth of Calvin Coolidge's National Conference on Outdoor Recreation. In 1925, the Trustees of Reservations organized a state symposium to establish cooperation among private and government bodies concerned about public lands, and a committee was appointed to identify and map existing areas. In 1928, the movement received formal stature when Governor Alvan T. Fuller appointed the Committee on Needs and Uses of Open Spaces. Chaired by Charles Bird, Jr., the delegation was staffed by Greeley, Fletcher, Chamberlain, Reynolds, Arthur Comey from the Boston Society of Landscape Architects, Conservation Commissioner William Bazeley, Joseph Lee from the Playground and Recreation Association of America, and a number of other representatives from key organizations. Mackaye was retained as a consultant and quickly set about identifying areas potentially valuable for public use. Although Chamberlain and Mackaye were old friends, it was Greeley who became intrigued by Mackaye's ideas after inviting him to speak at the Massachusetts Federation of Planning Boards' annual conference in 1927. The two subsequently collaborated with Reynolds on a plan to prevent "road towns" by enacting zoning legislation for state highways, and their work was published by the MFA in January 1928.[18]

Mackaye's work for Bird's committee made rapid headway, and a progress report was tendered in April 1928. He began by distinguishing two categories of open spaces: industrial and cultural. Forests grown for their timber or for protecting watersheds embodied the former; lands utilized primarily for recreation, the latter. State and town forests, he observed, exemplified both types. From there, he moved to the crux of his proposal, a system of cultural areas shaped as belts or "open ways" designed to check urban environments and offer escapes from civilization. Those in a primary class were sufficiently remote from population centers to constitute wild lands. They captured outstanding topographic features—mountain ridges, escarpments, canyons, and river valleys—or occupied unsettled lands such as pine barrens, glacial sand plains, and great swamps. Their purpose was to "form a series of breaks in the metropolitan deluge" and to separate "the flood waters of metropolitanism into separate basins . . . to avert their complete and total confluence." Secondary open ways were narrow belts along highways contrived to allow passage of motor traffic but not "road towns." These were crossed or "flanked" by primary ways (fig. 8.5).[19]

Mackaye regarded his plan as one for developing a psychological resource, namely, primeval environments. In it are beginnings of wilderness advocacy as a component of regional and urban planning in New England. In a separate writing for Bird's committee, he suggested more prosaic classifications for open spaces: public forests, water supply areas, and recreational spots. He also urged specific designation of public wilderness areas, or "conservation spaces," and devoted considerable energy to defining this category: "It may or may not be forested: the objective is to provide as far as possible a primeval environment, and this usually but not necessarily goes with a greater or less amount of forest growth. Water falls, gorges, and other natural features of marked scenic value are essentially wilderness features."[20]

Again, he identified two types: general and special. Large areas such as mountain slopes symbolized the first, and smaller outdoor laboratories for study of plant and animal life or geographic features were assigned to the second. In addition, he recommended extended vistas of surrounding country be included as part of these reserves. Mackaye credited Aldo Leopold as the inspired champion of wilderness areas, and their correspondence from two years before indicates that Leopold was pursuing a point of view rather than a piece

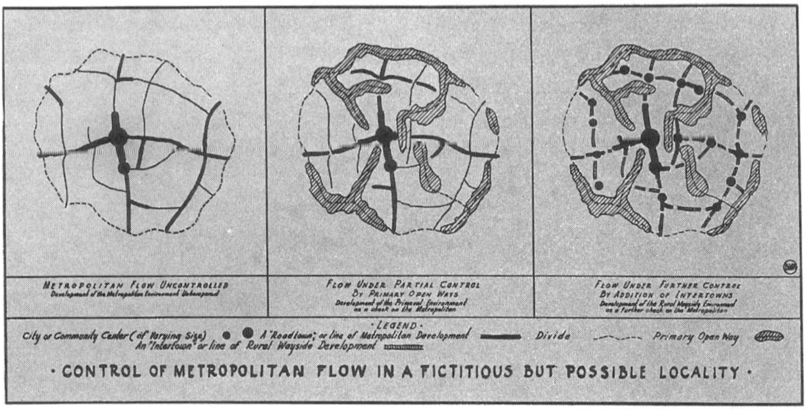

Fig. 8.5. Benton Mackaye's plan for "Control of Metropolitan Flow in a Fictitious but Possible Locality," as it appeared in his book, *The New Exploration: A Philosophy of Regional Planning*, published in 1928. Mackaye enlarged Howard's garden city plan to regional dimensions and offered the idea to the Committee on Needs and Uses of Open Spaces, established in Massachusetts that same year. The committee rejected Mackaye's scheme as impractical, largely because it required control of commercially valuable land along important transportation corridors. Dartmouth College Archives

of land. Mackaye grasped the significance inherent in this distinction and began to incorporate a wilderness philosophy into his plans. Ultimately, this wilderness point of view found its way into town planning and community structure in the shape of conservation areas.[21]

Bird's open spaces committee issued its initial report early in May 1929 and called for a statewide plan locating public lands according to some concrete (but unspecified) policy. This plan was to be preceded by a comprehensive survey identifying potential sites. Unfortunately, members hedged on Mackaye's imaginative solutions for controlling growth in population centers; nor was any elaboration given to his proposal for wilderness sanctuaries. Instead, emphasis was placed on more practical state-owned beaches, forests, and parks, and a new Division of Parks in the Department of Conservation was endorsed. Nevertheless, the committee did classify separate areas for wildlife conservation, wilderness camping, and foot trails. Mackaye's influence was also visible in their support for a "Bay Circuit," a series of connected reservations forming a semicircle around greater Boston and threaded together by a parkway. Years later, Mackaye was enlisted by the Trustees of Reservations to draft a plan for this circuit, and his report surfaced in published form in 1937.[22]

Although the committee as a body, perhaps fearing lukewarm public response, did not readily espouse Mackaye's package of novel ideas, a number of individual members were unwilling to dismiss them. In July 1929, he gathered the unabridged substance of his committee proposal into an article that appeared in *Landscape Architecture* under the title "Wilderness Ways." To illustrate his plan, Mackaye selected a belt of hill and mountain country extending south from Pack Monadnock in New Hampshire over the Watatic and Wachusett Mountains in central Massachusetts (fig. 8.6). Lesser peaks, among them Mt. Hunger and Bean Porridge Hill, added regional spice if not sustenance. Years later, he alluded to this territory fondly as "Mon-ta-chusett," an amalgamative word coined by Samuel Hopely, editor of the *Fitchburg Sentinel*. Mackaye used an existing foot path, the Sprague Trail from Wachusett to Watatic, as the spine of his wilderness way in Massachusetts, although a second passage, the Wapack Trail from Watatic to Monadnock, augmented this plan.

Mindful of the difficulty in translating abstract ideology into workable projects, he began with a struggle to define wilderness spaces. The resulting oration, awkward at times, symbolized a progressive statement of wilderness attitudes in very human terms. Scions of these themes ultimately helped to popularize local conservation areas. Much more than playgrounds, wilderness spaces personified "grounds for living" where primeval environments, actual or potential, elevated mere human existence to a higher level of appreciation for life itself. The metaphysical search for knowledge, as well as the more mundane inclination to just travel, lumped together under the term *wanderlust*, were satisfied by networks of trails leading from "somewhere to somewhere."

In more everyday language, Mackaye proposed state ownership of open land together with public acquisition or control of highway borders where his mile-wide zone intersected three routes leading to Boston. Anticipating negative reaction and possibly primed by his experience with the open-spaces committee, he justified interference with business interests for sake of recreational and cultural development. The two, industry and wilderness recreation, were incompatible and needed to be confined to their own spheres. In terms as relevant today as then, he confronted the battle between private and public concerns by intimating that future generations had a stake in the environment: "But 'who' are the people? Are they 'future' people as well as present people? And what do they 'ultimately' want?"[23]

If the results of his work for the open-spaces committee were disappointing to Mackaye, they did not adversely affect his friendships with New England's planners and conservationists. With Greeley and Stoddard, during the summer of 1930, he proposed a series of books and primers about regional planning and engineering to be published by the RPAA. That fall, both were invited to Hudson Guild Farm along with Fletcher, Reynolds, Hartman, and Walter Prichard Eaton. A year later, Greeley, Hartman, and Reynolds were elected RPAA members. At about the same time, Mackaye helped Reynolds with another

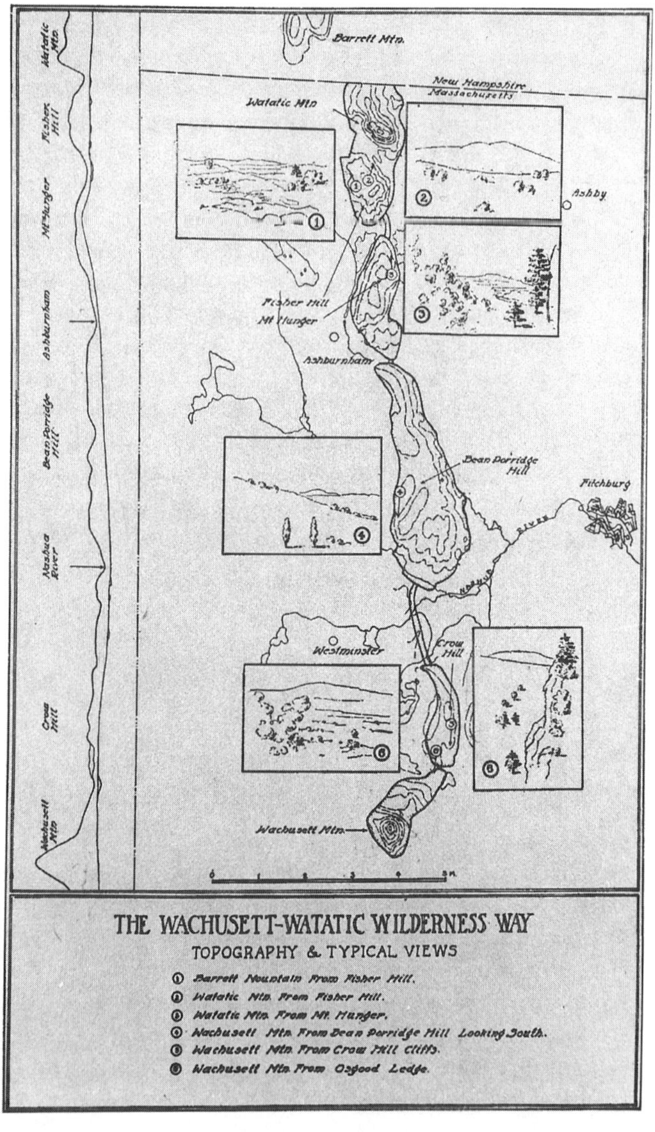

Fig. 8.6. Benton Mackaye's plan for a wilderness path between Mt. Monadnock in southwestern New Hampshire and Wachusett Mountain in Massachusetts, from an article he penned for *Landscape Architecture* in 1929, titled "Wilderness Ways." Although recreational forest uses were an important part of his plans, Mackaye focused primarily on the need for balance between natural and mechanical civilizations as the key to habitable communities. Footpaths through the forest primeval were simply one part of a much larger scheme, a means to a distant end. Dartmouth College Archives

MFA bulletin, this one a plan to save forests along highways, and penned an article, "Highway Approaches to Boston," for the Trustees of Reservations. At Mackaye's request, Reynolds and Hartman, as well as Stein, Mumford, and Henry Wright, all became charter members of the Wilderness Society in 1935.[24]

Mackaye's ability to capture the esteem of these influential men was a tribute to the fiber of his intellect and personality. Mumford called Mackaye a "Yankee Pan" and considered him a spiritual descendant of Thoreau. Although a spartan existence in Shirley Center may have offered a pointed contrast to the means of many who sought his counsel, he was well respected, and his opinions were clearly valued. More important, through his voluminous letters, plans, and published articles one can trace a progression of ideas, enlightened for the period, that had begun to circulate and that ultimately took root on community lands. In 1932, Reynolds consulted Mackaye about enlarging the MFA's focus to include parks as well as forests shortly before the association changed its name. Mackaye suggested the change be justified as a way of dealing with two classes of people affected by economic depression: "One class is strictly up against it—and looking for the next sandwich; the other class is pinched—and looking for ways of economizing and indulging in less expensive luxuries. Question:—Can a *forest* program be devised to help meet the first problem, and a *park* program to meet the second?"[25]

Reynolds scrawled a quick note back, inviting Mackaye to attend a forestry and planning conference where unemployment relief and park policies were to be addressed and town forests in Petersham, Athol, and Winchendon inspected. His remarks set a tone for the period: "Am short on office assistance, cash, spirit and ideas—Am think[ing] of moral bankruptcy—Where can I get a quart of pre-war?"[26] Five years later, when the MFPA appointed a committee to study park policy, Reynolds recruited Mackaye's support in assembling information about town lands suitable for public ownership around the state. Mackaye's circle of good friends extended to William Wharton, who served as MFPA president between 1936 and 1959. When Mackaye's presidency at the Wilderness Society overlapped, the pair held "Big Two" meetings at "The Empire" or at Wharton's home in neighboring Groton.[27]

For his part, Mackaye thrived on informal exchange of ideas among a wide range of loosely assembled groups. One such society, the Woodticks, is noteworthy because its reunions on Mt. Monadnock included New England's most influential policymakers in forestry and conservation (fig. 8.7). The group's timely gatherings between 1937 and the early 1950s also bridged an important period when Mackaye was most active in Wilderness Society affairs. These "Notorious Aggregations," as Mackaye comically portrayed their meetings, included Allen Chamberlain, Reynolds, Wharton, and Elmer D. Fletcher (Flooche), whose work with the U.S. Forest Service had been a driving force behind acquisition of the White Mountain National Forest. Others joined in,

Fig. 8.7. A gathering of the Woodticks at Wapack Lodge on Mt. Monadnock, May 25 and 26, 1946; (*left to right, front row*): Elmer Fletcher (Flooche), Benton Mackaye, John Foster, Harris Reynolds, and Karl Woodward; (*back row*): Austin Hawes, Kenneth Barraclough, William Wharton, and Larry Rathbun. Although loosely organized and somewhat whimsical in constitution, the group was allegiant to its yearly pilgrimages. Dartmouth College Archives

among them state foresters Harold O. Cook from Massachusetts, John Foster from New Hampshire, and Austin Hawes from Connecticut, as well as Yale Forestry School professor Herman H. Chapman. A number of New Hampshire foresters participated, including Philip Ayres and his successor, Larry Rathbun, from the SPNHF, New England Forestry Foundation forester Stanley Mesavage, Karl W. Woodward from the University of New Hampshire, extension forester Kenneth E. Barraclough, and Henry Ives Baldwin from Hillsboro. Lofty aspirations were most surely carried back down the mountain.[28]

As president of the Wilderness Society, Mackaye had ample opportunity to expand his already far-reaching sphere of influence. Still, he managed to keep New England's village environments in sharp focus, and his contributions foretold pivotal events for local conservation. Early in 1946, he dusted off his ideas for Bird's open-spaces committee and proposed Massachusetts legislation to stimulate community interest in spots of "aboriginal New England environment," particularly stream and ridge belts. His bill authorized the Trustees of Reservations to initiate projects by obtaining approval from the state's conservation commission and by subsequently acquiring title to selected lands. Custody of primitive areas would rest with the trustees, but costs of acquisition and management would be shared by the state. Mackaye delineated constraints against a host of intrusions, among them logging, grazing, farming, powerlines, and roads, but sanctioned unpaved footpaths opened and maintained by voluntary labor. Significantly, town organizations or clubs were encouraged to sponsor projects and to assume management duties after repayment of half of the acquisition costs, subject always to supervision by the trustees.[29]

Although nothing tangible materialized from this draft, Mackaye's fertile brain and spirited connections soon clicked again. In November 1947, he issued a memorandum to society members urging them to seek out and protect neighborhood patches of wildland. This missive was part of a continental effort by the society's council to build a wilderness area system encompassing lands of every size and type. Recalling his canoe voyages along the Squannacook River and plans for a wildland belt along its banks, he portrayed the operation as a little backyard job for every member. "If you as a 'lone wolf' would gather a crew and explore your bailiwick for a wildland area (some secret stream or marsh or ridge) and then *seek some local means* for getting it preserved, you would be taking part in a continent-wide campaign."[30]

Mackaye's wildland-patches letter was profound for its emphasis on wilderness areas as a distinct category of community lands. With the accent on ecology, these spaces were defined broadly to include every kind of wild landscape, not just forests. Its underscoring of the need for local means to preserve such areas also was laudable. In many ways, his letter was both similar in trait and comparable in importance to his proposal for an Appalachian Trail. Each synthesized ideas that already had begun to take shape but had not yet attained complete form, each was slightly vague beyond its core of an idea, each was bold in its vision for improving human environments, and each scattered seeds that bore fruit as unpredictable as plentiful. Combined with Reynolds's letters introducing forestry to the people, Mackaye's wilderness epistle completes a trilogy for town conservation begun in 1859, when Thoreau pondered forest parks and huckleberry patches in every town.

Replies from Wilderness Society members highlighted the importance of forming organizations capable of holding such lands in perpetuity. By 1949, Mackaye was headed in that direction, and in correspondence to Laurence Fletcher he asked whether legislation could help build paper museums around forests and swamps. In fairly awkward terms, he pictured an "artificial person" capable of protecting areas of primeval nature to the degree desired by ecologists. Copies were sent to Wharton and Greeley, and a more succinct memorandum was circulated to all three less than a week later. In it, Mackaye distinguished ecological from recreational wilderness areas, equating the former with yardsticks for measuring interactions of natural laws, a crucial step toward an understanding of conservation problems. Stimulating public appreciation of ecological values could be achieved "by visualizing the *biotic community* as a unit or element of the landscape, as distinguished from a mountain, or a river, or an artificially bounded area. Make the 'community' stick out (whether 'Old Field' or 'Forest Herb' or 'Northern Shrub' or 'Beech-Hemlock') as sharply as that there yonder lake."[31]

Mackaye's memorandum also considered political and legal methods for setting this policy into motion but did so in a somewhat timid manner. Public parks were handy spots for these landscape units, but he warned against confusion over scientific and recreational use and worried that distinctions could not always be guaranteed. In Massachusetts, the Trustees of Reservations represented a likely vehicle but were vulnerable to similar pressures. Private contracts, deeds, or wills offered a new approach, and Mackaye hinted that progress could be made along those avenues. However, he stopped barely short of considering that municipal or quasi-public bodies enabled by state law could own and protect swamps or meadows just as they did forests. Stimulus for that final ingredient did not ripen for seven more years, and it did not come from Mackaye. Still, this sequence of letters, memoranda, and draft legislation contained a nearly complete recipe for town conservation lands. Whether he depicted them as wilderness belts, biotic communities, aboriginal environments, wildland patches, or, as he had for the open-spaces committee in 1928, conservation spaces, Mackaye regarded them as ecological demonstration areas. He intuitively grasped their educational value for all matters of conservation and quite naturally sought community participation.

Without question, others were thinking and writing in like fashion during this period. The word *conservation* had by then become a familiar, all-inclusive term and had worked its way into the language of government policy at state as well as federal levels. A need for action at the community scene also had become apparent. In a 1950 issue of *Forest and Park News*, the MFPA urged every town to prepare a strategy for solving conservation problems and to apply sound remedies diligently. That same year, Wayland, Massachusetts, prepared a conservation land-use plan assembled by Allen Benjamin, briefly a field secretary for the MFPA. Groton compiled a similar document two years later through a collaborative effort by its town planner, the MFPA, and the New England Forestry Foundation. The appearance of local conservation commissions, the missing component, was simply a question of time.[32]

Mackaye's trail leading toward local conservation areas begins to fade at this point. He was destined, instead, to fill the role of prophet, and was probably no less content for it. The striking similarity between his vision of wildland patches and the approaching advent of town conservation spaces was no mere coincidence. Time and time again, with a knack for uncanny reliability, he blazed paths through uncharted territory and from there pointed the way to distant horizons. It happened with his survey of forests and watersheds in the White Mountains, a working plan for the Pennichuck Water Works in Nashua, themes for recreational use of forests, timber communities, regional and long-distance trail systems, wilderness belts, townless highways, and town conservation areas, to name but a few. All were creatively stitched together into a quilt of humanistic-environmental planning: community, regional, continental, and ultimately global.

A Secret Marsh

When scattered to the wind, seeds of thought sprout in unpredictable and sometimes unlikely places. So it was with the emergence of town conservation commissions in New England. In Ipswich, Massachusetts, town members sought a method for preserving an unspoiled marsh, and they turned to their state representative, John F. Dolan, for assistance. Dolan, a resident and former town clerk of Ipswich, did not know Mackaye, had not been a member of the Wilderness Society, and, although a member of the MFPA, was not an active one. However, he did sit on separate legislative committees for town affairs and conservation, and he was aware of a 1954 law that authorized towns to establish development and industrial commissions. With little fanfare or help from outside groups, Dolan borrowed the language from this enabling law, in many places word for word, and sponsored House Bill 2759.[33]

Dolan's bill enabled towns to create conservation commissions to safeguard natural resources and protect watersheds. Commissions were given responsibility for creating inventories of all open areas, specifically marshes, swamps, and other wetlands, and were encouraged to plan for their proper use as well as offer public education. Finally, commissions were authorized to accept and manage gifts of land, buildings, or money for their towns. The bill met little opposition and was enacted into law in March 1957. Ipswich citizens mustered support, organized a commission, and preserved their marsh.[34]

For several years, public awareness of the law remained spotty, and only a smattering of Massachusetts communities followed Ipswich's example. Currents of opinion suddenly shifted, and by 1964 more than two hundred communities had formed commissions. In part, this extraordinary turn of events was guided by a coalition of government and private groups throughout the state, notably the Department of Natural Resources, the MFPA, the Audubon Society, and the Massachusetts Federation of Planning Boards. Organizations with local chapters, such as the League of Women Voters, the Grange, and the Taxpayer's Association, also offered valuable assistance.[35]

Benefits of a strong local conservation movement to state environmental programs soon dawned on the Department of Natural Resources. They launched a survey early in 1960 to circulate information about the law and to establish communication between towns and state agencies. In April of that year, the department organized a conference in Cambridge to advance the theme "Solving Your Town's Natural Resource Problems." Delegates from all New England states attended, and the department produced a manual for local commissions. A year later, the Massachusetts Association of Conservation Commissions was formed by Stuart DeBard, an energetic member of the Massachusetts Federation of Planning Boards, and membership was offered to all commissions. In 1962, this newly cast organization collaborated with the MFPA to produce "A Guide for Town Forest Committees and Conservation Commissions" and recommended, among other things, that commissions be staffed with one or two members from town forest committees.[36]

Along another front, a series of legislative initiatives filled chinks in the enabling law and helped to insulate local enthusiasm against inclement opinion. A 1959 bill allowed towns to create a conservation fund to support their commissions. A year later, the state enacted its "Self-Help Conservation Program," designed, as its name implies, to assist communities in their efforts to address local environmental problems. The state's commissioner of natural resources was authorized to reimburse towns that had formed conservation commissions for up to half the cost of land acquisition and planning. This program was voluntary, and onus for seeking assistance rested on communities. A year later, commissions were given authority to purchase lands, including ability to acquire lesser estates such as easements, water or development rights, contractual obligations, and covenants.[37]

The movement for conservation commissions in Massachusetts swiftly spread beyond its borders, and by 1965 all New England states except Vermont had passed similar legislation: Rhode Island in 1960, Connecticut in 1961, New Hampshire in 1963, and Maine in 1965. The Rhode Island Audubon Society was responsible for migration of local commissions to that state, and Connecticut and New Hampshire enacted laws after their emissaries returned from the 1960 Cambridge conference. Initial headway in these states, measured by the number of commissions, was closely tied to the availability of financial incentives for towns and to technical support from state agencies.[38]

In Connecticut, author and conservationist William H. Whyte persuaded the state to furnish matching funds for commissions as part of a bond-supported initiative for local open spaces. This fledgling crusade also was nurtured by prompt formation of a state association of conservation commissions. Although New Hampshire's bill authorized the Department of Resources and Economic Development to adopt an assistance program, no funds were supplied. Nevertheless, money was made available through a private charitable trust, and awards were administered by the SPNHF. In both states, campaigns advanced steadily. Despite its headstart, Rhode Island's program faltered without a funding source and without coordination between towns and state agencies. Maine's legislation combined parks with conservation spaces and staffed commissions with elected, rather than appointed, officials. Their drive was slow to gain momentum until a 1969 amendment gave more autonomy to conservation commissions.[39]

No matter how influential proddings and handouts from government and private conservation groups may have been, credit for the movement's astonishing groundswell belonged to people and their towns. More Massachusetts communities formed conservation commissions in a brief half-decade than had created town forest committees during an era lasting a half-century. In Connecticut, the movement almost instantly surpassed that state's somewhat anemic town forest program. In other states, town commissions soon exercised jurisdiction over a wide variety of public lands, forests as well as wetlands and meadows. The movement's popularity smothered the remains of public interest in town forests, and woodlands left untended sometimes fell to the supervision of conservation commissions. Vermont finally passed enabling legislation in 1977, and its once vigorous municipal forest program was discontinued shortly after. By 1989, almost three thousand conservation spaces totaling more than 86,000 acres had been set aside in Massachusetts alone.[40]

Explanation for the movement's phenomenal surge can be found in New England's evolving conservation ethic and ever changing landscape. Conservation themes were not new to the region's towns, and one can observe the genesis of collective resource management in an assortment of seventeenth- and eighteenth-century practices—communal and public woodlots, for instance. More modern versions had been visible for almost a century in a progression of shapes, such as forest parks, town forests, watershed areas, open space reservations, and wildland patches. Twentieth-century planning furnished a glue to cement these and other pieces into a more comprehensible mosaic. Thus, when ideas about conservation commissions emerged, communities were thoroughly primed.

That alone, however, did not account for the movement's rapid and overwhelming acceptance. More significantly, commissions offered a ready outlet for people's concern about unfathomable change in their environments. Fast-paced technology, multiplying populations, creeping urbanization, and, imperceptibly, a shortening of time itself fostered a malaise of uncertainty. Control over incomprehensible forces seemed unattainable. Development, the physical embodiment of these evils, became a symptom and conservation the spiritual remedy. Unlike town forests, which demanded consistent management and patient vision, sanctuaries of wildland promised instant reward and gratification. Better yet, if public enthusiasm dwindled after its initial burst, lands acquired still remained relatively secure against development.

Several other comparisons to town forests warrant consideration. Both movements were immensely successful in the sense that each nourished grass-roots environmental awareness, and each was responsible for large endowments of land to the public domain. The value of these two elements in shaping habitable communities cannot be overstated, and their future worth, particularly that of publicly held land, will continue to increase. At the same time, both movements can be measured against more ambitious standards. Although town forests offered great potential for multiple use management and continue to do so, accomplishments in the areas of recreation and education must be tempered by a meager record for timber production. The goal of intense forestry management once displayed by European communities has, in most cases, remained elusive. Conservation commissions hoping to influence town planning on a broad scale, much as Mackaye's plans envisioned, not merely capture public-land fragments when opportunities arise, face a formidable task.[41]

. . .

Conservation areas administered by local conservation commissions represent a final category of New England's communal forests. Many writers have eloquently championed the cause of environmental ethics and have influenced thought in various tiers of our society. Few, however, have focused on the role of woodlands in shaping habitable communities as intently or with such humanism as did Benton Mackaye. Admittedly, his contributions to the origins of local conservation lands are only a small portion, albeit an important one, of a much larger context. The greatest part of that context can be found in separate classes of communal forests that have evolved in New England over three and one-half centuries. Recognition, and ultimately preservation, of these distinct forest types is crucial if we are to fully comprehend and utilize the timber lots and wildland patches currently owned by our towns.

Epilogue

Preserving the Forested Landscape of Community

The footpath through New England's history of town-owned forests has turned full circle. In a sense, communities have begun to reclaim expansive common lands given up long ago to individual ownership. Ironically, the random patterns of land division that characterized dispersal of seventeenth-century nucleated villages are being repeated, but the process is occurring in reverse. Where land splinters were once hewed from common domain and converted to severalty, public rights are now being reasserted through a similarly fragmented repertoire of local initiatives. Planning and zoning, conservation easements, purchase of development rights, and outright acquisition by organizations such as land trusts have created a complex and sometimes overlapping web of communal interests that have become part of town morphology. In some cases, too, state land-use laws give local governments an opportunity to participate in administrative decision making, adding a measure of oversight. All the while, woodland cover has steadily advanced over much of the region. Sustained by these trends, a course set almost a century ago by the movement for town forests has yielded prodigious return. Forested townscapes are now abundant (see, e.g., fig. 9.1). While character of use and relationship to community structure have evolved, as have methods of stewardship, an underlying bond remains the same: forests are a resource capable of substantial endowment to communities.

Struggles between public and private interests in land and its manner of use will continue to shape New England towns. The region's fertile history of town and forest has considerable present and future value in forums where current policy concerning land use is debated. For a start, the manifold benefits of forests to their communities for a period that spans four centuries justify policies calling for continued acquisition of woodlands for town use. Timber stands and patches of wildwood are an ancient and fitting part of community structure in the region, and this contiuum should be carried forward. There is

Fig. 9.1. The Frederick B. Knapp Town Forest in Duxbury, Massachusetts. Formerly part of the community's poor farm, the land now adjoins an extensive weald owned by the town and managed by its conservation commission. RLM, 1992

nothing to prevent towns from taking full advantage of existing town forest enabling laws, still valid in all New England states. Although state forestry offices no longer actively maintain municipal forest programs, technical assistance is generally available. The Agricultural Experiment Station at the University of New Hampshire in Durham has vigorously encouraged towns to utilize their forests to full purpose. Government foresters also participate in national and state urban and community forestry programs that can encompass municipal forests. Local conservation commissions and community or private land trusts are similarly equipped to acquire a broad range of landscapes, forests included. Purchase can take the form of complete title or be couched as lesser interests such as timber or development rights.

In recent years, property acquisition by community land trusts and conservation commissions has matured to a strategy of direct competition for valuable land resources with private development groups. Policy calls for conservation as an alternative preferable to that of development, and financial assistance is sometimes available from state and federal programs. Early in 1993, the small town of Granby, Vermont, located in the state's Northeast Kingdom, purchased land and timber rights, covering 1,640 acres of forest on Cow Mountain, from Champion International Paper Company. Potluck suppers and grants from a medley of participants, among them Vermont's Housing and Conservation Trust Fund, helped the eighty-odd town members amass sufficient purchase money. Title to a two-hundred-acre buffer zone around Cow Mountain Pond was taken by the town subject to a conservation easement that restricts cutting. Timber rights were acquired over the balance, and a local conservation commission was created to provide oversight. A large stipend was provided by the state, part of its allotment under the Forest Legacy Program administered by the U.S. Forest Service. That program allocates money to "state, regional, and other units of government" for purchase of forest lands and timber rights that otherwise might fall to nonforest uses such as development. Acquisition is through conservation easements or other mechanisms. Protected environmental values are identified, and in Granby's case, guidelines require sustained forest management with selective cutting methods and continued maintenance of nature trails and other recreational uses.[1]

Similar subsidies might easily be funneled to towns throughout New England whose members decide to purchase lands under existing town forest laws rather than conservation commission enabling acts. Where sustained-yield management and multiple use are objectives that parallel goals of retrieving land from the clutches of development, town forest laws are an eminently suitable mechanism. Ironically, federal participation comparable to that under the Forest Legacy Program never materialized during the heyday of town forests, despite the pleadings of Harris Reynolds and others. In part, the reason may be a sense of urgency that now compels policies of competition with private development by conservation groups. The history of federal policy toward local forestry—and ultimately its effect on the town forest movement—holds important lessons for today's policymakers. Where partnerships between federal, state, regional, and local entities can be forged, projects such as those in Granby become possible. A significant provision of the Forest Legacy Program requires state or regional agencies to adopt similar programs to gain eligibility. Concerns once voiced about federal interference in state affairs have thus been addressed by a method that allows broad state administration in a manner consistent with overall federal objectives.[2]

As these events illustrate, acquisition policies need not focus on merely ensuring the continued presence of forest cover. Public land is a valuable planning tool for controlling the direction of urban sprawl. Benton Mackaye's utopian vision of community and forest in physical harmony remains an ambitious but intriguing standard, possibly attainable in small villages such as Granby. Town forests in a number of communities—Woodbury, Vermont, and Mercer, Maine, for example—are located in close proximity to village centers and offer inviting opportunities. Along a slightly different tack, policy can assume the form of simply retaining woodlands that might otherwise might be sold. Watershed forests surrounding reservoirs that no longer supply community drinking water are a timely example.

Local management policies, far from consistent throughout the region, also stand to benefit substantially from improved understanding about the history of town-owned forests. For many reasons, management remains the single most challenging subject for future deliberation. A great number of such woodlands are underutilized, more than a few forgotten. Key to any policy is recognition that stewardship for utilitarian purposes has historically been a characteristic feature of these lands. In short, examples of the cultural legacy left by those who have worked New England's woods should be made visually apparent to town populations. By maintaining a publicly visible segment of this utilitarian character, we provide opportunity for greater insight about the cultural and natural history of forest landscapes and sharpen our perceptions about debate on current policy issues. Ultimately, this leads in turn to better decision making. Although the profitability of timber cropping on town forests has proved marginal at best, even demonstration forests provide a value that extends beyond simple economic concerns. Intense interest in this aspect of New England's cultural history lies just beneath the surface in many communities and needs only gentle prompting to become active. For example, a printed trail guide for the town forest in Warner, New Hampshire, refers to commercial as well as ecological properties of the tree species along its route. Plainly, this sort of demonstration evinces an intimate bond between cultural heritage and forest, one that can easily be made part of conservation policy. Moreover, drawing participation from diverse community sectors strengthens our sense of community and gives greater validity to democratic forums.

Multiple-use policies hold benefits that extend well beyond education about cultural and natural history and address recreational, aesthetic, conservation, and ecological objectives. In addition, each can be developed within the broader context of community planning, where a forum for competing interests can be established. When multiple uses are successfully accommodated, forests become a meeting ground where diverse community segments are drawn into dialogue and where resolution of opposing interests is possible. Public involvement in these debates stimulates, in turn, greater emphasis on community well-being. Diligent management of town-owned forests to sustain multiple use, including crops of timber, will be critical to the success of any future policy. In the past, fickle town administration has often neglected the timber stand improvement required to produce a paying crop of trees. Public participation in conservation areas has been more consistent, partly because the specter of unwanted change is pervasive and partly because mangagement is not as time-consuming, expensive, or exacting. Successful management policy will depend on creating a structure in which enthusiasm similar to that voiced for conservation of undeveloped spaces can be enlarged to encompass multipe-use management, in other words, a structure where highly efficient use of public property—by another definition, private property for a defined group—becomes possible.[3]

A sophisticated coalition of groups representing cultural, natural, recreational, and commercial elements of local woodlands might be able to sustain that energy. Interpreting the cultural history of New England's town-owned forests will require crossing several disciplines—primarily, forestry, urban and village history, and historic preservation. Formal methods for identifying and evaluating the significance of historic resources rest with the last. Although such a coalition has yet to materialize, potential benefits appear to be substantial. Surprisingly, at precisely the same time that conservation commission enabling laws were being passed in New England, confrontation with change was taking place in the built environment. Beginning with Rhode Island in 1959 and followed by Massachusetts in 1960 and Connecticut in 1961, the region's states began adopting enabling laws for municipal historic districts. Together, historic district commissions and conservation commissions furnished a blanket of local protection for both built and natural environments. Federal legislation, the National Historic Preservation Act, followed in 1966 and created the National Register of Historic Places, which provides a means for formally recognizing historic resources. In more recent years, historic landscapes have been considered eligible for listing on the national register.

As a discipline, historic preservation does certain things very well. Most important, it emphasizes the reflective image of society, past as well as present, mirrored by our cultural environment. Individual pieces of this environment are vivid learning tools, capable of translating subjects of hugh dimension and complexity into simple visual form. Where natural and cultural environments blur, however, interpretation and management policies sometimes become difficult, opening a door for participation by other disciplines. The value in continuing sustained-yield management on town-owned forests as policy consistent with the cultural integrity of these lands is clear from a historical perspective. Similar management policies favored by forestry professionals will thus be strengthened. Moreover, historic preservation groups have very recently placed an emphasis on the role of cultural resources in creating livable communities, a distant echo of themes expounded by Benton Mackaye and others during the 1920s and 1930s. This trend is a further enticement for the creation of a collaborative network of resource groups, such as town forest committees, park boards, conservation commissions, and historic preservationists. If a broad coalition of conservation groups can be established, people from diverse fields of interest will learn from one another and, through compromise, achieve a larger gain. When that happens, towns will have moved a step closer toward developing political consensus about prudent management of resources.

Potential policy benefits also extend beyond local borders. Forests have afforded an endowment of incalculable value to New England. Almost every fragment of the region's history, from art and literature to commerce and industry to urban and village settlement, is bound to woodland culture. In communal forests, one can observe a microcosm of the manifold benefits, traditions, ideals, and shortcomings that characterize policy toward New England forests as a whole. Communities hold answers to key questions about habitable living environments. Forests, as both natural and cultural resources, should play a critical role in any humanistic working plans.

Finally, this work's focus on New England should not be interpreted to suggest that communal forests have not made similar contributions in other regions. Spanish settlements in the southwestern United States initially employed extensive common lands, and New Englanders transported their culture to places such as the Western Reserve in Ohio during the nineteenth centry. School lots and other large areas of public land were allotted by the Public Land Survey during settlement of midwestern and western territories. The history of utopian societies in America is a richly diverse one, and many relied on communal ownership of property. The town forest movement was national in scope and became immensely popular in states such as New York, Pennsylvania, Ohio, Michigan, and Minnesota. Forest plantations sown to protect watersheds are ubiquitous, and forest parks exist in an assortment of unlikely places, as do local conservation spaces. Unquestionably, these topics invite further exploration, and it is hoped that this study will encourage precisely that.

Notes

Chapter 1. From the Common Wood (Notes to Pages 6–46)

1. P. J. Fowler, *Late Prehistory*, in *The Agrarian History of England and Wales*, ed. Joan Thirsk, vol. 1, pt. 1.
2. For a discussion of the influence of Roman settlement in Britain, see A. L. Rivet, *Town and Country in Roman Britain*; S. Applebaum, "Roman Britain," in *The Agrarian History of England and Wales, A.D. 43–1042*, ed. H. P. R. Finberg, vol. 1, pt. 2; and F. Seebohm, *The English Village Community*, 278.

   For a review of the historical theories offered to explain origins of the open-field farming system in England and in particular the weaknesses and gaps in these theories, see C. S. and C. S. Orwin, *The Open Fields*, 1–20. See also Joan Thirsk, "The Common Fields," *Past and Present* 29 (1964):3–25.
3. Orwin and Orwin, *Open Fields*, 30–62.
4. Ibid. See also Seebohm, *English Village Community*; Thomas E. Scrutton, *Commons and Common Fields*; Howard Levi Gray, *English Field Systems*; W. G. Hoskins and L. Dudley Stamp, *The Common Lands of England and Wales*; W. E. Tate, *The English Village Community and the Enclosure Movement*; and J. A. Yelling, *Common Field and Enclosure in England 1450–1850*. Probably the most thorough discussion of field systems in a recent work can be found in Alan R. H. Baker and Robin A. Bultin, eds., *Studies of Field Systems in the British Isles* (1973).
5. Oliver Rackham, *Ancient Woodlands*, 3–4. Rackham makes this observation in his comprehensive treatment of the forest history of eastern England. He convincingly argues that less than half of Anglo-Saxon England was forested and that by the eleventh century that amount had decreased to 15%. He also disputes the long-standing opinion that shipbuilding and iron industries devastated England's forests, suggesting that relentless clearing for cultivation was the primary culprit. His work also contains an excellent bibliography. For less recent discussions of these issues, see H. C. Darby, "The Clearing of the English Woodlands," *Geography* 36 (May 1951):71–83.
6. Rackham, *Ancient Woodlands*, 3–5, 137–60.
7. Scrutton, *Commons*, 5–18; Rackham, *Ancient Woodlands*, 174–75; H. P. R. Finberg, "Anglo-Saxon England to 1042," in *The Agrarian History of England and Wales*, vol. 1 pt. 2, pp. 431–48; and Sir Frederick Pollock and Frederic W. Maitland, *The History of English Law*, 2d ed., 1:25–63.

   The origins and period of the phrase "by hook or by crook" are far from clear, and written evidence does not appear until the fourteenth century. See Ihor Evans, ed., *Brewers Dictionary of Phrase and Fable*, 1981 ed., 564, and *Oxford English Dictionary*, 7:365.
8. Gray, *English Field Systems*, 107–10, 137, 303–4, 350–56; for Kent, see K. P. Whitney, *The Jutish Forest: A Study of the Weald of Kent from 450 to 1380 A.D.*, 13, 31, 56–60, 78–82.

9. Rackham, *Ancient Woodlands*, 7; and David Grayson Allen, *In English Ways*.
10. Finberg, "Anglo-Saxon England," 479–82.
11. For a discussion of Norman land tenure, see Pollock and Maitland, *English Law*, 1:64–110, 232–382; for treatment of royal forests, see John Manwood, *A Treatise of the Forest Laws*; John C. Brown, *The Forests of England and the Management of Them in Bye Gone Times*; and Margaret Ley Bazeley, "The Extent of the English Forest in the Thirteenth Century," *Transactions of the Royal Historical Society* 4 (1921):140–172.
12. Manwood, *Treatise*; Brown, *Forests of England*, 14–16; Rackham, *Ancient Woodlands*, 175–99.
13. Young, *The Royal Forests of Medieval England*, 114–34.
14. Brown, *Forests of England*, 215–19; Bazeley, "Extent of English Forests," 146–50; Rackham, *Ancient Woodlands*, 179. Rackham estimates that in the thirteenth century the total physical area of forests amounted to 1 million acres, roughly 3% of England, but the legal area of forest jurisdiction, royal and private, covered at least 33%. Young is more conservative at 25%.

    For the Forest Charter, see *Statutes of the Realm*, 9 Hen. 3, 1225 (Charta Foresta).
15. Brown, *Forests of England*, 48–49; Rackham, *Ancient Woodlands*, 119–20, 155, 180–85; Scrutton, *Commons*, 3, 42–43; H. L. Bennett, *Life on the English Manor: A Study of Peasant Conditions, 1150–1400*, 49–60.
16. H. L. Bennett, *Life on the English Manor*, 49–60; Joan Thirsk, "Enclosing and Engrossing," in *The Agrarian History of England and Wales*, ed. Joan Thirsk, 4:200–55.
17. Act of 1543, *Statutes of the Realm*, 34–35 Henry 8, Stat. c. 3; Rackham, *Ancient Woodlands*, 147–48.
18. Thirsk, "Enclosing," 200–255.
19. Tate, *English Village*, 44–73, 91–127; see also J. A. Yelling, *Common Field*.
20. Melville Egleston, *The Land System of the New England Colonies*, 4–13; Roy H. Akagi, *The Town Proprietors of the New England Colonies*, 6–9. A good summary of the laws affecting title to land in colonial New England can be found in James Sullivan's *The History of Land Titles in Massachusetts*, published in 1801. The Massachusetts Bay Company received a royal charter in 1629 confirming a prior grant and acquired jurisdiction over what is now Massachusetts, southern New Hampshire, and Maine, as well as the Plymouth Colony.
21. Egleston, *Land System*, 27–28; Charles M. Andrews, *The River Towns of Connecticut: A Study of Wethersfield, Hartford, and Windsor*, 48–52, 63–66; and Akagi, *Town Proprietors*, 10, 30–38.
22. Herbert B. Adams, "Village Communities of Cape Anne and Salem," 80; Egleston, *Land System*, 41–42; Andrews, "River Towns," 68–71; Ann Bush Maclear, *Early New England Towns: A Comparative Study of Their Development*, 148, 167; and Akagi, *Town Proprietors*, 85–104, 110–14. For a more recent evaluation of town morphology, see John Reps, *The Making of Urban America*, 115–46.
23. David G. Allen, *In English Ways*, 30–36. See also Kenneth A. Lockridge, *A New England Town: The First Hundred Years: Dedham, Mass., 1676–1736*; Sumner

Chilton Powell, *Puritan Village: The Formation of a New England Town*; and Michael Zuckerman, *Peaceable Kingdoms: New England Towns in the Eighteenth Century*.

24. For a discussion of diversity in New England's precolonial landscape, see William Cronon, *Changes in the Land*, 26–34. Town records provide satisfactory evidence of land divisions and various categories of common land. However, traditions of woodsmanship and their effect on the duration of common land are far more difficult to establish due to the absence of written records. This topic represents a challenging but worthy domain for future research.
25. Egleston, *Land System*, 34–39; and Akagi, *Town Proprietors*, 103–4. (Original 17th- and 18th-century spellings retained in quotations throughout.)
26. Braintree Records, 4 (9-10-1646).
27. Dedham Town Records, 145 (11-23-1656).
28. Cambridge Town Records, 126 (12-14-1658).
29. Reading Town Records, 30 (2-18-1667).
30. Dorchester Town Records, 191 (1-10-1672/3).
31. Brookline, Mass.: *Muddy River and Brookline Records*, 51 (1-28-1678).
32. Farmington Town Votes, 63 (12-15-1691). The town borders Hartford and Wethersfield to the west and was founded in 1645, roughly a decade after its two neighbors.
33. Salem Town Records, 34 (1636).
34. Dorchester Town Records, 112–13 (2-10-1667).
35. Plymouth Town Records, 1:142 (10-14-1674).
36. Braintree Town Records, 359 (11-14-1757).
37. Providence Town Records, 2:21 (3-14-1662)
38. William Wood, *New England's Prospect*, 38.
39. Meetinghouse Swamp has retained this title and is part of Lynn Woods, established in 1888 as one of the earliest examples of a town-owned forest park in New England.
40. Wood, *New England's Prospect*, 38; Cambridge Town Records, 242 (1-13-1678).
41. Ipswich Town Records, 89 (9-16-1652).
42. Medfield Town Records, vol. 1, n.p. (8-26-1663).
43. Weymouth Town Records, 1:109 (11-26-1666).
44. Plymouth Town Records, 142 (10-14-1674).
45. Salem Town Records, 119 (10-30-1670).
46. Andover Town Records, 351 (1-1-1676/7), 382 (3-2-1690), and 383 (3-7-1691/2); these records are unpublished and are located at the Andover Historical Society. The town was located inland on the Merrimack River and chartered in 1646.
47. Providence Town Records, 3:68 (2-12-1665).
48. Saybrook, Conn.: *Records of the Proprietors of Oyster River Quarter*, 1:8 (9-3-1672).
49. Akagi, *Town Proprietors*, 105–6.
50. Muddy River and Brookline Records, 35 (3-25-1657).
51. Medfield Town Records, vol. 1, n.p. (12-22-1663).
52. Reading Town Records, 29 (12-3-1666), 33 (2-7-1669).

53. Cambridge Town Records, 149 (12-8-1663).
54. Dedham Town Records, 122 (1-7-1665). In that case, adjoining land had already been divided.
55. Cambridge Town Records, 126 (12-18-1658).
56. Ibid., 158 (2-11-1683).
57. Ibid., 94 (12-9-1651).
58. Wenham Town Records, 45 (11-7-77); see also 34 (11-2-1670) for restrictions against cutting on the town's Cedar Swamp.
59. Reading Town Records, 34 (2-6-1670), 207 (5-9-1727), 228 (5-21-1733).
60. Taunton Proprietors Records, 1:34 (11–25/26–1711).
61. Braintree Town Records, 236 (3-2-1740).
62. Saybrook, Conn.: *Records of the Proprietors* . . . , 1:215 (1769).
63. Adams, "Village Communities," 49; Egleston, *Land System*, 41–42; Andrews, *River Towns*, 63; Herbert L. Osgood, *The American Colonies in the Seventeenth Century*, 1:441; Maclear, *Early New England Towns*, 93; and Akagi, *Town Proprietors*, 110–14.
64. See Topsfield Town Records, 8 (3-19-1666/7).
65. Adams, "Village Communities," 15, 49; Egleston, *Land System*, 41–42; Andrews, *River Towns*, 63; Osgood, *American Colonies*, 1:441; Maclear, *Early New England Towns*, 93; Akagi, *Town Proprietors*, 110–14. Adams indicates that settlement gravitated toward attractive planting land and frequently occurred in open spaces formerly habituated by Indians. This theme also has been argued in Charles J. Hogan, "Puritans and Meadows: The Interplay of English Culture and the American Environment in the Massachusetts Bay Colony, 1630–1660."
66. New Haven Town Records, 103 (12-3-1651).
67. Ibid., 114–15 (12-10-1651).
68. Ipswich Town Records, 102 (3-1-1656).
69. Ibid., 24 (3-18-1697), 53 (3-14-1700).
70. Cambridge Town Records, 168 (2-3-1667); 197 (12-10-1671); 257 (12-27-1681); 131 (3-14-1660).
71. Salem Town Records, 122 (10-11-1643), 116 (9-23-1670).
72. Topsfield Town Records, 7 (3-7-1664). The village was settled as early as 1639, originally part of the Ipswich grant, and was chartered as a separate town in 1650.
73. Hingham Town Records, 88 (3-7-1679).
74. Marshfield Town Records, 1:172 (1-14-1682). Groton inhabitants proffered similar encouragement and set like restrictions for the establishment of iron works at Masabog Pond; see Groton Town Records, 98 (5-21-1688). Sandwich proprietors, in 1712, voted to lease all pine trees on the town common for production of turpentine; see Sandwich Proprietors' Records, 1:47, 55 (10-2-1712).
75. New Haven Town Records, 338 (3-8-1658).
76. Cambridge Town Records, 126 (11-20-1658).
77. Reading Town Records, 35 (2-4-1671).
78. Weymouth Town Records, 1:88 (12-19-1660).
79. Dedham Town Records, 9–10 (1660).

80. Carl Bridenbaugh, "Yankee Use and Abuse of the Forest in the Building of New England, 1620–1660," 11–25.
81. William Cronon, *Changes in the Land*, 116–18; John Stilgoe, *The Common American Landscape, 1630–1850*, 170–82.
82. Norwalk Town Records, 43 (12-29-1653). The town was chartered in 1651.
83. New Haven Town Records, 109 (2-9-1651).
84. Weymouth Town Records, 1:55 (3-24-1651).
85. Medfield Town Records, vol. 1 (1649–1700).
86. Charles F. Carroll, *The Timber Economy of Puritan New England*, 75–97.
87. Providence Town Records, 3:8 (1-27-1661), 66 (10-27-1665).
88. Ipswich Town Records, 75 (8-31-1649); see also Marshfield Town Records, 1:32 (1-30-1650), where the town prohibited removal of oak beyond its borders, and Plymouth Town Records, 224 (3-13-1692), where cutting oak from the commons for any purpose, whether for cordwood or bark, was strictly prohibited.
89. Carroll, *Timber Economy*, 75–97.
90. The practice—known as "driving a piece"—of notching a succession of small trees to cushion the fall of valuable timber such as pine, which was used for ships' masts and which shattered easily on impact, may also have been employed. However, town records consistently evince control of cutting on common land, and it is unlikely that such wasteful practices were extensive. See W. Cronon, *Changes in the Land*, 111; and J. Wilcox Brown, "Forest History of Mt. Moosilauke," *Appalachia* 24 (1958):26.
91. New Haven Town Records, 323 (Aug., 1657).
92. Guilford Town Records, book B, 16 (1643).
93. Warwick Town Records, 72 (1-3-1652); see also Norwalk Town Records, 42 (12-29-1653).
94. Weymouth Town Records, 1:65–66 (3-7-1654).
95. Reading Town Records, 88 (1-29-1693). Again, use of the term *lopping* points to English practice.
96. Salem Town Records, 1:14 (12-15-1635).
97. Ibid., 1:31 (1-11-1636). However, in 1658, selectmen seemed most concerned about orderly clearing and required those who cut fuelwood on the town common to cut all wood, leaving only timber trees, with a penalty of 10 shillings for every offense, half to the town and half to the informer; see ibid., 219 (9-21-58).
98. Watertown Town Records, 2 (9-23-1635).
99. Ipswich Town Records, 31 (1639).
100. Cambridge Town Records, 94 (12-9-1651).
101. Providence Town Records, 2:54 (11-27-1650).
102. Ibid., 3:102–3 (12-14-1681).
103. Cambridge Town Records, 168 (1-20-1667).
104. Manchester Town Records, 8 (10-6-1658).
105. Plymouth Town Records, 224 (3-13-1692).
106. Providence Town Records, 157 (11-9-1685).
107. Cambridge Town Records, 170 (4-22-1668).

108. Salem Town Records, 112 (5-5-1670).
109. Braintree Town Records, 407 (3-13-1766); 507 (3-14-1780).
110. Bridenbaugh, "Yankee Use," 33.
111. For example, at a town meeting in Ipswich in 1659, town members complained about being overburdened by increasing numbers of dwelling houses. They ordered that all houses subsequently erected would acquire neither rights to common land nor privileges of common pasture, wood, or timber without express consent. See Ipswich Town Records, 113 (3-15-1659).
112. Duxbury Town Records, 199 (5-17-1703).
113. Edna Scofield, "The Origin of Settlement Patterns in Rural New England," 659; Akagi, *Town Proprietors*, 190–91, 219–29.
114. Egleston, *Land Systems*, 23; Scofield, "Origin of Settlement," 659–61; Akagi, *Town Proprietors*, 189–91.
115. Andrews, *River Towns*, 75–81; Maclear, *Early New England*, 29–32. Transitional towns also developed new, linear patterns of land division, often where geographical factors such as rivers were influential. See Scofield, "Origin of Settlement," 661–62. Title disputes over conflicting patents and royal grants issued by the English Crown contributed to political uncertainty, and overlapping charters were issued for many towns. See Maclear, *Early New England*, 31; Scofield, "Origin of Settlement," 657; and Akagi, *Town Proprietors*, 183–85, 230–87.
116. Akagi, *Town Proprietors*, 177, 293.
117. Ibid., 293. Akagi considers the system of common pasturage to have died out by the opening of the eighteenth century, but Andrews (*River Towns*, 71) indicates that common meadows were still employed near the close of the nineteenth century. Moreover, Herbert Adams, in "Village Communities," 76–77, discusses the remaining common land in Salem, which totaled approximately 300 acres at the time of his writing in 1882. Philip J. Greven, *Four Generations: Populations, Land and Family in Colonial Andover, Massachusetts*, provides a thoroughly documented examination of the disappearance of common lands in Andover. For a discussion of the division of lands by range and township, see James L. Garvin, "The Range Township in Eighteenth Century New Hampshire," 48–62. See also Scofield, "Origin of Settlement," 661–63.
118. For examples of early 17th-century common grants in Massachusetts, see Topsfield Town Records, 189 (1-18-15), 191 (12-6-1715), and 196 (1-7-1718); Bridgewater Town Records, 103 (3-11-1717). In Connecticut, see Lyme Town Meeting Book, 1:163 (12-24-1708). For discussion of timber trespass, see William B. Weeden, *Economic and Social History of New England, 1620–1789*, 62; Andrews, "River Towns," 52; and Akagi, *Town Proprietors*, 147–51.
119. Manchester Town Records, 114–15 (11-21-1706).
120. Lunenburg Town Records, 45 (6-5-1728).
121. Braintree Town Records, 322 (3-28-1753); 355 (3-28-1757).
122. Warner Proprietor Records, 122 (1790).
123. Warner Proprietor Records, 17–23 (3-26-1766) and 123 (11-13-1793).
124. Weymouth Town Records, 1:77 (3-2-1659).

125. Medfield Town Records, vol. 1, n.p. (2-6-1670); Topsfield Town Records, 7 (3-7-1664).
126. Lyme Records, Town Meeting Book, no. 1, 23 (1-6-1686).
127. G. Wheeler and H. Wheeler, *History of Brunswick, Topsham, and Harpswell, Maine*, 521–27.
128. Wheeler and Wheeler, *History of Brunswick*, 526–27; "Brunswick's Experiment in Forestry," *Lewiston (Maine) Journal*, June 10, 1901.
129. Austin Cary, "Forestry on the Town Common," *Brunswick (Maine) Telegraph*, February 26, 1902.
130. For a general review of America's utopian societies, see Mark Holloway, *Heavens on Earth*.
131. Ibid., 53–79; E. D. Andrews, *The People Called Shakers*; David R. Starbuck, "Canterbury Shaker Village: Archeology and Landscape," *The New Hampshire Archeologist* 31 (1990):1–5.
132. E. Andrews, *The Community Industries of the Shakers*.
133. For a comprehensive review of the ideological and social makeup of villages, see P. Brewer, *Shaker Communities, Shaker Lives*. For Sabbathday Lake, see Alan M. Hall, "The Shaker Response to Industrialism: A Case Study of the Sabbathday Lake Shaker Community." See also Sister Aurelia's Diary, March 11, 1896.
134. Starbuck, *Canterbury Shaker Village*, 25; Blinn, *A Historical Record* 120, 123, 161, 162.
135. Blinn, *Historical Record*, 2:250–66; Blinn, *Church Record*, 1:114, 120.
136. Blinn, *Church Record*, 1:195, 201. For blister rust control, see J. H. Foster to Arthur Bruce, Elder (1-9-24) and Cooperative Agreement (6-28-23).
137. Sales occurred in 1891, 1894, 1902, 1907, 1908, 1909, and 1926. For example, see Agreement (12-23-02) between Canterbury Shaker Trustees and John T. Dodge to cut standing timber, not including pine, on Patten Lands for three years, with authority to construct a steam mill.
138. "A Historical Narrative of the Rise and Progress of the United Society of Shakers, Enfield, N.H.," 51, 63, 94–96, 99, 100, 109, 110. See also "H. Cumings, "A Sketch in the Life of Caleb M. Dyer," *Enfield (New Hampshire) Advocate*, (December 30, 1904), "Some Early Industries and Inventions . . . of the Shakers," *Enfield (New Hampshire) Advocate*, March 31, 1905; and "A Day Amongst the Pines," *Enfield (New Hampshire) Advocate*, January 10, 1908.
139. Cumings, "A Day Amongst the Pines."
140. Hall, "Shaker Response," 171; Sister Elsie McCool, "A Brief History of the Shaker Mills, Sabbathday Lake, Maine," *Shaker Quarterly* 3 (fall 1963):99–102; "Church Family Journals," vol. 3, (2-10-86, 5-14-1887).
141. Hall, "Shaker Response," 171; Delmer Wilson's Diary 1890, general entries; Church Family Records, vol. 3 (1918), vol. 5, 312–14 (1903).
142. J. D. Lyman, "Omar Pease Plantation," in *Annual Report of the Rhode Island Board of Agriculture* (1900):9–10.

Chapter 2. Sylvan Public Lands (Notes to Pages 47–92)

1. Maurice Beresford, *New Towns of the Middle Ages: Town Plantations in England, Wales and Gascony*, 79–135.
2. Ibid., 164–66; citing "Calendar of Patent Rolls" (1284), 133.
3. Ibid., 222–23.
4. James Steves Curl, *The Londonderry Plantation 1609–1914*, 22–26; T. W. Moody, *The Londonderry Plantation 1609–1641*, 33–35.
5. Curl, *Londonderry Plantation*, 33–36, 247; Moody, *Londonderry Plantation*, 62–83. The Deputation Commission of the Fishmongers Company, organized to survey their proportion, traveled the countryside and noted the great tracts of bog in which large quantites of wood and timber were found.
6. *An Historical Narrative of the Irish Society*, London, 1865, app. 3, p. 63. The Charter of Coleraine, similar in content, was also executed in 1613.
7. Curl, *Londonderry Plantation*, 40–71, 124–26.
8. Sidney and Beatrice Webb, *English Local Government from the Revolution to the Municipal Corporations Act: The Parish and the County*, 9–15; Colin Platt, *The Parish Churches of Medieval England*, 1–12; Eleanor Trotter, *Seventeenth Century Life in the Country Parish*, 23; Toulmin Smith, *The Parish: Its Powers and Obligations at Law*, 214–26; Warren O. Ault, *The Self-Directing Activities of Village Communities in Medieval England*; Sedley Lynch Ware, "The Elizabethan Parish in Its Ecclesiastical and Financial Aspects," *Johns Hopkins University Studies in Historical and Political Science*, 26 nos. 7–8 (1908), 312–95; John Godfrey, *The English Parish, 600–1300*, 17–56.
9. Webb and Webb, *English Local Government*, 9–15; Platt, *Parish Churches*, 1–12, 55–64.
10. Platt, *Parish Churches*, 55–64, 88–91; Godfrey, *English Parish*, 60; Smith, *Parish*, 103.
11. Trotter, *Seventeenth Century Life*, 36.
12. See Acts and Resolves, Province of the Massachusetts Bay, vol. 2, chap. 1, p. 99 (5-28-1718), and vol. 2, chap. 6, p. 181 (11-2-1720). See also Laws of New Hampshire, vol. 2 (Province Period—1702 to 1745), p. 155 (5-11-1714). Few studies address the influence of parish settlements on the formation of splinter towns in New England. Methods of determining boundaries, practices of land management, and matters of community organization beckon further inquiry.
13. Topsfield Town Records, 8 (3-6-1665).
14. Marshfield Town Records, 64 (2-10-1657), 88 (7-6-1663), 111 (8-5-1667), 115 (4-27-1668), 150 (2-18-1667). See also Reading Town Records, 53 (2-6-1681); the town voted that house and land bought for the minister should be for the inhabitants of Reading and their heirs forever and for the use of the ministry. See also Norwalk, Connecticut, Town Records, 89 (5-23-1695).
15. Ipswich Town Records, 172 (11-22-1711).
16. Rowley Town Records, 129 (2-25-1662).
17. Reading Town Records, 48 (2-2-1679), 110 (3-2-1701).

18. Cambridge Town Records, 215–16 (12-9-1673), 259 (1-8-1682), 296 (9-9-1691).
19. Weymouth Town Records, 214 (3-7-1697).
20. Plymouth Town Records, 1:296 (3-16-1701), 2:27 (5-9-1709), 2:260 (12-29-1728).
21. Topsfield Town Records, 236 (2-23-1723).
22. Manchester Town Records, 40 (3-19-1744), 113–15 (12-22-1766).
23. Weymouth Town Records, 299, (3-14-1727). In 1838, the north parish successfully petitioned the Massachusetts legislature to sell its woodlands and expend the proceeds on construction of a parsonage house. See Massachusetts Acts and Resolves, chap. 26 p. 323 (1838).
24. Brookline, Mass. *Muddy River and Brookline Records*, 145 (3-7-1737), 199–202 (5-18-1761), 208 (5-19-1763), 598–603 (4-7-1834), 635–36 (11-1836), 182 (3-2-1846). For information concerning the salary of Rev. John Pierce, see Brookline, Mass.: "Records of the First Parish Church," 99 (1796).
25. Simon G. Griffin, *A History of the Town of Keene*, 16–18, 104–6, 293, 462–63, 514; Keene *City Reports*, 16 (1857), 29 (1861), 24 (1864).
26. William P. House, "Timber Report—City of Keene's Woodlots," (May 1, 1979); Thomas C. Lacey, "Forest Inventory and Management Plan—Minister's Lot—City of Keene. Land, Keene, N.H." (March 7, 1985).
27. *State of New Hampshire: Biennial Reports of the Forestry Commission*, 76–79 (1923–24); John F. Rowe, *Newington, New Hampshire*, 78–83.
28. Rockingham County Registry of Deeds, 79:57 (7-19-1765); Newington Town Records, Extract Book (1786, 1799); Early Parish and Town Records, 335, 351 (1712–1821). See also W. Dennis Chesley, "Summary of 1978 Archival Research, Old Parsonage Site, Newington, N.H.," 18–22 (1-16-1981).
29. Chesley, "Summary," 23 (1-16-1981). Mr. Chesley's work organizes primary data from a variety of sources, among them Newington Town Records, Early Parish Records, and a portion of the very extensive but largely uncataloged collection of the Newington Historical Society. See also *State of New Hampshire: Biennial Reports*, 79 (1923–24).
30. Newington, N.H.: *Early Parish and Town Records* (10-22-1787); Newington Town Records, vol. 4 (4-13-1866); Clerk Minutes, 267 (1868); Chesley, "Summary," 22–24; Newington Annual Town Reports, 16 (1912), 14 (1920), 5 (1924). See also Rowe, *Newington*, 78–83; *State of New Hampshire: Biennial Reports*, 76–79 (1923–24); Nelson C. Brown, "The First Community Forest," *American Forests* 44 (September 1938): 406–8.
31. Newington, N.H.: Newington Annual Town Reports, 16–17 (1915).
32. Danville, N.H.: *Danville Parsonage Committee Records*. Unpublished writing on Danville's town forest includes the following: Clarence Collins, "A Few Facts Gathered from the Records of the Parsonage Committee of the Town of Danville," n.d.; and Ernest O. Buhler, "Acres of Diamonds," (11-21-38). Copies of both articles are kept with the Parsonage Committee records.

Published writing includes *State of New Hampshire: Biennial Reports* (1927–28), 144–46; and (1949–50), 47; Nelson C. Brown, "First Community Forest," 20–21; J. P. Barrett, D. S. Linden, J. H. Gove, R. A. Langevin, and J. M. Brien,

Timber Values of Town Forests; H. I. Mauran, L. Alexander, K. W. Barba, and J. P. Barrett, *Recreational and Education Opportunities in New Hampshire's Town Forests*; Warren F. Hale, "Town Forests in New Hampshire," 526; Thomas N. Stern, "Community Forests and Parks in Pennsylvania," 110, 113a, 113b.

33. *State of New Hampshire: Biennial Reports* (1925–26), 74–75. Roxanna S. Richardson, "History of Northwood, N.H.," (master's thesis, University of New Hampshire, 1946), 12–20. Portions of the school lot have been leased and summer camps constructed.
34. Records of the New Hampshire Division of Forests and Lands—Concord. Richardson, *History of Northwood*, 13, 18. *Concord Daily Monitor*, January 11, 1949. New Hampshire Forestry Commission: *Biennial Reports* (1947–48): 51. Timber sales occurred in 1909, 1927, and 1949.
35. See correspondence from Henry A. Howard, chairman of the Board of Selectmen, to Warren F. Hale, assistant state forester, 11-11-1926, records of the New Hampshire Division of Forests and Lands—Concord. See also *N.H. Forestry Commission: Biennial Reports* (1925–26): 80; G. H. Collingwood, "The Making of a Town Forest," *American Forests and Forest Life*, 36 (September 1930): 587–90; and Hale, "Town Forests," 525–28. See also Isaac W. Hammond, ed., *Town Papers: Documents Relating to Towns in New Hampshire*, 12:56, 13:199.
36. See Woodland Inventory and Conservation Report, Records of the State Agricultural Extension Service, Haverill, N.H.
37. Massachusetts Resolves (1838) chap. 25, p. 323.
38. Massachusetts Statutes (1836), title 8, chap. 20, secs. 1, 7, 8, 18, 23, 30, 41, 43. Episcopal ministers were excluded from the requirement of consent to conveyances by town, parish, or religious society.
39. Works Progress Administration, *Rhode Island: A Guide to the Smallest State*, 341.
40. Providence Town Records, 3:35 (5-9-1663).
41. Ipswich Town Records, 45 (9-13-1642); n.p. (1-11-1650); n.p. (1-16-1650); n.p. (4-15-1782).
42. A similarly named water boundary in Worcester, Massachusetts, is described in W. Cronon, *Changes in the Land*, 66.
43. Hanover Town Records, 90 (3-18-1790).
44. Caleb Stark, *History of the Town of Dunbarton*, 144–45.
45. See "Plan of Land in Northern Part of Washington Co., Part of Incorporated as Houlton in 1831," an unsigned plan of land in Houlton, Maine, dated 1820, MSA. See also plan titled "Gore of Land Granted to Lincoln Academy (at Damariscotta Pond)" located near Waldoboro, Maine, dated 1813, also at MSA.
46. Pomfret Deed Records, 6: 105, 107 (1-14-1818). See also Edward D. Andrews, "The County Grammar Schools and Academies of Vermont," 166–67, and records of Royalton Academy, Vermont Historical Society, Montpelier.
47. Richardson, *History of Northwood*, 30–34. The school's name was changed to Coe's Northwood Academy in honor of a local benefactor, Ebenezer Coe, whose family donated several woodlots. It became Coe-Brown Northwood Academy in 1939.
48. Matthew Kelty, "The Development and Productivity of Hemlock-Hardwood Forests in Southern New England," 83.

49. See *Cilley v Cayford*, Smith's N.H. Reports 150 (1807).
50. Walter T. Bogart, *The Vermont Lease Lands*, 295–303; Laws of Vermont (1937), chap. 65.
51. Bogart, *Vermont Lease Lands*, 55, 299, 317. See also Vermont Resource Management, "Forest Management Plan—Cross and Glebe Lots, West Windsor" (9-30-86), in records of the West Windsor Town Clerk.
52. *Lemington v Stevens*, 48 Vt. Reports 38 (1875).
53. Austin H. Wilkins, *Ten Million Acres of Timber*, 7–21; Lee M. Schepps, "Maine's Public Lots: The Emergence of a Public Trust," 219–30.
54. Wilkins, *Ten Million Acres*, 7–21; Schepps, "Maine's Public Lots," 219–30.
55. Schepps, "Maine's Public Lots," 226–30.
56. Sir George Nicholls, *A History of the English Poor Law*, 62–106, 149–93, 174–204. A committee appointed to study that country's poor laws issued a report in 1817 and considered a plan to establish parochial farms as a means of affording work. The plan was based on a successful example in the county of Kent, but its influence on practice in New England is not clear.
57. Marshfield Town Records, 1:15 (4-10-1645).
58. Watertown Records, 70 (1-8-1660), 88 (9-1-1666). See also Plymouth Town Records, 1: 3–4 (7-16-1638).
59. Records of the Colony of Rhode Island, vol. 1 (1636–1663), 184 (1647).
60. Laws of New Hampshire, vol. 1 (Provincial Period), chap. 39, pp. 254–55.
61. Ibid., chap. 17, pp. 218–21. In 1672, a Connecticut law mandated similar welfare; see Connecticut Statutes, Title 130 (1672, rev. 1702), *Public Statute Laws of the State of Connecticut*, bk. 1, 552–553.
62. A 1680 New Hampshire law established that persons residing in a town for more than three months were to be considered inhabitants and permitted towns to warn before that period had elapsed. See Laws of New Hampshire, vol. 1 (Acts of the Assembly in Portsmouth, 1680–81), p. 36.
63. See Acts and Laws of the State of Connecticut (1796), 343.
64. See Laws of the Commonwealth of Massachusetts, vol. 2 (2-26-1794), 619–20; Laws of New Hampshire, vol. 5 (First Constitutional Period) (2-15-1791), 691; Laws of Rhode Island and Providence Plantation (1798), 348–49.
65. Maine Revised Statutes, chap. 24, sec. 7 (1857), 251.
66. For a general discussion of poor farms in Vermont, see Steven R. Hoffbeck, " 'Remember the Poor' (Galatians 2:10): Poor Farms in Vermont," *Vermont History* 57 (fall 1987): 226–40. An 1865 plan of Foster's Town Asylum Farm by P. and P. T. Howard has been recorded in the Foster Town Clerk's Office, "Plans," 1:41.
67. Leander Cogswell, *History of Henniker*, 100.
68. Henniker, N.H., *Annual Reports* (1856), 10. See also *Rice v Wadsworth*, 27 *Foster's New Hampshire Reports* 104 (1853), for a discussion of the town's ability to control proceeds from the sale of its ministry and school lands. Newington, N.H., is another example of a town that has converted public lands to use as a poor farm. Few other examples appear to exist.
69. Groton, Mass., *Annual Reports* (1859), 13; (1869), 18; (1872), 15–16; (1877), 27; (1880), 26; (1881), 29; (1885), 23; (1888), 34; (1889), 33; (1890), 33; (1891), 33.

70. Petersham, Mass., *Annual Reports* (1900), 13; (1902), 13; (1903), 12; (1908), 16; (1909), 17; (1910), 27.
71. Pepperell Town Reports (1883), 23; (1884), 22–23; (1885), 23; (1886), 23; (1887), 28; (1888), 26; (1891), 27; (1892), 25. For example, 5½ cords of wood sold in 1883 produced a return of $28. Barre's poor farm produced a crop of walnuts in 1848; see Barre, Mass., *Annual Reports* (1848), 8. See also D. Hamilton Hurd, *History of Worcester County, Mass.*, 360.
72. Sidney Bull, *History of the Town of Carlisle*, 53. The sale yielded $2,025.
73. William Little, *History of Weare, 1735–1888*, 387–9. Weare, N.H., *Annual Reports* (1873), 7; (1874), 6–7; (1875), 6; (1898), 3; and (1900), 43. Between 1854 and 1891, town records consistently show supplies of lumber and wood kept at the farm although it is unclear whether entries represent new inventory.
74. Concord, N.H., *Annual Reports* (1855), 30–31; (1857), 24; (1858), 25; (1859), 29; (1863), 44; (1864), 52; (1866), 45; (1868), 41; (1870), 51; (1883), 96–98. See also *Annual Report of the Board of Water Commissioners* (1875), 48.
75. Newmarket, N.H., *Annual Records* (1881), 16; (1909), 17.
76. See G. A. Hubbard, "The Municipal Forest in Fitchburg, Mass.," *American City* 24 (February 1921): 121–24.
77. Lucy C. Kellogg, *History of the Town of Bernardston*, 279.
78. Warner, N.H., *Town Reports* (1857), 3; (1858), 4–5; (1865), 5–6; (1867), 4; (1873), 2; (1875), 4; (1880), 14; (1883), 15. Deed of sale dated 4-5-1890, recorded 4-8-90, Merrimack County Registry of Deeds. See also Walter Harriman, *The History of Warner, New Hampshire*, 353–57.

    See Harrison H. Harriman, *Survey Notes* (10-4-1847 and 10-19-1852), Pillsbury Free Library, Warner, N.H. Harriman's records, as well as county deed records, are frustratingly silent regarding both acquisition and sale of this parcel. In addition, town records and poor-farm records also make no reference to its use. However, the land is located less than a mile from the former location of the poor farm.
79. Harris A. Reynolds, "The Town Forest in Massachusetts," *Bulletin of Massachusetts Forest and Park Association* 159 (December 1936): 14–15.
80. Athol, Mass., *Annual Reports*, (1922), 82–83; (1923), 71; (1924), 86–87; (1925), 74–75. Petersham *Annual Reports* (1900), 13; (1902), 13; (1908), 16; (1910), 10–11. See also Records of the Petersham Town Forest at offices of Harvard Forest, Petersham, Mass.
81. Westfield, Mass., *Annual Reports* (1914), 144. See also. U.S. Forestry Service, *Field Handbook of Community Forests*, Leaflet 18.
82. Groton *Annual Reports* (1922), 11–13. See also Nelson C. Brown, "Community Forests Come of Age," *American Forests* 45 (January 1939): 16–19.
83. Harris A. Reynolds, "The Town Forest in Massachusetts," *Bulletin of Massachusetts Forest and Park Association* 159 (December 1936): 8. See also Harris A. Reynolds, *Report on the First Conference of Massachusetts Town Forest Committees*.

Chapter 3. Community of Interest (Notes to Pages 95–130)

1. Henry D. Thoreau, *The Maine Woods*.
2. Michael Williams, *Americans and Their Forests*, 146–89, 331–52. For the devastation to America's forests by agriculture, see William Cronon, *Changes in the Land*, 114–20, and Howard S. Russell, *A Long Deep Furrow*, 170–76, 384–93. For New England, see Lloyd Irland, *Wildlands and Woodlots*, 28, 64–80.

   For the impact of railroads on American forests, see N. H. Egleston, "Report of the Chief of the Forestry Bureau," 203–4. What railroad companies didn't use for construction or waste in the cutting, they often burned when sparks from passing locomotives ignited bordering forests.
3. Williams, *Americans*, 160–89, 201–30. On the practice of nineteenth-century lumbering and its impact on forest floor, see Irland, *Wildlands*, 28–35, 64–80; see also George Ahern, *Deforested America*.
4. Williams, *Americans*, 396–99.
5. Timothy Dwight, *Travels in New England*, 106; Samuel T. Dana, *Forests and Range Policy*, 73–75.
6. "American Forest Trees," *New England Farmer* 15 (May 10, 1837): 348–49, (May 17, 1837): 353–54, (May 24, 1837): 361–62, and (May 31, 1837): 369–70, at 369. The article initially appeared in *North American Review* 45 (1837): 334–61. The 1833 article was also entitled "American Forest Trees" and appeared in *New England Farmer* 11 (n.s. vol. 2) (November 14, 1833): 140.

   The work by A. Micheaux and his son was published in Paris under the full title *North American Sylva: A Description of the Forest Trees of the United States, Canada and Nova Scotia with a Description of the Most European Forest Trees*, by A. F. Micheaux.
7. George P. Marsh, *Man and Nature*, 201.
8. Ibid., 201, 202. See also Lewis Mumford, *The Brown Decades*, 78.
9. John Warder, "Report on Forests and Forestry," 16.
10. Ibid., 10.
11. American Association for the Advancement of Science, Forestry Committee, "Cultivation of Timber and the Preservation of Forests," 15, 24. The author, Franklin Hough, relied extensively on sources gathered by George Perkins Marsh and repeated the latter's example of the French forest of Montant, owned by the town of Labruguiere, which exemplified the possibilities of forest reclamation through public education. See Marsh, *Man and Nature*, 208.
12. See *Proceedings of the American Forestry Congress*, Cincinnati, April 1882, and Montreal, August, 1882, 8–20. See also *Montreal Herald*, August 23, 1882, 2.
13. The Mark Society's influence on village structure in both England and New England has been widely argued and has stimulated volumes of writing. Whatever effects Germanic society may have had upon New England's early communal forests, they were steeped in centuries of English tradition. A summary of the work that considers the influence of German village form on New England's first towns—notably the writing of A. Johnston, H. B. Adams, and C. M. Andrews—

can be found in A. S. Eisenstadt, *Charles Mclean Andrews: A Study in American Historical Writing*, 14–20. See also C. F. Adams, "The Genesis of the Massachusetts Town," *Proceedings of the Massachusetts Historical Society*, January 1982.

A summary of works that discuss the influence of German village form and agrarian tradition on English practice, including the work of W. Stubbs, F. W. Maitland, P. Vinogradoff, F. Seebohm, and H. L. Gray, can be found in C. S. and C. S. Orwin, *The Open Fields*. See also Erwin Nasse, *On the Agricultural Community of the Middle Ages and Inclosures of the Sixteenth Century in England*.

A brief discussion of German scholars who have considered these issues, including A. Meitzen and G. Hanssen, appears in F. Seebohm, *The English Village Community*. For a more current analysis of German investigation, see Joan Thirsk, "The Common Fields," *Past and Present* 29 (1964): 3–25, and T. Z. Titow, "Medieval England and the Open Field System," *Past and Present* 32 (December 1965): 86–102.

14. Franz Heske, *German Forestry*, 69–70; E. A. Gutkind, *Urban Development in Central Europe*. Vol. 1, *International History of City Development*, 89–99. See also: B. E. Fernow, *A Brief History of Forestry in Europe, the United States and Other Countries*, 35, and Alan Mayhew, *Rural Settlement and Farming in Germany*, 28.
15. Gutkind, *Urban Development*, 71–85; B. E. Fernow, *Brief History*, 31–32, and Heske, *German Forestry*, 70–71.
16. Gutkind, *Urban Development*, 97; Fernow, *Brief History*, 23–36.
17. Fernow, *Brief History*, 39; Gutkind, *Urban Development*, 129; and Heske, *German Forestry*, 71–72.
18. A. E. J. Morris, *History of Urban Form. Before the Industrial Revolution*, 109–11; Rolf Hager, ed., *The Zahringen New Towns*. See also New Brandenberg, founded in 1248, where settlers were given land both inside and outside city walls.
19. Heske, *German Forestry*, 71–73, 265–66; Mayhew, *Rural Settlement*, 48; Fernow, *Brief History*, 39. In 1368, for example, the city of Nuremberg sowed large burn areas with pine, spruce, and fir.
20. Lowood, "The Calculating Forester," in *Quantifying Spirit in the 18th Century*, 315–42; see also Fernow, *Brief History*, 41–91.
21. Lowood, "Calculating Forester," 315–42.
22. Heske, *German Forestry*, 94–7, 264–67.
23. Hosmer, "City, Town and Communal Forests," *Lumber World Review* 42 (May 10, 1992): 27–30.
24. Meister, *Die Stadtwaldungen von Zurich*, 54–67; Hager, *Zahringen New Towns*. Meister's work is ambiguous on the distinction between *Forst* and Sihlwald. At one point, he makes it clear that the two are separate; at other times, both seem to be included as one.
25. Meister, *Stadtwaldungen*, 54–72; Hosmer, "City, Town," 27–30.
26. Meister, *Stadtwaldungen*, 73–97.
27. Pinchot, "The Sihlwald," pts. 1–3, *Garden and Forest* (July 30, 1890): 374, (August 6, 1890): 386, (Aug. 13, 1890): 397–98. See also Hosmer, "City, Town,"

27–30; H. Reynolds, "Notes on European Town Forests, (1913)," from files of the Environmental League of Massachusetts.

28. Pinchot, "The Sihlwald," *Garden and Forest* (July 30, 1890): 374; (Aug. 6, 1890): 386; and (Aug. 13, 1890): 397–98. See also Hosmer, "City, Town," 27–30.
29. Hosmer, "City, Town," 27–30.
30. For Villigen, see "Economic History of a German City Forest," *Forest Worker* 6 (September 1930): 19–20. For Baden-Baden, see H. Reynolds, "Notes on European Town Forests (1913). For Forbach, see H. Reynolds, "Town Forests—a European Idea by Which America Might Profit," *American City* 11 (October 1914): 310–14. For Freiburg, see E. Meeman, "Green Spaces in German Cities," *Rotarian* 45 (November 1934): 16–18, 58–60.
31. Andrew D. Rodgers, *Bernhard Eduard Fernow*, 14–26, 51–79, 106–10.
32. Ibid., 122–29, 177–83, 207–9, 305–6.
33. Bernard E. Fernow, "Report upon the Forestry Investigations of the U.S. Department of Agriculture," 1877–98, 208. Rodgers, *Bernhard Eduard Fernow*, 61. Similar information appeared in Fernow's lectures at Yale in 1904–1905 on the history of forestry and in his book, *A Brief History of Forestry in Europe, the United States and Other Countries*, published in 1913.
34. Bernhard E. Fernow, "Communal Forests," *Garden and Forest* 3 (July 16, 1890): 349.
35. Ibid.
36. Sylvester Baxter, "The Lynn Public Forest," *Garden and Forest* (October 30, 1889): 526–27. See also "The Lynn Public Forest" (June 24, 1891): 290, and Wilhelmine Seliger, "German Forests as I Remember Them" (May 4, 1892): 214–15, in the same journal.

    Bernhard E. Fernow, "Report of Chief of Division of Forestry," in *Annual Report of the Secretary of Agriculture*, 1889. See specifically subheading "Forestry Interests in the United States," 290–91.
37. Harold T. Pinkett, *Gifford Pinchot: Private and Public Forester*, 20–30. W. A. Buchanan to B. E. Fernow, 5-23-93; B. E. Fernow to W. A. Buchanan, n.d.; U.S.F.S. Collection, National Archives.
38. National Academy of Sciences, "Report of the Committee Appointed by the National Academy of Sciences upon the Inauguration of a Forest Policy for the Forested Lands of the United States," May 1, 1897. See also Samuel T. Dana, *Forest and Range Policy*, 98–118.
39. Samuel P. Hays, *Conservation and the Gospel of Efficiency: The Progressive Conservation Movement, 1890–1920*.
40. Gifford Pinchot, "The Sihlwald," *Garden and Forest* (July 30, 1890): 374, (August 6, 1890): 386, (August 13, 1890): 397–98. While in France, Pinchot also wrote a report about the communal forest at Vandoeuvres that demonstrated coppice management; see Pinchot, *Breaking New Ground*, 12.
41. Gifford Pinchot, "Forest-policy Abroad," *Garden and Forest* (January 1, 1891): 8–9, (January 14, 1891): 21–22, (January 21, 1891): 34–35.
42. Samuel T. Dana, "Extent to Which Foreign Methods of Forest Administration are Suited to Conditions in the United States," in *Report of the National Conser-*

vation Commission, 375–89. See also Raphael Zon, "Foreign Sources of Timber Supply," 280–370.

43. In southwestern New Hampshire, for example, sash and door mills in Keene and surrounding towns began paying much higher prices for pine after extensive cutting by match companies had depleted the region's standing timber. See "State Pine Problem," *Boston Morning Herald*, March 9, 1902.
44. Ralph Widner's edited anthology, *Forests and Forestry in the American States*, contains a good review of state forestry programs, especially their origins.
45. George F. Talbot, "Forest Planting and Municipal Ownership of Forest Lands," in *First Annual Report of the Forest Commissioner of the State of Maine*, 1891.
46. "An Act for the Establishment of a Forestry Commission," Laws of New Hampshire, 1893, chap. 44, 36–37.
47. Ralph C. Hawley and Austin Hawes, *Forestry in New England*, 381–404.
48. "Forest Tree Planting in Massachusetts," *Garden and Forest* (January 2, 1889): 9–10.
49. Robin Winks, *Frederick Billings: A Life*, 293–300. See also J. R. Simmons, "The Older Forest Plantations in Massachusetts: Conifers," *Bulletin of the Massachusetts State Forester*, 1915; and Harold B. Kempton, *The Planting of White Pine in New England*, 12–24.

    Kempton describes yearly plantings of white pine and Norway spruce seedlings conducted on worn-out farms owned by Isaac Adams in Moultonboro, New Hampshire. Although unsuccessful as a commercial venture when the expense of planting was measured against the value of timber grown, Adams's experiment established that an acceptable rate of growth could be achieved through planting from natural seedlings. Kempton indicates that the Moultonboro plantation was begun in 1868 and concluded in 1882. However, correspondence from Adams's son to Gifford Pinchot in 1899 indicates the period of planting did not start until the spring of 1874. See series of letters, Isaac Adams to Gifford Pinchot and Henry Graves, all written in February 1899, U.S.F.S. Collection, National Archives.
50. See Jacob Manning to the Division of Forestry, 8-31-1899, U.S.F.S. Collection, National Archives. See also "A Rhode Island Forest," June 8, 1895, from an unidentified newspaper (Forest History Society Collection).
51. M. Richard Applegate, *Massachusetts Forest and Park Association: A History, 1898–1973*. The group's other founding members were Joseph Newell, its first president; James H. Bowditch; Myron S. Dudley; D. Blakely Hoar; Warren H. Manning; Jacob U. Pierce; and Walter C. Wright.
52. Harris A. Reynolds, "A Citizens' Movement to Reduce the Tax Rate," *Bulletin of the Massachusetts Forestry Association* 104 (May 1913).
53. Paul E. Bruns, *A New Hampshire Everlasting and Unfallen*, 1–39; see also William P. House, "Forty Years of Forestry," *Appalachia* 23 (December 1940): 203–11. Among its founding members were then-governor Frank W. Rollins, future-governor Nahum J. Bachelder, and Joseph T. Walker, whose father, Joseph B. Walker, had been an early champion of forestry conservation in New Hampshire.
54. See archival collection, Society for the Protection of New Hampshire Forests, Special Collections Library, University of New Hampshire. Ayres prepared a

handwritten index of the society's activities that provides a wealth of important data.

See also Philip W. Ayres, "State Aid to Town Forests—a Suggestion," in *Seventh Annual Report. Society for the Protection of New Hampshire Forests, 1909*, 32–33; "Forestry in New Hampshire," *Eighth Annual Report of the Society for the Protection of New Hampshire Forests, 1910*, 19–20.

55. Harris A. Reynolds, "The First Quarter Century of the Town Forest in Massachusetts," *Bulletin of the Massachusetts Forest and Park Association* 163 (December 1939): 7. The date of publication of Maxwell's articles by the society is uncertain. Ayres's index indicates 1914 and conflicts with Reynolds's statement that the Massachusetts group reprinted the articles in 1912. Reynolds also linked the publication to the passage of Massachusetts's municipal forest law in 1913.
56. "Get an Acre and Live On It," *Maxwell's Talisman* 6 (March 1906): 1; "Forestry and Tree Culture," *Maxwell's Talisman* 8 (September 1907): 43.
57. Maxwell based his statistics on a report by a German forester from Donaueschingen whose last name was Jaeger. Jaeger's report had found its way to Pennsylvania's State Forestry Reservation Commission, where Maxwell procured it.
58. John L. Cowan, "Forest to Pay City's Taxes," *Technical World Magazine*, November 1911, pp. 316–19. Max Watson was named city forester.
59. At their convention in Harrisburg in 1910, Pennsylvania's foresters listened to a presentation about the benefits of community-owned forests and considered examples at Lock Haven and Lewiston. See Forrest H. Dutlinger, "Municipal Ownership of Forests," *Forest Leaves* 12 (June 1910): 137–38. See also Charles Barnard, "Communal Forests," *The Rural New Yorker* 49 (July 5, 1890): 430–31.
60. Raphael Zon, "Communal Forests," in *Cyclopedia of American Agriculture*, 4:144–48; Arthur B. Recknagel, "County, Town and Village Forests," *The Cornell Reading Courses. Lesson for the Farm* 2 (May 15, 1913): 143–56.
61. A number of local water utilities sought assistance, among them Middletown and Waterbury, Connecticut. For Middletown, see A. K. Chittenden to the Middletown Water Co., 10-11-05; for Waterbury, see P. H. Danaher, 2-3-06; for Nashua, see Thomas Sherrard, 6-6-06; for Bar Harbor, see L. E. Spdycke to Gifford Pinchot, 10-6-06; all in U.S.F.S. Collection, National Archives.
62. "Brunswick's Experiment in Forestry," *Lewiston (Maine) Journal*, June 10, 1901 (Forest History Society Collection); "An American Town Forest," *The Forester* 6 (May 1900): 116; "The First Town Forest," *Forest and Stream* 54 (May 5, 1900): 341.
63. Frank Heyward, Jr., "Austin Cary—Yankee Peddler in Forestry," pts. 1 and 2, *American Forests* 61 (May 1955): 29–30, 43–44; (June 1955): 28–29, 52–53.
64. Austin Cary, "The Town Common," *Brunswick, (Maine) Telegraph*, March 2, 1901; "Forestry on the Town Common," *Brunswick (Maine) Telegraph*, February 26, 1902 (Forest History Society collection).
65. T. H. Root, "Westmore Church Forest, *Green Mountain State Forest News* 6 (June 1930): 3. See Deed dated 1-19-1900, from C. O. Gill to The First Congregational Church of Westmore, recorded 2–7-1900, Westmore Town Clerk.
66. Resolution of First Congregational Church of Westmore, 8-7-1904; Thomas H.

Sherrard to John T. Ritchie, 10-11-04; certification by Trustees of the First Congregational Church of Westmore, 7-23-34; John E. Nutting, Acting Vermont Conference Minister, to James E. Hildebrandt, 10-10-79, Records of the Vermont Conference of the United Church of Christ, Inc., successor to Vermont Domestic Missionary Society, Burlington.

67. B. E. Fernow, "Fundamentals of Forestry," in *Special Bulletin of the Forestry Association of Vermont*. Fernow's address was so well received that the president of the forestry association, W. J. Van Patten, printed and circulated more than a thousand copies throughout the state.
68. B. E. Fernow, "A Plan Adequate to Meet Our Needs for Wood Timber," *Forestry Quarterly* 11 (September 1913): 307; reprinted in the Society for the Protection of New Hampshire Forests, *12th Annual Report*, 1913, 35–43.
69. Massachusetts Laws of 1882, chap. 255, sec. 1–8, adopted May 25, 1882. See Senate Bill No. 1, 1882, for the text of Governor Long's address.
70. "The New Forestry Law of Massachusetts," *The American Journal of Forestry* 1 (April 1883): 304–7.
71. Sylvester Baxter, "The Lynn Public Forest," *Garden and Forest* 2 (October 30, 1889): 526–27.

    Massachusetts Laws of 1913, chap. 564, sec. 1–7, approved April 26, 1913. Laws of New Hampshire, 1913, chap. 27, sec. 1–4, approved March 14, 1913.
72. Vermont Public Laws of 1915, no. 24, sec. 1–5, approved April 1, 1915.
73. "Local Branch Forestry Associations," in *Tenth Report of the Society for the Protection of New Hampshire Forests*, (1911): 33; "Municipal Forests," in *Biennial Report of the Forestry Commission for the State of New Hampshire* (1913–14): 53–55.
74. Harris A. Reynolds, "Establishment of Foresters in Towns and Cities through State Forestry Associations," *American City* 9 (November 1913): 412–14.
75. Harris A. Reynolds, "An Opportunity for the Young Landscape Architect," *Landscape Architecture* 4 (October 1913–July 1914): 47–51.
76. Ibid., 50.

Chapter 4. Forestry for the People (Notes to Pages 132–166)

1. H. A. Reynolds, "Forest Revival in New England," *Forest Leaves* 15 (October 1916): 169–73. Reynolds made very similar predictions about the importance of town forests to the conservation movement in another article that year, "Town Forests and the Lincoln Highway," *American Forestry* 22 (March 1916): 174–75.
2. See H. A. Reynolds, "Town Forest Contest," *MFA Bulletin* 112 (June 1, 1914); "Seventeenth Annual Report of the MFA," *MFA Bulletin* 115 (December 10, 1914); "Town Forest Contest of 1916," *MFA Bulletin* 116 (October 1, 1915); "Eighteenth Annual Report of the MFA," *MFA Bulletin* 117 (1915).
3. P. Bunker, "A Town Forest in America," *Forest Quarterly* 13 (March 1915): 4–7; G. Hubbard, "The Municipal Forest in Fitchburg, Mass.," *The American City* 24 (February 1921): 121–24; "The First Town Forest," *American Forestry* 21

(November 1915): 1042; H. O. Cook, "The Fitchburg Town Forest," *American Forestry* 29 (October 1923): 592–93.

4. J. Hancock, "John Nolen: The Background of a Pioneer Planner," in *The American Planner, Biographies and Recollections*, 37–57. See also E. Engst and H. T. Hickerson, "Urban America: Documenting the Planners," Department of Manuscripts and University Archives, Cornell University, 1985.
5. J. Murphy, "The Town of Walpole Establishes a Communal Forest of 200 Acres," *American City* 15 (August 1916): 144–53; "First Town Forest Celebrates Fiftieth Anniversary," *Forest and Park News* 3 (spring 1966): 5; "Town Forests," *MFA Bulletin* 119 (1916): 6–7. See also Walpole Town Planning Committee, "Town Forests," in *Town Planning for Small Communities*, 101–24.

   See also J. Nolen to C. S. Bird, 9-5-14; John Nolen to H. A. Reynolds, 10-28-14; and H. A. Reynolds to J. Nolen, 10-29-14; all in John Nolen Collection, Cornell University Archives. See also G. Carlisle to C. S. Bird, 4-25-16, records of the Environmental League of Massachusetts.
6. W. Colton, "Organization and Management of Town Forests," *MFA Bulletin* 115 (1914).
7. For lists of early New Hampshire town forests, see "Town Forests," *Tenth Report of the SPNHF, 1911*: 33–35; and New Hampshire Forestry Commission, *Biennial Report 1915–16*: 95–97.
8. New Hampshire Forestry Commission, *Biennial Report 1915–16*, 95–97; *Biennial Report 1919–20*, 65–67; *Biennial Report 1921–22*, 78–81.

   See also "Contract and Agreement" between W. B. Hastings, Acting State Forester, and William H. Lund, 3-5-18; A. F. Hildreth to V. A. Beede, Assistant Forester, 3-14-16; both in records of the New Hampshire Division of Forests and Lands, Concord.
9. W. F. Hale, "The Warner Town Forest," *New Hampshire Forests* 4 (December 1927): 12–13; New Hampshire Forestry Commission, *Biennial Report 1919–1920*: 65–67; *1921–1922*: 78–80. See also Capt. L. H. Chandler to E. C. Hirst, State Forester, 3-1-19, records of New Hampshire Division of Forests and Lands, Concord.
10. J. W. Toumey, "County, City, Town and School Forests," *American Forestry* 22 (July 1916): 428–29.
11. "Why Massachusetts Needs Town Forests," *MFA Bulletin* 132 (October 1921).
12. "Town Forests," 68th Cong., 1st sess., S. Doc. 151; H. A. Reynolds, "Town Forests," *Playground* 18 (July 1924): 234–36, 258.
13. C. Pack, "Foreword," in H. A. Reynolds, *Town Forests. Their Recreational Value and How to Maintain Them*, 5.
14. Ibid., 14.
15. Reynolds, "Town Forests," 236.
16. "The Town Forest," *American Forestry* 22 (November 1916): 689.
17. P. Buttrick, *Town Forests and Parks for Connecticut*.
18. C. L. Pack, "Forests–State and Town," *Nature Magazine* 5 (February 1925): 117–18; H. A. Reynolds, "Town Forests and Farming," *World Agriculture* 6 (July 1926): 440.

19. Massachusetts Acts and Resolves, 1920, chap. 604, pp. 668–69; see specifically sec. 5. See also "Why Massachusetts Needs Town Forests," *MFA Bulletin* 132 (1921); H. A. Reynolds, "Town Forests—the Massachusetts Plan," *MFA Bulletin* 137 (January 1924); H. A. Reynolds, "Town Forests in Massachusetts," *MFA Bulletin* 145 (January 1, 1927); "Wood-Using Industries of New Hampshire," *New Hampshire Forests* 6 (December 1929): 12, 14.
20. MFA, "26th Annual Report," *MFA Bulletin* 138 (1923); H. Reynolds, "Town Forests in Massachusetts—Report of Progress," *MFA Bulletin* 139 (November 1924); "Summarized Report of the Third Annual Conference of Town Forest Committees," *MFA Bulletin* 150 (1930). For information on Petersham's town forest, see committee reports, 1930–1938, records of Harvard Forest, Petersham, Mass.
21. Groton, Mass., *Annual Reports* (1922): 11–13, (1935): 73–74; H. S. Ripley, "Forest Management Plan for the Groton Memorial Town Forest" (July 1952), records of the New England Forestry Foundation, Boston.
22. "Second Conference of Town Forest Committees," *MFA Bulletin* 147 (November 8, 9, 1929): 4; H. A. Reynolds, "The Russell Town Forest," *MFA Bulletin* 170 (1946); MFA, "Town Forests," *Forest and Park News* 10 (December 1946): 1, 3.
23. New Hampshire Forestry Commission, *Biennial Reports* (1913–14): 55; (1923–24): 63–79; (1925–26): 69–73; (1929–30): 75–80.For an example in *New Hampshire Forests*, see "Winchester Given Town Forest," 4 (June 1927): 9. Most of New Hampshire's woodland records for town forests are located in the Department of Resources and Economic Development, Division of Forests and Lands, Concord. Material is also available at Fox State Park, Hillsborough.
24. New Hampshire Forestry Commission, *Biennial Reports* (1913–14): 55; (1923–24): 63–79; (1925–26): 69–73. See also "An Act Relating to Public Forest Lands," *Laws of New Hampshire* (1929), chap. 74, pp. 87–88. For a discussion of Foster's paper, see "Massachusetts Town Forest Committees Meet," *New Hampshire Forests* 5 (December 1928): 1–2.
25. New Hampshire Forestry Commission, *Biennial Reports* (1923–24): 76–79; (1925–26): 74–75; (1927–28): 144–46. For Northwood, see Report of the Northwood Forestry Committee (3-10-25), New Hampshire Division of Forests and Lands, Concord. For Danville, see records of the Danville Parsonage Committee, Danville. For Winchester, see "Winchester Given Town Forest," *New Hampshire Forests* 4 (June 1927): 9. See also New Hampshire Forestry Commission, *Biennial Reports* (1927–28): 146–47. For Richmond, see New Hampshire Forestry Commission, *Biennial Reports* (1925–26): 79–80, and "Unusual Town Forestry Work of Richmond, New Hampshire," *Forest Worker* 3 (May 1927): 4.
26. New Hampshire Forestry Commission, *Biennial Reports* (1927–28): 151, (1947–48): 53.
27. New Hampshire Forestry Commission, *Biennial Reports* (1927–28): 147–48.
28. F. E. Garvin to Arthur Winslow, 9-1-29; Arthur Winslow to Frank E. Garvin, 9-4-29; records of the New Hampshire Division of Forests and Lands, Concord. See also Deed, Winslow Corporation to Town of Dunbarton, 1-7-30, and Resolution, Board of Selectmen and Town Forest Committee, 3-9-37, both in records of the New Hampshire Division of Forests and Lands, Concord. See also J. Fos-

ter, "Dunbarton Town Forest," *New Hampshire Forests* 7 (June 1930): 9; New Hampshire Forestry Commission, *Biennial Reports* (1929–30): 77–78.

29. Vermont Commissioner of Forestry, *Biennial Reports* (1926): 25; (1930): 41–43. Public Laws of Vermont (1915): chap. 24, (1917): chap. 26. The forty-acre requirement was finally eliminated in 1973; see Public Acts of Vermont (1973 Adj. Session), no. 148.
30. For an example of Vermont's reports, see Vermont Commissioner of Forestry, *Biennial Report* (1922–24): 37–38.
31. Public Acts of Vermont (1945), no. 86; Public Acts of Vermont (1951), no. 74.
32. R. Ross, *Town Village and City Forests in Vermont*, Vermont Forest Service Publication no. 29 (December 1924); K. Flint, "Forestry as a Municipal Undertaking," *The Vermont Review* 2 (July–August 1927): 41–44.
33. Public Laws of Vermont (1915): chap. 24; (1927): no. 11; (1929): no. 13; Vermont Resolutions (1933), no. 232: 292. For Billings's letter, see "Town Forests," *Green Mountain State Forest News* 2 (February 1926): 1. For nursery activity, see "Forest Planting," *Green Mountain State Forest News* 3 (December 1929): 1. For Vermont Forestry Association activity, see K. Flint, "Forestry as a Municipal Undertaking," *The Vermont Review* 2 (July–Aug. 1927): 41–44; also, "Vermont Association Offers Trees for Municipal Forests," *Forest Worker* 2 (January 1926): 10.
34. "Municipal Forest of Essex Junction," *Green Mountain State Forest News* 2 (March 1926): 1; "Essex Jct. Village Forest Pays," *Green Mountain State Forest News* 6 (February 1930): 3. See also U.S. Forestry Service, "Essex Junction, Vermont," in *Field Handbook of Community Forests*, L-12 (1939). Several smaller towns, lacking incorporated cities within their boundaries (Chester and Northfield, for instance), also established watershed forests for their villages during this same period.
35. "Town Forests," *Green Mountain State Forest News* 2 (April 1926): 3; "Municipal Forests of Vermont," *Green Mountain State Forest News* 3 (July 1927): 9; Vermont Forestry Commissioner, *Biennial Report* (1932): 55–57; and "Municipal Forests," *Green Mountain State Forest News* 4 (October 1926): 6; see also records of the Vermont Forest Service, Pittsford District Office.
36. Calais, Vt., *Annual Reports* (1926): 19–20, (1927): 11–12, (1928): 17–18, (1929): 11–12, (1930): 19; "Report of Town Forester," *Green Mountain State Forest News* 2 (April 1926): 4. See also deed dated 4-3-43, records of Calais Town Clerk, vol. 23, p. 5.

    For St. Johnsbury, see Clare B. Johnson, *History of St. Johnsbury*; Vermont Forest Service, "Municipal Forests," *Green Mountain State Forest News* (October 1926): 4.

    See also Plan of the Lyndonville Forest Reserve, also known as the Morrill Pasture, by W. F. Smith, dated 1930; Records of the Lyndon Town Clerk, Lyndonville, Vt. For a discussion of this land, see Vermont Commissioner of Forestry, *Biennial Report* (June 30, 1930): 41–43.
37. Public Laws of Maine (1927), chap. 33; Public Laws of Rhode Island (1929): chap. 1389; Public Laws of Connecticut (1939), supp. sec. 152e.
38. Maine Forest Commissioner, *Biennial Reports* (1891): 51–60; (1921–22): n.p.

39. Ibid. (1929–30): n.p. For Fryeburg, see "Fryeburg Builds New School with Town Forest Money," *Green Mountain State Forest News* 6 (September 1930): 3. For Old Town, see "Town Farm Becomes City Forest," *New Hampshire Forests* 4 (September 1927): 9. See also plan entitled *Brunswick Town Common*, an unsigned forest type map dated 1926, Brunswick Department of Public Works.
40. For Farmington and Ellington, see T. Hewes, "The Farmington Town Forest," *Connecticut Woodlands* 4 (September 1939): 53–54; C. Bickford, *Farmington in Connecticut*, 388; and Connecticut Park and Forest Commission, *Annual Report* (1928): 104. For Ellington, see also deed dated 2-28-27, records of Ellington Town Clerk, vol. 39, pp. 454–57.

    For Salisbury's Barrack-Matiff Town Forest, see deed dated 11-4-05, recorded in vol. 41, p. 21, and deed dated 9-2-05, recorded in vol. 39, p. 648; see also a plan titled *Barackmatiff Town Forest* by Kiefer and Quaile in 1955; records of the Salisbury Town Clerk. For Salisbury and Newtown, see Buttrick, *Town Forests and Parks*, 15.

    For each town, see records of the Connecticut Department of Environmental Protection, Division of Forestry, Hartford.
41. Deed dated 2-18-27, records of Norwich Town Clerk, vol. 2, p. 133. See plan titled *The Solomon Lucas Memorial Woods*, referred to as the Norwich Town Forest, by Chandler and Palmer in 1927, records of the Norwich Town Clerk.
42. Rhode Island State Board of Agriculture, *Annual Reports* (1933): 14, (1938): 88.
43. H. A. Reynolds, "Some Indirect Benefits from Town Forests," *The Illustrated Canadian Forest and Outdoors* 20 (October 1924): 602.
44. J. W. Toumey, "Town Forests," *American Forestry* 28 (February 1922): 96, 113.
45. R. S. Hosmer, "City, Town and Communal Forests," *Lumber World Review* 42 (May 10, 1922): 27; and "Town Forests and Community Chests," *American Forestry* 29 (March 1923): 155–57.

    N. C. Brown, "Possibilities of Municipal Forestry in New York," *Bulletin of the New York State College of Forestry at Syracuse University*, 14 (June 1914): 1–19.

    G. Pratt, "Communal Forests for America," *Town and County Edition of American City* 22 (June 1920): 563–66.

Chapter 5. Salesman's Samples (Notes to Pages 171–199)

1. H. A. Reynolds, *Report on the First Conference of Massachusetts Town Forest Committees* (January 10, 1929). Percival S. Ridsdale, editor of the American Tree Association's publication, *Nature Magazine*, offered similar commentary.
2. "Second Conference of Town Forest Committees, Westfield and Russell Town Forests, November 8 and 9, 1929," *MFA Bulletin* 147, n.d.; "Summarized Report of the Third Annual Conference of Town Forest Committees," *MFA Bulletin* 150, n.d.
3. M. Applegate, *Massachusetts Forest and Park Association, A History*, 20–21; see also letter to Town Forest Committees from H. A. Reynolds, 11-21-33, records of the Environmental League of Massachusetts. See also Massachusetts State For-

ester, *Annual Reports*, 1930–39. For example, numbers ranged from a high of 490,000 in 1930 to a low of 214,500 in 1934, with 375,000 about average from 1930 to 1939.

4. H. A. Reynolds, "The Town Forest in Massachusetts: A Twenty-Two Year Record," *MFPA Bulletin* 159 (December 1936). Other MFPA publications included "Forest Weeding with Special Reference to Young Natural Stands in Central New England," and "Improvement Cutting and Thinning as Applied to Central New England Hardwoods."
5. For Groton, see U.S. Forest Service, *Field Handbook of Community Forests*, (1939): L-20. For Norton, see "Reports of Reforestation Committee," in Norton, Mass., *Annual Reports* (1929): 19–21, (1930): 46–49, (1933): 21–23, (1936): 30–32. Norton's committee was chaired for years and years by Gertrude Cornish Milliken.
6. W. F. Hale, "Warner Holds First Town Forest Meeting," *New Hampshire Forests* 7 (December 1930): 13, 16.
7. "Co-operation with Towns in Forest Management," *Forest Notes* 2 (January 1938): 2–3; New Hampshire Forestry and Recreation Commission, *Biennial Report* (1937–38): 25–30.
8. "Working Plans for Town Forests," *Green Mountain State Forest News* 6 (December 1930): 4; "Silvicultural Operations on Essex Junction Village Forest," *Green Mountain State Forest News* 7 (January 1931): 1; "Care of the Forest," *Green Mountain Forest News* 8 (February 1932): 1.
9. For New Hampshire, see "Cooperation with Towns in Forest Management," *Forest Notes* 2 (January 1938): 2–3; see also a forest type map for woodland belonging to the town of Litchfield by H. K. Baldwin, c. 1938, WPA Project No. 1323, NHDF-C; for Vermont, see "Unemployment and Municipal Forestry," *Green Mountain State Forest News* 7 (November 1931): 1; and Vermont Commissioner of Forestry, *Biennial Report* (June 30, 1932): 55–56; for Connecticut, see Records of the Division of Forestry, Connecticut Department of Environmental Protection.
10. "Town Forests Endorsed by U.S. Forest Service," *Forest and Park News* 3 (January 1939): 1, 4; H. A. Reynolds, "Town Forest in Massachusetts." Concord, Plainville, Hanover, Woburn, and West Bridgewater were among the fifteen new towns to create forests during this period.

    New Hampshire Forestry and Recreation Commission, *Biennial Report, 1937–38*, 25–30; for Vermont, see Vermont Forestry Commissioner, *Biennial Report, 1934*, which indicates that 43 municipalities had established forests; for Connecticut, see reports of the State Park and Forest Commission.
11. "Formulating a Forest Policy," *The Timberman* 21 (February 1920): 1–2.
12. 43 U.S. Statutes 653, 68th Cong., 1st sess., (June 7, 1924), chap. 348, sec. 4.
13. H. A. Reynolds, *Town Forests*, 78. For example, see L. C. Everard, "Town Forests," *Forest Worker* 1 (September 1925): 27–28.
14. U.S. Forest Service Collection, National Archives. See Central Files of the Division of State and Private Forestry, 1913–44, boxes 21, 22.
15. U.S. Secretary of Agriculture, *A National Plan for American Forestry* (Washington, D.C.: GPO, 1933). See also E. H. Clapp, "The Philosophy of the Copeland

Report," unpublished memorandum, U.S. Forest Service Collection, National Archives, "Copeland Report Data, 1923–33."

16. "Copeland Report Data, 1923–33," U.S. Forest Service Collection, National Archives.
17. U.S. Dept. of Agriculture, *A National Plan for American Forestry* (Washington, D.C.: GPO, 1933). See specifically, C. R. Tillotson, "Community Forests," 843–47. Tillotson's report identified town forests nationwide, but in the case of New England his tallies seem to rely more on biennial reports from forestry commissions than on the questionnaires. Numbers are as follows: Connecticut, 23; Maine, 8; Massachusetts, 90; New Hampshire, 82; and Vermont, 42.
18. F. Roosevelt, "A Debt We Owe," *The Country Home* 54 (June 1930): 12–14. See also Edgar B. Nixon, ed., *Franklin D. Roosevelt and Conservation, 1911–1945*, 71–76.
19. "Nelson Courtland Brown Retires," *Yale Forest School News* 39 (October 1951): 69–70.
20. See "Nelson Brown Scrapbook," Terence J. Hoverter College Archives, Franklin Moon Library, State University of New York College of Environmental Science and Forestry, Syracuse, N.Y.
21. N. C. Brown, "Possibilities of Municipal Forestry in New York," *Bulletin of the New York State College of Forestry* 14 (June 1914).
22. N. C. Brown, "Municipal Forestry," *American Forestry* 28 (December 1921): 777–82. For a chronological listing of Brown's articles, see Bibliography.
23. N. C. Brown, "Community Forestry: A Neglected Phase of the American Forestry Program," *Journal of Forestry* 36 (July 1938): 687–94; "The First Community Forest," *American Forests* 44 (September 1938): 406–8; "Community Forests Come of Age," *American Forests* 45 (January 1939): 16–19, 46–47.
24. N. C. Brown, "Progress in Community Forests," *Journal of Forestry* (December 1939): 25–28; "Community Forests—a New Idea in the American Forestry Program," *Southern Lumberman* 161 (December 15, 1940); H. A. Reynolds, "The First Quarter Century of the Town Forest in Massachusetts," *MFPA Bulletin* 163 (December 1939). See also "Town Forests Endorsed by U.S. Forest Service," *Forest and Park News* 3 (January 1939): 1, 4.
25. U.S. Forest Service, *Community Forests in the United States*, B-26, n.d. (probably 1939); see also "Memorandum to United States Forest Service" (questionnaire circulated in 1941), records of the Central File of the Division of State and Private Forestry, 1913–44, U.S. Forest Service Collection, National Archives. Silcox reported preliminary estimates in 1939 showing more than 1,500 forests in twenty-seven states on roughly 3 million acres. See also letter from Earl S. Pierce to all regional foresters, 2-14-38, and Earl S. Pierce to A. C. Cline, 2-14-38, same collection.

    See also F. A. Silcox, "The Community Forest as a Public Asset," *The Kiwanis Magazine*, March 1939, pp. 142–43, 188; and U.S. Forest Service, *Report of the Chief of the Forest Service* (1939): 17. A later definition included YMCA forests.
26. N. C. Brown, "Program Started for Community Forests, Neighborhood Woods and Town Forests," *Forestry News Digest, Southern Issue* (May 1938): 29.

27. New Hampshire Forestry Commission, *Biennial Report* (1927–28): 144–46. See also E. O. Buhler, "Acres of Diamonds," unpublished manuscript dated 11-21-38, and W. F. Hale, "Town Forest Helps Pay Cost of Preaching," undated, unsigned, unpublished manuscript, both in records of the New Hampshire Division of Forests and Lands, Concord and Hillsborough.

    See also E. O. Buhler to A. H. Collins, 11-23-38; A. H. Collins to E. O. Buhler, 11-28-38; E. O. Buhler to A. H. Collins, 12-1-38; W. F. Hale to C. M. Collins, 12-13-18; all in records of the New Hampshire Division of Forests and Lands, Concord and Hillsborough. See also N. C. Brown, coll., *Community Forests* (1939), 20–21.

    At first, Buhler estimated yields as high as $5.19 per acre for a thirty-year period, including interest income from the parsonage fund. Later calculations for a hundred-year period were lower, $4.45. However, that amount could have been higher had a share of profits been used to improve the forest.
28. W. P. House, "Danville Town Forest. Report on Visit to Danville 12-2-37," and "Danville Town Forest, 2-13-38," both located in SPNHF Collection, Special Collections Department, University of New Hampshire.
29. *Community Forests in the United States*, undated but probably issued in late 1939, is typical. See also U.S. Forest Service, *Field Handbook of Community Forests*, 1939; *Community Forests—a Program for Clubwomen*, undated but probably compiled prior to 1945; and *Community Forests: A Bibliography*, 1939.
30. Claude R. Wickard to Franklin D. Roosevelt, 11-12-41, in E. B. Nixon, ed., *Franklin D. Roosevelt and Conservation 1911–1945*, 538–39.
31. U.S. Forest Service, Collection, National Archives, Records of the Central File of the Division of State and Private Forestry 1913–44, box 24.
32. G. Duthie to H. A. Reynolds 10-19-48, SAF Collection, Forest History Society, box 110, folder 434.
33. *U.S. Statutes at Large*, 52 (1938): 1452.
34. For Silcox's report, see "Summary of Recommendations Presented by the Forest Service on February 16, 1940, to the Joint Congressional Committee on Forestry with respect to A Forest Program for the United States," records of the National Forest Products Association, Forest History Society.

    For the Fulmer Act, see 49 Stat. 963 (August 29, 1935).
35. *Forest Lands of the United States*, Report of the Joint Committee on Forestry, 77th Cong., 1st sess., March 24, 1941, H. Doc. 323.

    See also C. Wickard to Franklin D. Roosevelt, 5-12-41; Franklin D. Roosevelt to Secretary Wickard, 5-14-41; E. H. Clapp to John H. Bankhead, 10-10-41, all in U.S. Forest Service Collection, National Archives, Records of the Office of the Chief (Clapp), RG 200, box 12.
36. Memorandum from Wayne Coy to Franklin D. Roosevelt, 5-16-42, in Nixon, *Franklin D. Roosevelt*, 549, 553.
37. H. A. Reynolds, "The Place of Community Forests in Conservation," pts. 1 and 2, *Forest Leaves* 31 (January–February, March–April 1941): 1–2, 7–8, 12–14, 16–18.
38. "Report of the Committee on Community Forests," *Journal of Forestry* 40 (February 1942): 112–17.

39. Ibid.; see also 42 (May 1944): 348–51.
40. Ibid., 40 (February 1942): 112–17; 44 (November 1946): 1005–7.
41. H. Clepper to H. A. Reynolds, 9-1-48; memorandum from H. Clepper, 10-19-48; and N. C. Brown to H. Clepper, 11-5-48; all in SAF Collection, Committee on Community Forests, Forest History Society, Box 110, folder 434. For reference to SAF sections, see "Report of the Committee on Community Forests," *Journal of Forestry* 41 (May 1943): 332–34; 47 (February 1949): 132–34; and 47 (December 1949): 1009–10.
42. "Report of the Committee on Community Forests," *Journal of Forestry* 47 (February 1949): 132–34; "Report of AFA Board of Directors," 1-28-49, AFA Collection, Forest History Society, box C2.

    See H. A. Reynolds to S. L. Frost, 2-18-49 and 3-10-49, and reply from S. L. Frost to H. A. Reynolds, 3-11-49; both in AFA Collection, Forest History Society.
43. "Report of the Committee on Community Forests," *Journal of Forestry* 49 (February 1951): 132–34.
44. H. A. Reynolds to H. Clepper, 7-6-51, SAF Collection, Committee on Community Forests, Forest History Society. See also "Report of Committee on Community Forests," *Journal of Forestry* 50 (February 1952): 150–51.
45. Memorandum from G. Duthie to SAF committee members Brown, Reynolds, Charles Geltz, E. V. Jotter, G. S. McIntire, J. W. Myers, Jr., and William A. Sylvester, undated. Duthie's memo was almost certainly written in April 1952, shortly after his appointment by the AFA the previous month. SAF Collection, Committee on Community Forests, Forest History Society, box 110, folder 434.

    See also "Report of Committee on Community Forests," *Journal of Forestry* 51 (January 1953): 48–49, and G. Duthie, "The Community Forest," *American Forests* 59 (May 1953): 14–15, 36–37. At the time of Duthie's census, communities had established 3,226 county, municipal, school, and other local forests on almost 4½ million acres of land.
46. Memorandum from Henry Clepper, 1-19-54, SAF Collection, Committee on Community Forests, Forest History Society, box 110, folder 434.
47. H. A. Reynolds, "Town Forests—a Neglected Opportunity," *Journal of Forestry*" 37 (May 1939): 364–67. For yield prediction, see H. A. Reynolds, "The First Quarter Century of the Town Forest in Massachusetts," *MFPA Bulletin* 163 (December 1939): 9.
48. T. Noel Stern, "Community Forests and Parks in Pennsylvania," Ph.D. dissertation, University of Pennsylvania, 1945. See also "A Community Forests Program," *Forest Leaves* 29 (October 1939): 1–2, 19–21; and "Pennsylvania's Community Forests," *Conservation* 7 (May–June 1941): 31–32.
49. Letter to Spencer H. Palmer, N.Y. State College of Forestry, 2-24-33, probably from A. C. Cline; see also "The Management of a Small Forest for Sustained Yield: The Petersham Town Forest," undated but probably 1926. Both in records of Petersham Town Forest, Harvard Forest, Petersham, Mass.
50. For reference to water boards, see "Town Forests Endorsed by Forest Service," *Forest and Park News* 3 (January 1939): 4.
51. "Town Forests in the Northeast," *Forest and Park News* 4 (November 1940): 3.

52. "Town Forest Idea Sweeping the Nation," *Forest and Park News* 5 (November 1941): 1, 4.
53. M. Richard Applegate, *Massachusetts Forest and Park Association: A History*, 12–13, 40. See also *New England Forestry Foundation*, undated pamphlet issued by the NEFF (prior to 1954). As of that date, management centers were located in Massachusetts, New Hampshire, Maine, and Vermont.

    For NEFF plans in Massachusetts, see "A Survey of the Town Forests in Massachusetts," *MFPA Bulletin* 171 (Nov. 1949), 4–5, and records of the NEFF. For New Hampshire plans, see records of the NEFF. Later, in a much different capacity, Mesavage authored an article, "Community Forestry as a Substitute for Illegal Mining," *Journal of Forestry* 39 (June 1941): 569–70, which encouraged towns in northern Pennsylvania to employ bootlegging miners in local forests.
54. Applegate, *Massachusetts Forest and Park Association*, 30–32; Petersham Town Forest committee, *Report of the Town Forest Committee*, 1938. See also H. A. Reynolds, "Bringing Forests to the People," *MFPA Bulletin* 165 (November 1941); at the time of that publication, 124 towns had established forests on 37,430 acres and had planted 7,654,814 trees. A trust fund to advance town forests was given to MFPA in 1938 by Mrs. Harriett L. Hemenway in memory of her husband, August Hemenway.
55. "Town Forests Aiding in Wood Fuel Campaign," *Forest and Park News* 6 (December 1942): 1, 4; "War Demonstrates Value of Town Forests," *Forest and Park News* 7 (November 1943): 1, 4;
56. "Town Forestry Ideal Post-War Employment," *Forest and Park News* 8 (November 1944): 1, 3; H. A. Reynolds, "A Survey of the Town Forests in Massachusetts," *MFPA Bulletin* 171 (November 1949). Miner's report recognized that sale or acquisition of lands consistently played a factor in the total number and acreage of town forests.
57. N.H. Forestry and Recreation Commission, "Town Forests," *Biennial Report* (1941–42): 38; "Town and City Forests," *Biennial Report* (1949–50).
58. NEFF, "Report of Woodland Examination (Louden)," SPNHF Collection, Special Collections Department, University of New Hampshire; "Chopping Bee Scheduled at Hollis Town Forest," *Forest Notes* 15 (December 1951): 30; "Hollis Town Forest Sets Good Example," *Forest Notes* 36 (November 1952): 12–13; "Plan for Better Town Forests," *Forest Notes* 15 (October 1951): 10. For Walpole and Dover, see New Hampshire Forestry and Recreation Commission, *Biennial Report* (1949–50): 46–47.
59. Vermont Forest Service, *Biennial Report* (1949–50): 20; (1957–58): 26–28.
60. J. Wilkinson to Sanborn Partridge, 2-2-66, Vermont Forest Service, Pittsford District Office. Both Royalton and Danville, Vermont, have fire district forests classified under the state's municipal forest law.
61. B. Miner, "A Town Forest for Troy," *American Forests* 50 (November 1944): 552–54.
62. E. Keenan, "Good Forestry Practices Awards for Municipal Forestry," unpublished manuscript, records of Department of Forests, Parks and Recreation, District Office, Pittsford, Vt.
63. See T. G. Gregoire, J. E. Carroll, J. P. Barrett, C. J. Mitchell, and J. H. Gove, *A*

Survey of the Town Forest Resources in New Hampshire (1980), and J. Barrett and S. Baumann, "Overlooked and Underappreciated," *Forest Notes* 190 (November–December 1991): 20–22. See also J. Dirkman, "Town Forests: Their Place in Forestry in Maine"; and "Green Acres Land Acquisition Act of 1964," Rhode Island Acts and Resolves, January sess., 1964, chap. 174.

Chapter 6. Old Pasture to Young Pine (Notes to Pages 203–230)

1. G. P. Marsh, *Man and Nature*, 176–79, 195–96.
2. Ibid., 215.
3. F. Starr, "American Forests: Their Destruction and Preservation," 230.
4. American Association for the Advancement of Science, Forestry Committee, "Cultivation of Timber and the Preservation of Forests," in *Reports of the Committees of the House of Representatives*, 43rd cong., 1st sess., 1873–74, H. Doc. 259 serial 1623, 20, 79, 81–83, 88.
5. American Association for the Advancement of Science, Forestry Committee, "Report upon Forestry for 1877," 221–83. For reasons unknown, this report was not included in the annual report of the Department of Agriculture.
6. J. A. Warder, "Forestry and Its Needs," paper delivered to the American Forestry Association meeting, Washington, D.C., February 19–21, 1878.
7. N. H. Egleston, "Report of Chief of the Forestry Bureau" (1884), 157–58.
8. "The Forestry Congress," *Cincinnati Enquirer*, April 1, 1882 (Forest History Society collection).
9. *Proceedings of the American Forestry Congress*, Cincinnati (April 1882), Montreal (August 1882), Boston (1885), Denver (1886). William Saunders chaired Section C, and M. C. Read presented "Forest Reserves at the Head Waters of the Ohio."
10. A. Rodgers, *Bernhard Eduard Fernow*," 96–98; Fernow's article appeared in the May 1884 issue of *Forestry Bulletin*, pp. 8–9. See also Fernow, "Report of Chief of Forestry Division" in *Annual Report of the Commission of Agriculture, 1886*; see specifically subheading "Climate Influence," 152–53.
11. B. E. Fernow, "Report of Chief of Forestry Division," in *Annual Report of Commissioner of Agriculture, 1888*; see specifically subheading "Forest Influences," 602–18.
12. Ibid.
13. B. Fernow, "Report of Chief of Forestry Division," in *Annual Report of the Secretary of Agriculture, 1889*; see specifically subheading "Influence of Forests on Water Supplies," 297–330; "Forests and Floods," *Garden and Forest* (January 1, 1890): 9–10; "Notes on the Climatic Influence of Forests," *Garden and Forest* (March 29, 1893): 147–48.
14. Fernow, "Forest Influences," U.S. Dept. of Agriculture, Bulletin No. 7.
15. National Academy of Sciences, *Report of the Committee Appointed by the National Academy of Sciences upon the Inauguration of a Forest Policy for the Forested Lands of the United States to the Secretary of the Interior, May 1, 1897*.

16. J. W. Toumey, "The Relation of Forests to Stream Flow," in *Yearbook of the United States Department of Agriculture. 1903*, 279–88. See also S. P. Hays, *Conservation and the Gospel of Efficiency. The Progressive Conservation Movement, 1890–1920*, 22–27, for a discussion of the role of western states in conservation of water resources.
17. P. P. Wells, "In the Forest Service Law Office," *Forest History* 16 (April 1972): 23–29. See also R. Hawley and A. Hawes, *Forestry in New England*, 374–78.
18. "The Forest Cover on the Watershed Examined by the Geological Survey in the White Mountains, New Hampshire," in M. O. Leighton, A. C. Spencer, B. McKay [*sic*], and others, *The Relation of Forests to Stream Flow*, U.S. Department of Interior, Geological Survey, Administrative Report, 1913.
19. New Hampshire Laws (1881), chap. 117.
20. N. H. Egleston, "Report of Chief of Forestry Division," in *Annual Report of the Commissioner of Agriculture, 1885*, 192–95.
21. A. Messer, "Preserving Old Forests vs. Reforesting Newly Cut or Waste Land," *The New England Farmer*, January 5, 1901.
22. G. Swain, "The Influence of Forests upon the Rainfall and upon the Flow of Streams," *Journal of the New England Water Works Association* 1 (March 1887): 11–26.
23. *Journal of the New England Water Works* 22 (December 1908): 389–408, 25 (June 1911): 243–46), 30 (March 1916): 64–85, 31 (June 1917): 247–55.
24. F. W. Rane, "The Reforestation of Watersheds for Domestic Supplies," *Journal of the New England Water Works Association* 25 (June 1911): 234–42. Reprinted in *Eighth Annual Report, The State Forester of Massachusetts, 1911*, 58–61; reprinted in *Bulletin of the Massachusetts Forestry Association*, n.d., n.n.
25. J. W. Toumey, "Effects of Deciduous and Coniferous Forests on Rainfall and Run-off," *Journal of the New England Water Works Association* 40 (March 1926): 40–43. See also C. E. Behre, "Are Further Studies Needed on the Relation of Forests to Water Supply in New England?" *Journal of the New England Water Works Association* 46 (June 1932): 170–83.
26. For an example of a small system, see George W. Lane, "Plan of Land—Northfield and Tilton, N.H. Water Works," 1888, New Hampshire State Archives, Concord.
27. *Journal of the New England Water Works Association* 10 (September 1895): 55–63, 15 (March 1901): 272–87. *Records of the Hartford Water Commissioners* 8 (December 17, 1912): 108.
28. Comparative review of watershed forestry among New England states is a topic that summons further inquiry and promises results that extend into the topics of water conservation and quality.
29. R. C. Hawley, "Forest Planting for Water Companies," *Proceedings of the Connecticut Forestry Association, 1909–11*, Publication no. 7; and *Guide to the Maltby Tract of the New Haven Water Company* (New Haven, Conn.: n.p., 1920). J. W. Toumey, "Forestry In Relation to Public Water Supplies," *Journal of N.E.W.W.A.* 31 (June 1917): 247–55; U.S. Forest Service, *Field Handbook of Community Forests*, L-27.

30. In October 1900, the letter to Graves somehow landed on the desk of Overton Price at the U.S. Forest Service's office in Washington. Graves, who left Gifford Pinchot's ranks to become director of Yale's program that year and who took James Toumey with him, put off his visit to Middletown and probably referred the matter to Price. Possibly confused about who was doing what, the Forest Service responded with a copy of *Circular 21* but not until 1905. See R. N. Jackson to H. S. Graves, 8-11-1900 and 9-26-1900; A. K. Chittenden to Middletown Water Co., 10-11-05. U.S. Forest Service Collection, Office of State and Private Cooperation, Division of Timber Management, 1896–1908, National Archives.

    Kornel Bailey, "Forest History, 1901–1921," unpublished account of entries in ledger book labeled "Forestry–Mt. Higby," records of the Middletown Water Department. See also *Middletown City Report* (1904): 46–47.
31. Bailey, "Forest History." See also Connecticut Forest and Park Association, "Forestry at Mt. Higby Reservoir," 75th Anniversary Field Meeting, October 17, 1970, which contains a short history of the forest.
32. L. W. Goodrich, "Forestry," *Journal of the New England Water Works Association* 24 (September 1910): 345–52. See also "Water Department Plans for Future," *Hartford Telegram*, October 14, 1901, records of the Forest History Society; and records of the Hartford Board of Water Commissioners, vol. 4 (7-19-04): 305, vol. 7 (4-11-12): 294, vol. 8 (7-16-12): 45–46, vol. 8 (9-17-12): 64–65, and vol. 8 (12-17-12): 106. Records are now held by the Metropolitan District Commission.
33. P. H. Danaher to U.S. Department of Agriculture, 2-3-06, and reply dated 2-10-06, U.S. Forest Service Collection, Office of State and Private Cooperation, Division of Timber Management, 1896–1908, National Archives. See also P. Buttrick, *Town Forests and Parks for Connecticut*, 16.
34. New Hampshire Forestry Commission, *Biennial Report of the Forestry Commission, 1925–26*, 81–83. Concord Town Reports (1875): 48, (1895): 176, (1907): 374, (1908): 178, (1913): 270–71.
35. T. Sherrard to C. Lyford, 6-5-06, U.S. Forest Service Collection, Office of State and Private Cooperation, Division of Timber Management, 1896–1908, National Archives. See also B. Mackaye, "A Working Plan for the Woodlands of the Pennichuck Water Company, Hillsborough County, N.H.," Mackaye Collection, Dartmouth College Archives, box 178, folder 15.
36. *Journal of the New England Water Works* 24 (September 1910): 357–64.
37. W. F. Sullivan to John F. Stark, 6-23-10, records of Pennichuck Water Works.
38. Massachusetts Forest Service, F. W. Rane State Forester, H. O. Cook, assistant in charge, *Forest Working Plan for Land Belonging to the City of Fall River on the North Watuppa Watershed.*
39. "To Plant Pines of City Watershed," *Municipal Journal and Enquirer*, November 25, 1908. See also U.S. Forest Service, "Westfield, Massachusetts," *Field Handbook of Community Forests*, L-18.
40. Fred Park, Superintendent of Hanover Water Works Co., to Warren F. Hale, State Forester, 10-31-90. Records of the New Hampshire Division of Forests and Lands, Concord.
41. *Annual Reports of the Board of Water Commissioners* (1913): 10, (1914): 5–8;

Manchester Water Works Annual Reports (1929): 6–7, (1930): 8–9. In 1937, University of New Hampshire forestry professor Kenneth Barraclough drafted a ten-year management plan. Shortly after completion of repairs to the mill, SPNHF held its annual meeting at the forest, and members observed demonstrations showing the difference in value between pruned and unpruned timber.

42. *Manchester Water Works Annual Reports* (1933): 16–17, (1934): 16–17, (1935): 18–19, (1936): 12, (1937): 10–12, (1940): 11–12, (1944): 22–23, and (1971): 28–29; "Forestry Report, 1937–1946," *Manchester Water Works Annual Reports* (1946): 1–10.

    See also K. E. Barraclough, A. J. Christie, and C. S. Herr, "Management Plan of Manchester Water Works Forest," *Manchester Water Works Annual Report* (1937) 46–77; and A. J. Christie, "An 8000-Acre Watershed Forest Pays Its Way," *Public Works* 91 (June 1960): 88–89. For example, 1 million board feet of timber were marketed in 1971.
43. "Montpelier Water Works," in *Financial Statement of the Village of Montpelier*, 1884: 13–17. *Annual Report, City of Montpelier*, 1913: 85–87. *Annual Report, City of Barre* (1912): 107–12, (1914): 115–20. Between 1931 and 1933, Vermont's forest service prepared timber-stand maps for forests belonging to Montpelier, Bellows Falls, Essex Junction, and Rutland, the last by Wilbur Bradder.
44. "Revised Charter, the Laws and By Laws of Bellows Falls Village Corporation together with Rules and Regulations of the Water and Fire Departments, and Board of Health, Inc." (1891), 15–17. "Rutland Vermont," in U.S. Forest Service, *Field Handbook of Community Forests*, 1939: L-14; Warranty Deed dated 8-18-1896, recorded 8-18-1896 in vol. 9, p. 465, Mendon Town Clerk's Office, Henry T. Davis, grantor, City of Rutland, grantee.
45. U.S. Forest Service, "Essex Junction, Vt.," in *Field Handbook of Community Forests*, L-12. For Chester, see "Plan of Reservoir Property, Chester, Vermont," by L. N. Farnum in 1900, Chester Town Clerk. For Rhode Island watersheds, see *Annual Report, Rhode Island State Department of Agriculture* (1929): 54.
46. R. Hawley, "Forestry Operations of the New Haven Water Company," *Journal of the New England Water Works Association* 61 (September 1947): 235–37; "General Forestry Practices," *Journal of the New England Water Works Association* 66 (March 1952): 100–103.
47. Hawley, "Forestry Operations," 236–37.
48. Warren E. Archey and David R. Miller, "Water Conservation Begins in the Forest," *Journal of the New England Water Works Association* 105 (March 1991): 34–41.

Chapter 7. Fountains of Timber (Notes to Pages 231–274)

1. H. D. Thoreau, *Journal*, ed. B. Torrey, 386–87 (October 15, 1859).
2. William Cullen Bryant, *Poems*, 2:207.
3. A. J. Davis, *Rural Residences*.
4. R. W. Emerson, *The Method of Nature*, 127.

5. H. D. Thoreau, "Chesuncook," in *The Maine Woods*, 167–68.
6. Adirondack Park Commission, *Annual Report* 1899. See also "Our Forest Lands," *New York News*, March 19, 1899 (Forest History Society collection).
7. J. Muir, "A Wind Storm in the Forests of the Yuba," *Scribner's Monthly* 17 (November 1878): 55–59.
8. L. and G. Waterman, *Forest and Crag*, an informative and well-written history of hiking in the Northeast.
9. Ibid., 183–94, 476.
10. Allen Chamberlain, "The Club's Reservations," *Appalachia* 10 (April 1904) and "Exploration and Forestry," *Appalachia* 12 (July 1909): 81–85 (Report of the Councillors for autumn of 1908). See also A. Chamberlain to Gifford Pinchot, 1-21-07; O. Price to A. Chamberlain, 1-23-07; A. Chamberlain to O. Price, 2-27-07; and O. Price to A. Chamberlain, 2-28-07, all in U.S. Forest Service Collection, Office of State and Private Cooperation, Division of Timber Management, 1896–1908, National Archives.
11. *Forest and Crag*, 333–35, 351–66.
12. Larry Anderson, "Benton Mackaye and the Art of Roving," *Appalachia* 46 (December 15, 1984): 85–104. Paul Bryant, "The Quality of the Day: The Achievement of Benton Mackaye," 33–37.
13. B. Mackaye, "Recreational Possibilities of Public Forests," *Journal of the New York State Forestry Association*, 3 (October 1916): 4–10, 29–31.
14. G. Cranz, *The Politics of Park Design: A History of Urban Parks in America*, 24–27. Cranz has classified nineteenth-century urban parks as pleasure grounds.
15. American Institute of Architects, Committee on Town Planning, *City Planning Progress in the United States*, ed. G. B. Ford.
16. Ministerialrath F. Muhl, "The Communal Forests of the Grand Duchy of Hesse," pts. 1 and 2, *The Forester* 3 (June 1, 1897): 70–73, 3 (July 1, 1897): 83–85. *The Forester* was first published by the New Jersey Forestry Association in 1895 and began publication under the auspices of the American Forestry Association in January 1898.
17. N. H. Egleston, *Arbor Day: Its History and Observance*, 14–19.
18. Ibid., 12–19; "Arbor Day," *Cincinnati Enquirer*, April 28, 1882 (Forest History Society collection).
19. Massachusetts Laws of 1882, chap. 255, sec. 1–8. See also S. Baxter, "The Lynn Public Forest, *Garden and Forest* 11 (October 30, 1889): 526–27.
20. Metropolitan Park Commission, "Draft National Register Nomination," 2–14.
21. Ibid., *Annual Report* (January 1895): 30.
22. Ibid., "Draft National Register Nomination," 16–18.
23. Medford Historical Society, *Round about Middlesex Fells*, 5–12. "New Forestry Law of Massachusetts," *The American Journal of Forestry* 1 (April 1883): 304–7. Metropolitan Park Commission, "Draft National Register Nomination," 178–80.
24. For Blue Hills, see Metropolitan Park Commission, *Annual Report* (1900): 7, 37, (January 1919): 13; Metropolitan District Commission Annual Reports, (February 1922): 15; (February 1925): 6. For Middlesex Fells, see MPC Annual Report

(January 1905): 42–44. See also Metropolitan Park Commission, "Draft National Register Nomination," 16–17, 178–80.

25. Metropolitan Park Commission, "Draft National Register Nomination," 16–41, 178–94.
26. E. S. and H. H. Grant, *The City of Hartford 1784–1984*, 132–34.
27. Lynn, Mass.: *Records of the Town Meetings of Lynn*, vol. 1, 1691–1701/2, (March 8, 1694): 31, (October 6, 1698): 55; vol. 2, 1701–1717 (April 15, 1706): 37.

    See also Keith N. Morgan, Elizabeth H. Cushing, and Katherine Boonin, "Historic Landscape Report: Lynn Woods, Lynn, Mass." This report contains a comprehensive history of Lynn Woods, portions of it based on annual reports of the Board of Park Commissioners, which are available at the Lynn Historical Society. It also provides a useful guide to other primary materials.

    See also S. Baxter, "The Lynn Public Forest," *Garden and Forest* 2 (October 30, 1889): 526–27.
28. Lynn Semi-Centennial Celebration Committee, *City of Lynn* (Lynn; Mass.: 1900), 152. Alonzo Lewis and James R. Newhall, *History of Lynn*, 455.

    For a discussion of Cyrus Tracy in Morgan et al., "Historic Landscape," see pp. 7–14.
29. James R. Newhall, *History of Lynn 1864–1893*, 2:90; S. Baxter, "The Lynn Public Forest," *Garden and Forest* 2 (October 30, 1889): 526–27; K. Morgan et al., "Historic Landscape," 15–17, 18–21.
30. Frederick Law Olmsted to Sylvester Baxter, 8-20-89, Lynn Historical Society. See also Morgan, et al., "Historic Landscape," 32–44.

    See also S. Baxter, *Lynn's Public Forest. A Handbook Guide to the Great Woods Park in the City of Lynn* (Boston: Author's Mutual Publishing Co., 1891).
31. K. Morgan et al., "Historic Landscape," 45–51. The Harris map is noteworthy because it shows the earliest part of Lynn Public Forest above Breed's Pond and surrounding Dungeon Rock.
32. B. Fernow, "Communal Forests," *Garden and Forest* 3 (July 16, 1890): 349; "The Lynn Public Forest," *Garden and Forest* 4 (June 24, 1891): 290.

    K. Morgan et al, "Historic Landscape," chaps. 9–12. Following the report prepared by Morgan, Cushing, and Boonin, renewed efforts to maintain the park were begun, much of the urban debris removed, and interpretation of the park as a historic urban forest emphasized.
33. "Report of the Park Commissioners," *Waltham Annual Reports* (1894): 245–47. See also report for 1893, 265–72.
34. Ibid., (1905): 235–38, at 237–38; see also reports for the following years: (1900): 335–36, (1902): 217–23, (1903): 337–46. Improvements include a ski slope, game farm, and playground, all clustered near the park's northern perimeter. An air force laboratory sits atop one summit, a fire tower on another.
35. M. C. Robbins, "Forest Park, Springfield, Massachusetts," *Garden and Forest* 4 (December 2, 1891): 566–67. See also A. Marston, "Landscape History of Forest Park, Springfield, Massachusetts," 41–42.
36. Springfield Park Commissioners' Report (1898): 17–22; (1922): 13; and Marston,

"Landscape History," 29–30, 42. In 1922, park commissioners reported that 9,000 white pine, 700 spruce, 500 maples, 100 sycamores, and 100 golden willows were planted during the preceding year.

37. Francis Sullivan, "Coggshall Park," (1936), unpublished manuscript, Fitchburg Department of Parks. H. O. Cook, "The Fitchburg Town Forest," *American Forestry* 29 (October 1923): 592–93. See also "Annual Report of Park Commissioners," in Fitchburg, Mass., *Annual Reports* (1913): 334, (1916): 7.
38. Susan Blake, "A Communal Forest for Andover," *The Forester* 3 (November 1, 1897): 132. Juliet Mofford, *AVIS: A History in Conservation*, 22–33.
39. C. Taylor and G. MacLean, *History of Great Barrington*, 1676–1882 (Part 1), and 1882–1922 (Part 2), 506–7. See deed from Edward F. Searles dated 5-7-1894, recorded 5-28-94 in book 170, 430, Berkshire County Registry of Deeds, Southern District.
40. A. Annett and A. E. Lehtinen, *History of Jaffrey*, 511–13. New Hampshire State Forester, "Report of the State Forester to the City of Franklin Park Commission," January 15, 1915, New Hampshire Division of Forests and Lands, Concord.
41. Keene History Committee, *Upper Ashuelot: A History of Keene, New Hampshire*, 133, 142, 189, 243, 492.
42. C. B. Nelson, *History of Stratham*, 58–69. See also "A Brief History of Stratham Hill Park," Stratham Historical Society. Edward Tuck, then a resident of Paris, donated the land in memory of his father, Amos Tuck, who had been a prominent citizen in neighboring Exeter. A Portsmouth sparmaker was hired to fell and shape a flagstaff for the park and was given his choice from a town-member's cherished lot of old-growth pine. With considerable publicity, the staff was raised in time for Old Home Day in 1908.
43. Executor's Deed from the estate of John E. Hubbard, dated 8-29-1901, recorded 9-23-01, in book 20, 491, Montpelier Town Clerk.

    See also "Reports of Board of Park Commissioners, Montpelier, Vt., *Annual Reports* (1908): 73–75, (1909): 75–76, (1910): 72–73, (1911): 71–72, (1916): 153–54.
44. "Report of East Rock Park Commissioners," in *New Haven City Yearbook* (1880 and 1881): 291–93, (1885): 277–80. "Report of Commission of Public Parks," in *New Haven City Yearbook* (1889): 289–92, (1892): 311–14. See also Elizabeth M. Brown, *New Haven. A Guide to Architecture and Urban Design*, 33, 57. For Orenaug Park, see Julia M. Strong, ed., *The Town and People—Woodbury, Conn.*, frontispiece.
45. "Report of Park Commissioners," in Meriden, Conn., *Municipal Register* (1899): 198; (1900): 206, (1901): 163–65, (1904): 199–201, (1940): 107.
46. Waterman, *Forest and Crag*, 431–41.
47. W. M. Ellicott, "Forests as Pleasure Parks," *Art and Progress* (November 1910): 13–20.
48. F. Roth, "Woods as Parks," *Parks and Recreation* 5 (September–October 1921): 16–19.
49. Ibid.
50. "Memorandum for State Forester on Field Examination," November 26, 27, 28,

n.y. (probably 1917), New Hampshire Division of Forests and Lands, Concord. See also New Hampshire Forestry Commission, *Biennial Report* (1917–18): 72–73. John Nolen's collaboration with Boston forester George Carlisle on Walpole's town forest is another example.

51. Shurtleff's papers, together with an extensive collection of glass lantern slides, are held by the Massachusetts Historical Society under the name Shurcliff. Brief reference to Shurtleff's work for the SPNHF is made in Philip Ayres's index, contained in the organization's archives at Special Collections, University of New Hampshire, Durham. See also "Dr. Arthur H. Nichols Plan of Land in Cornish, New Hampshire," by A. Shurtleff, dated December 1920, Sullivan County Registry of Deeds, Newport.
52. See two letters from A. Shurtleff to A. B. Hastings, both dated 2-28-18, New Hampshire Division of Forests and Lands, Concord.
53. New Hampshire Forestry Department, "Plan for Development of Moody Park," dated 3-5-18, New Hampshire Division of Forests and Lands, Concord.
54. Will of Henry V. Shattuck, Cheshire County Registry of Probate, no. 2089.
55. Will of Albert Annett, Cheshire County Registry of Probate, no. 868. See also letter by George Duncan (Town Clerk) titled: "Notes on Children's Woods, Jaffrey," summer 1956. SPNHF Archives, Special Collections, University of New Hampshire, Durham.
56. Last Will and Testament of Joseph Battell, recorded 10-26-1915 in vol. 38 of Deeds, p. 38, Middlebury Town Clerk.
57. J. R. Jackson, *History of Littleton, New Hampshire*, 2:13; J. H. Colby, *Littleton. Crossroads of Northern New Hampshire*, 522.
58. New Hampshire Forestry Commission, *Biennial Report* (1947–48): 53.
59. "Third Annual Report of Board of Park Commissioners," in *Journal of the Proceedings of the Court of Common Council, City of Norwich* (1909): 3. See also the following journal entries: (1905/06): 372–73, (1906/07): 287–90, (1908/09): 37–8; (1922/23): Appendix, (1925/26): 1–2.
60. "A Guide for Town Forest Committees and Conservation Commissions," *MFPA Bulletin* 179 (1962); "A Survey of the Town Forests in Massachusetts," *MFPA Bulletin* 171 (1949).
61. For Skowhegan, see L. H. Coburn, *Skowhegan on the Kennebec*, 967–71. For Walpole, see New Hampshire Forestry Commission *Biennial Report*, (1949–50): 45–46, and Leslie Hubbard, "Walpole Town Lands and Forest," unpublished summary of town-owned lands, available at the town clerk's office.

Chapter 8. Working Plans (Notes to Pages 278–300)

1. Walpole Town Planning Committee, "Town Forests," in *Town Planning for Small Communities*, 101–24.
2. C. L. Pack, "The Community Forest and the Community Plan," *City Planning* 2 (October 1926): 278–81.
3. H. A. Reynolds, "The Town Forests in Massachusetts," *A Planning Forum* 3

(October–November, 1939): 34–36. See also "Massachusetts Wants More Public Beaches and Town Forests," *The American City* 56 (August 1941): 91.

4. H. A. Reynolds, "Town Forests and the Lincoln Highway," *American Forestry* 22 (March 1916): 174–75.

   N. Brown, "Community Forestry: A Neglected Phase of the American Forestry Program," *Journal of Forestry* 36 (July 1938): 687–94; and "Community Forests Come of Age," *American Forests* 45 (January 1939): 16–19, 46–47.

5. C. Stein, "The Regional Planning Association of America," 6-8-23, Cornell University Manuscripts and Archives, Stein Collection, box 8, folder 12.
6. "The Regional Planning Association of America," 6-8-23; box 8, folder 12; see also "Regional Plan Assoc.," 9-5-23, box 8, folder 19; both in Cornell University Manuscripts and Archives, Stein Collection.
7. E. Howard, *Garden Cities of Tomorrow*; F. Osborn, *Green-belt Cities*, 32–33.
8. B. Mackaye, "The RPAA Era, A Reminiscence," unpublished manuscript, Cornell University Manuscripts and Archives, Stein Collection, box 8, folder 19.
9. P. Bryant, "The Quality of the Day: The Achievement of Benton Mackaye," 94–101; See also B. Mackaye, "Some Social Aspects of Forest Management," *Journal of Forestry* 16 (February 1918): 210–14, and 75th birthday letter to friends, Dartmouth Archives, Mackaye Papers, box 188, folder 32.
10. B. Mackaye, *Employment and Natural Resources*, 126–28. See also "Memorandum to Prof. Artemus Packard," 2-10-34, Cornell Univ. Manuscripts and Archives, Stein Collection, box 2903, folder 74.
11. Mackaye, *Employment and Natural Resources*, 116–23. See also T. Adams, "Town Planning, Housing and Public Health," in *Seventh Annual Report, Commission of Conservation, Canada*, (1916): 117–37.
12. Bryant, "Quality of the Day," 119, 130. RPAA Program Committee (Mumford, Chase, Mackaye, Stein), "Report to the Executive Committee," 6-12-23, Cornell University Manuscripts and Archives, Stein Collection, box 28, folder 4.
13. B. Mackaye, "An Appalachian Trail: A Project in Regional Planning," *Journal of the American Institute of Architects* 9 (October 1921): 325–30.
14. Ibid.
15. B. Mackaye, "A New England Recreation Plan," *Journal of Forestry* 27 (December 1929): 927–30.
16. B. Mackaye, "The RPAA Era, A Reminiscence," unpublished manuscript, Cornell University Manuscripts and Archives, Stein Collection, box 8, folder 19.
17. C. Stein [attributed to], "The Regional Planning Association of America, Past and Future," unpublished manuscript, Cornell University Manuscripts and Archives, Stein Collection, box 8, folder 12.
18. B. Mackaye, "The RPAA Era, A Reminiscence," unpublished article, Cornell University Manuscripts and Archives, Stein Collection, box 8, folder 19; and 75th birthday letter, Dartmouth Archives, Mackaye Papers, box 188, folder 32. See also H. Reynolds, "Zone the State Highways. The Lesson of the Mohawk Trail," *MFA Bulletin* 146 (January 16, 1928).
19. B. Mackaye, "Progress Report," unpublished manuscript, 4-4-28, Dartmouth Archives, Mackaye Papers, box 183, folder 67.

20. B. Mackaye, "Suggested Classification of Open Spaces," Dartmouth Archives, Mackaye Papers, box 183, folder 66.
21. See A. Leopold to B. Mackaye, 2-3-26, Dartmouth Archives, Mackaye Papers, box 166, folder 12.
22. C. Bird and L. Fletcher, *Report of Governor's Committee on Needs and Uses of Open Spaces*, 1929. See also B. Mackaye, *The Bay Circuit*.
23. B. Mackaye, "Wilderness Ways," *Landscape Architecture* 14 (July 1929): 237–49, at 247.
24. B. Mackaye, W. R. Greeley, and W. L. Stoddard, "Confidential Memorandum on Regional Engineering Series," box 15, folder 29; B. Mackaye to C. Stein, 8-1-30, box 15, folder 29; RPAA invitation list, 10-19-30, box 8, folder 12; RPAA minutes, 11-14-31, box 8, folder 12; all in Cornell University Manuscripts and Archives, Stein Collection.

    See also H. A. Reynolds, "Help Preserve Our Roadside Beauty," *MFA Bulletin* 151, n.d. (1930 or 1931), and an unsigned, undated typewritten draft, Dartmouth Archives, Mackaye Papers, box 214, folder 32; and B. Mackaye, *Highway Approaches to Boston: A Wayside Situation and What to Do About It*," Trustees of Public Reservations Bulletin no. 2, 1931.

    For Wilderness Society, see B. Mackaye to R. S. Yard, 4-1-35, Dartmouth Archives, Mackaye Papers, box 168, folder 3.
25. B. Mackaye to H. Reynolds, 9-26-32, Dartmouth Archives, Mackaye Papers. See also L. Mumford, "Citation to Benton Mackaye, Receiving the Conservation Award, 1954," Cornell University Archives, Stein Collection, box 15, folder 29.
26. H. Reynolds to B. Mackaye, 10-4-32, Dartmouth Archives.
27. H. Reynolds to B. Mackaye, 3-30-37, Dartmouth Archives, Mackaye Papers, box 168, folder 14. B. Mackaye, 75th birthday letter, 3-6-54, Dartmouth Archives, Mackaye Papers, box 188, folder 32.
28. See B. Mackaye's copy of Allen Chamberlain's book, *The Annals of Grand Monadnock*, for handwritten records of attendance. Mackaye's book is in the Appalachian Trail Conference archives. See also Mackaye's diary, 1-28-37, Dartmouth Archives, Mackaye Papers.
29. B. Mackaye, "A Proposed Bill. To authorize a system of wildland areas in the Commonwealth of Massachusetts, and for other purposes," 1-31-46, Dartmouth Archives, Mackaye Papers, box 187, folder 49.
30. B. Mackaye, memorandum to Wilderness Society members, 11-9-47, Dartmouth Archives, Mackaye Papers, box 187, folder 61.
31. B. Mackaye to L. Fletcher, 6-29-49, box 171, folder 2; and memorandum from B. Mackaye to F. Egler, L. Fletcher, W. Greeley, and W. Wharton, 7-5-49, box 187, folder 83; both in Dartmouth Archives, Mackaye Papers.
32. "Township Conservation," *Forest and Park News* 14 (October 1950): 2; "Wayland Breaks New Ground," *Forest and Park News* 14 (December 1950): 3. See also MFPA, "A Conservation Land Use Plan for the Town of Groton Massachusetts," *MFPA Bulletin* 175 (October 1952).
33. Acts and Resolves of Massachusetts (1954), chap. 297. House Bill 2759 was initially introduced as Senate Bill 388. Mr. Dolan continues to live in Ipswich.

34. Acts and Resolves of Massachusetts (1957), chap. 223.
35. A. J. W. Scheffey, *Conservation Commissions in Massachusetts*, 30–34.
36. Ibid., 35–43; MFPA and Massachusetts Association of Conservation Commissions, "A Guide for Town Forest Committees and Conservation Commissions," *MFPA Bulletin* 179 (1962).
37. Acts and Resolves of Massachusetts, 1959, chap. 208; 1960, chap. 517; 1961, chap. 258. See also Charles H. W. Foster, "A Massachusetts Self-Help Conservation Program," *Forest and Park News* 24 (October 1960): 3–4.
38. Public Laws of Rhode Island (1960, January sess.), chap. 203; Connecticut Public Acts (1961), no. 310; New Hampshire Laws (1963), chap. 36-A; Maine Public Laws (1965), chap. 203.
39. W. Duddleson, "Idea on the Move," in Scheffey, *Conservation Commissions*, 166–93. Maine Laws (1969), chap. 394.
40. Laws of Vermont (1977, Adj. sess.), no. 250. See memorandum from Roy S. Burton to James Billings, 2-1-79, noting the passing of the state's municipal forest program; Vermont Division of Forest, Parks and Recreation, Waterbury.

    See also Peter Westover and David Litterer, *Conservation Areas of Massachusetts*, Massachusetts Society of Municipal Conservational Professionals, 1989, p. 91.
41. In *Conservation Commissions*, Scheffey and Duddleson chronicle the movement's origins and first decade. Scheffey concludes by arguing that commissions must ultimately prove themselves through resolution of discord caused by a full span of human values and social needs. In doing so, he suggests a point of departure for future inquiry.

Epilogue (Notes to Pages 303–305)

1. The Forest Legacy Program was created by the Food, Agriculture, Conservation and Trade Act of 1990, Public Law 101-624, 11-28-1990; see specifically Title 12, sec. 1217. The portion of that act that incorporates the Forest Legacy Program and other forestry matters was subtitled "Forest Stewardship Act of 1990." See *U.S. Code*, vol. 16, sec. 2101.

    For Granby, see Melissa Tarkington, "Tiny Granby Celebrates Its Big Land Purhcase," *Sunday Rutland Herald and Sunday Times Argus*, January 17, 1993.
2. See *U.S. Code*, vol. 16, sec. 2103c.
3. See Daniel W. Bromley, "Property Rights as Authority Systems: The Role of Rules in Resource Management," and John W. Bruce and Louise Fortmann, "Property and Forestry," both in *Emerging Issues in Forest Policy*, ed. Peter N. Nemetz, 453–96. Bromley suggests that effective resource management by communities in developing countries is, in part, dependent on a visible social structure that defines appropriate conduct toward resources. Where that structure has broken down, external authority systems are often responsible. He implies that creation of reciprocal communal obligation and benefit, unfettered by control outside the community, is an essential ingredient in this structure.

Bibliography

This bibliography is organized into two major divisions, American topics and European topics. Each is subdivided into three separate categories according to subject matter: forest and conservation history, urban and village history, and communal forests. Three additional American categories, "Town Records and Annual Reports," "Map and Survey Collections," and "Archival Collections and Other Records," also are treated separately to accommodate more detailed explanation. Works concerning American communal forests are lengthy in number and are arranged into subdivisions, with additional explanation offered at that point.

A sincere effort has been made to avoid confusion caused by overlapping subject matter. Where that occurred, sources were assigned a category according to the purpose served by a particular citation. For example, although Lewis Mumford's work, *The Brown Decades*, is hardly a study of forest or conservation history, he makes insightful observations about George Perkins Marsh's *Man and Nature* and thus contributes to that category. One area of discussion, that of watershed forestry, has posed some difficulty. Articles addressing the general topic of forestry practice on watershed lands have been included in forest and conservation history. Those dealing primarily with specific examples of municipal watershed forests have been placed in the sections on communal forests. References to works in the categories of forest and conservation history and urban and village history for both American and European materials are far from exhaustive. They do, however, supply a starting point for those who are interested in further inquiry.

American Forest and Conservation History

Ahern, George P. *Deforested America*. Washington, D.C.: GPO, 1929.

American Association for the Advancement of Science, Forestry Committee. "Cultivation of Timber and the Preservation of Forests." *Reports of the Committees of the House of Representatives*, 43rd Cong., 1st sess., 1873–74, serial 1623, report no. 259. Washington, D.C.: GPO.

"American Forest Trees." *New England Farmer* 11, n.s., 2 (November 14, 1883): 140.

"American Forest Trees." Parts 1–4. *New England Farmer* 15 (May 10, May 17, May 24, May 31, 1837): 348–49, 353–54, 361–62, 369–70. Originally published in *North American Review* 45 (1837): 334–61, and based on Daniel Brown, *Sylva Americana*.

Anderson, Larry. "Benton Mackaye and the Art of Roving." *Appalachia*, 46 (December 15, 1984): 85–104.

Applegate, Richard M. *Massachusetts Forest and Park Association: A History: 1898–1973*. Boston: MFPA, 1974.

Archey, Warren, and David R. Miller, "Water Conservation Begins in the Forest." *Journal of the New England Water Works Association* 105 (March 1991): 34–41.

Behre, Edward C. "Are Further Studies Needed on the Relation of Forests to Water Supply in New England?" *Journal of the New England Water Works Association* 46 (June 1932): 170–83.

Bird, Charles S., Jr., and Laurence Fletcher. *Report of Governor's Committee on Needs and Uses of Open Spaces*. Boston: n.p., 1929.

Bridenbaugh, Carl. "Yankee Use and Abuse of the Forest in the Building of New England: 1620–1660." *Proceedings of the Massachusetts Historical Society* 89 (1977): 3–35.

Brown, William R. *Our Forest Heritage: A History of Forestry and Recreation in New Hampshire*. Concord: New Hampshire Historical Society, 1958.

Browne, Charles C., and Howard B. Reed, Jr., eds. *Visions, Toil and Promise: Man in Vermont's Forests*. St. Johnsbury, Vt.: Fairbanks Museum and Planetarium, 1985.

Bruns, Paul E. *A New Hampshire Everlasting and Unfallen*. Concord: SPNHF, 1969.

Bryant, Edward S. "Practical Forestry for Water Works." *Journal of New England Water Works Association* 25 (June 1911): 243–46. Reprinted by the MFA as an undated, unnumbered bulletin.

Bryant, Paul. "The Quality of the Day: The Achievement of Benton Mackaye." Ph.D. diss., University of Illinois, 1965.

Bryant, William Cullen. "The Antiquity of Freedom." In *Poems*, vol. 2. New York: D. Appleton, 1854–1876.

Buttrick, Philip L. "Forest Growth on Abandoned Agricultural Land." *Scientific Monthly* 5 (July 1917): 80–91.

Carroll, Charles F. *The Timber Economy of Puritan New England*. Providence, R.I.: Brown University Press, 1974.

Clapp, Sidney K. "Detail Methods Used in the Forestation of Watersheds." *Journal of New England Water Works Association* 30 (March 1916): 64–68.

Clepper, Henry, ed. *Origins of American Conservation*. New York: Ronald Press, 1966.

"Comprehensive Forest Policy a Vital Necessity." *The Timberman* 20 (July 1919): 1.

Cook, Harold O. *Fifty Years a Forester*. Boston: MFPA, 1961.

Coolidge, Philip T. *History of the Maine Woods*. Bangor, Maine: Furbush-Roberts Printing Co., n.d.

Cox, Thomas P. *This Well-Wooded Land*. Lincoln: University of Nebraska Press, 1985.

Cronon, William. *Changes in the Land: Indians, Colonists and the Ecology of New England*. New York: Hill and Wang, 1983.

Dana, Samuel T. "Extent to Which Foreign Methods of Forest Administration Are Suited to Conditions in the United States." In *Report of the National Conservation Commission*. 60th Cong., 2nd sess., 1908–1909. S. Doc. serial 5397, 5398, 5399. Washington, D.C.: GPO, 1909.

———. *Forest and Range Policy: Its Development in the United States*. New York: McGraw-Hill, 1956.

Davis, Alexander Jackson. *Rural Residences*. 1837. Reprint, New York: Da Capo Press, 1980.

Egleston, Nethaniel H. "Report of the Chief of the Forestry Bureau." In *Report of the Commissioner of Agriculture, 1884*, 137–180. House Executive Documents, 48th Cong., 1st sess., 1884–85, vol. 32, ser. 2212. Washington, D.C.: 1884.

———. "Report of the Chief of the Forestry Bureau." In *Report of the Commissioner of Agriculture, 1885*, 183–206. House Executive Documents, 48th Cong., 2nd sess., 1884–85, vol. 34, ser. 2309. Washington, D.C.: GPO, 1885.

———. *Arbor Day: Its History and Observance*. Washington, D.C.: GPO, 1896.

Ellicott, William M. "Forests as Pleasure Parks." *Art and Progress* (November 1910): 13–20.

Emerson, Ralph Waldo. "The Method of Nature." In *Nature: Addresses and Lectures*. Boston and New York: Houghton Mifflin, 1876. Reprint, Cambridge, Mass.: Riverside Press, 1903.

Fernow, Bernhard E. "Report of the Chief of the Forestry Division." In *Annual Report of the Commissioner of Agriculture, 1886*, 149–226. House Executive Documents, 49th Cong., 1st sess., 1885–86, vol. 37, serial 2405. Washington, D.C.: GPO, 1886.

———. "Report of the Chief of the Forestry Division." In *Annual Report of the Commissioner of Agriculture, 1888*, 597–642. House Executive Documents, 50th Cong., 1st sess., 1887–88, vol. 32, serial 2564. Washington, D.C.: GPO, 1889.

———. "Report of the Chief of the Forestry Division." In *Annual Report of the Commissioner of Agriculture, 1889*, 273–330. House Executive Documents, 51st Cong., 1st sess., 1889–90, vol. 17, serial 2732. Washington, D.C.: GPO, 1890.

———. *Report upon the Forestry Investigations of the U.S. Department of Agriculture, 1877–98*. House Documents, 55th Cong., 3rd sess., 1898–99, vol. 71, serial 3813. Washington, D.C.: GPO, 1899.

———. "Forests and Floods." *Garden and Forest* 3 (January 1, 1890): 9–10.

———. *Forest Influences*. U.S. Department of Agriculture, Forestry Division, bulletin no. 7. Washington, D.C.: GPO, 1893.

———. "Notes on the Climatic Influence of Forests." *Garden and Forest* 6 (March 29, 1893): 147–48.

———. *Fundamentals of Forestry*. Burlington, Vt.: Free Press Association, 1904.

Fletcher, E. D. "Harris A. Reynolds: Father of Town Forests." *Journal of Forestry* 43 (May 1945): 377–78.

"Forest Tree Planting in Massachusetts." *Garden and Forest* 2 (January 2, 1889): 9–10.

"Forest Trees." *New England Farmer* 1 (April 14, 1849): 139.

"Formulating a Forest Policy." *The Timberman* 21 (February 1920): 29–30.

Goodrich, L. W. "Forestry." *Journal of the New England Water Works Association* 24 (September 1910): 345–64.

Gould, Ernest M., and Carl H. Reidel. "The Yankee Forest." *Journal of Forestry* 77 (September 1979): 558–66, 586.

Graves, Henry S. "Lumber Exports and Our Forests." *The Timberman* 20 (July 1919): 36, 81, 83–84.

Graves, Henry S., and Cedric H. Guise. *Forest Education*. New Haven, Conn.: Yale University Press, 1932.

———. "When Forestry Education Began." *American Forests* 55 (February 1949): 18–19, 37, 40.

Hawley, Ralph C., and Austin F. Hawes. *Forestry in New England*. New York: John Wiley & Sons, 1912.

———. "Forest Planting for Water Companies." *Proceedings of the Connecticut For-*

estry Association, Publication No. 7 (1909–11). Hartford, Conn.: Hartford Printing Co., n.d.

———. "General Forestry Practices." *Journal of New England Water Works Association*. 66 (March 1952): 100–103.

Hays, Samuel P. *Conservation and the Gospel of Efficiency: The Progressive Conservation Movement, 1890–1920*. Cambridge, Mass.: Harvard University Press.

Heyward, Frank. "Austin Cary: Yankee Peddler in Forestry," pts. 1 and 2. *American Forests* 61 (May, June 1955): 29–30, 43–44, 28–29, 52–53.

House, William P. "Forty Years of Forestry: A Sketch of the Society for the Protection of New Hampshire Forests." *Appalachia* 23 (December 1940): 203–11.

Irland, Lloyd C. *Wildlands and Woodlots: The Story of New England's Forests*. Hanover, N.H.: University Press of New England, 1982.

Joint Committee on Forestry. "Forest Lands of the United States." In *House Reports*, 77th Cong., 1st sess., 1941–42, vol. 2, serial 10553, report no. 323. Washington, D.C.: GPO, 1941.

Kellogg, Royal S. "Forest Conservation." *Journal of New England Water Works Association* 22 (December 1908): 389–408.

Kelty, Matthew J. "The Development and Productivity of Hemlock-Hardwood Forests in Southern New England." Ph.D. diss., Yale University, 1984.

Kempton, Harold B. *The Planting of White Pine in New England*. U.S. Department of Agriculture, Bulletin No. 45. Washington, D.C.: GPO, 1903.

Kinney, J. P. *Forest Legislation in America prior to March 4, 1789*. Ithaca, N.Y.: Agricultural Experiment Station of the New York State College of Agriculture at Cornell University, 1916. Bulletin No. 370.

———. *The Development of Forest Law in America*. New York: John Wiley & Sons, 1917.

Lyman, J. D. "Omar Pease Plantation." In *Annual Report of the Rhode Island Board of Agriculture* (1900): 9–10.

Mackaye, Benton. "The Forest Cover on the Watershed Examined by the Geological Survey in the White Mountains, New Hampshire." In M. O. Leighton, et al., *The Relation of Forests to Stream Flow*. Washington, D.C.: U.S. Department of Interior, Geological Survey, 1913.

———. "Recreational Possibilities of Public Forests." *Journal of the New York State Forestry Association* 3 (October 1916): 4–10, 29–31.

———. *Employment and Natural Resources*. U.S. Department of Labor, Office of the Secretary. Washington, D.C.: GPO, 1919.

———. "An Appalachian Trail: A Project in Regional Planning." *Journal of the American Institute of Architects* 9 (October 1921): 325–30.

———. *The New Exploration: A Philosophy of Regional Planning*. New York: Harcourt, Brace, 1928.

———. "A New England Recreational Plan." *Journal of Forestry* 27 (December 1929): 927–30.

———. "Wilderness Ways." *Landscape Architecture* 19 (July 1929): 237–49.

———. *The Bay Circuit*. Boston: The Massachusetts Trustees of Public Reservations, 1937.

———. *From Geography to Geotechnics*. Edited by Paul T. Bryant. Urbana: University of Illinois Press, 1968.

McNeel, W. "School Forests and Why They Grow." *The National Waltonian* 1 (May 1934): 4–7.

Marsh, George P. *Man and Nature; Or, Physical Geography as Modified by Human Action*. Edited by David Lowenthal. 1864. Reprint, Cambridge, Mass.: Belknap Press of Harvard University Press, 1965.

Mason, David T. "Pinchot, Cary, Greeley: Architects of American Forestry." *Forest History* 5 (Summer 1961): 6–8.

Massachusetts Forest and Park Association. *A Conservation Land Use Plan for the Town of Groton Massachusetts* (MFPA Bulletin No. 175), 1952.

Merrill, Perry H. *A History of Forestry in Vermont, 1909–1959*. Montpelier, Vt.: State Board of Forests and Parks, 1959.

Messer, A. "Preserving Old Forests vs. Reforesting Newly Cut or Waste Land," *The New England Farmer*, January 5, 1901, n.p. See Forest History Society, news clippings collection, under the subheading "Regions—U.S."

Metropolitan Park Commission. "Draft National Register Nomination—Metropolitan Park System." Manuscript, Boston, April 23, 1991.

Micheaux, Andre Francois. *The North American Sylva; or, a Description of the Forest Trees of the United States, Canada, and Nova Scotia*. Philadelphia: Rice, Rutler & Co., 1865.

Muir, John. "A Wind Storm in the Forests of the Yuba." *Scribner's Monthly* 17 (November 1878): 55–59.

Mumford, Lewis. *The Brown Decades: A Study of the Arts in America, 1865–1895*. New York: Harcourt, Brace, 1931.

Nash, Roderick. *Wilderness and the American Mind*, 3rd ed. New Haven, Conn.: Yale University Press, 1982.

National Academy of Sciences. *Report of the Committee Appointed by the National Academy of Sciences upon the Inauguration of a Rational Forest Policy for the Forested Lands of the United States, 1896*. 55th Cong., 1st sess., 1897. S. Doc. vol. 5, serial 3562. Washington, D.C.: GPO, 1897.

Nemetz, Peter N., ed. *Emerging Issues in Forest Policy*. Vancouver: UBC Press, 1992.

New England Regional Planning Commission. "A Forestry Policy for New England." In *Forestry Organizations in New England*. New England Regional Planning Commission Bulletin No. 51 (June 1940).

Newton, Norman T. *Design on the Land: The Development of Landscape Architecture*. Cambridge, Mass.: Harvard University Press, 1971.

Nixon, Edgar B., ed. *Franklin D. Roosevelt and Conservation, 1911–1945*. Washington, D.C.: GPO, 1957.

Olmsted, Frederick L., Jr. "Landscape Problems in the Improvement of Spot Pond Reservoir, Metropolitan Water Works." *Journal of New England Water Works Association* 15 (March 1901): 272–87.

Penick, James Jr. *Progressive Politics and Conservation: The Ballinger–Pinchot Affair*. Chicago: University of Chicago Press, 1908.

Pinchot, Gifford. *Breaking New Ground*. New York: Harcourt, Brace, 1947.

Pinkett, Harold T. *Gifford Pinchot: Private and Public Forester*. Urbana: University of Illinois Press, 1970.

Plass, William T., and Stephen G. Boyce. "Aesthetic Forests for Urban Areas." *Parks and Recreation* 2 (October 1967): 42, 73, 90.

"Public Forests and Public Parks." *Garden and Forest* 5 (November 16, 1892): 542–43.

Rhode Island State Department of Agriculture, Bureau of Forestry. *Planting Idle Land Pays*. N.p.: Author, 1928.

Rodgers, Andrew D. *Bernhard Eduard Fernow: A Story of North American Forestry*. Princeton, N.J.: Princeton University Press, 1951.

Russell, Howard S. *A Long Deep Furrow: Three Centuries of Farming in New England*. Hanover, N.H.: University Press of New England, 1976.

Sabine, Lorenzo. "The Forest Lands of Maine: Report of the Land Agent of the Commonwealth of Massachusetts Before the Legislature, January 10, 1844." *North American Review* 58 (April 1884): 299–335.

Salmond, John A. *The Civilian Conservation Corps, 1933–1942*. Durham, N.C.: Duke University Press, 1967.

Scheffey, Andrew J. W. *Conservation Commissions in Massachusetts*. Washington, D.C.: Conservation Foundation, 1969. (Supplementary report by William J. Duddleson)

"The School Forest Idea Growing." *Nature Magazine* 12 (July 1928): 69.

Simmons, J. R., and F. W. Rane. *The Older Forest Plantations in Massachusetts*. Boston: Wright & Potter Printing Co., 1915.

Smith, David C. *A History of Lumbering in Maine, 1861–1960*. Orono: University of Maine Press, 1972.

Starr, Frederick, Jr. "American Forests: Their Destruction and Preservation." In *Report of the Department of Agriculture, 1865*, pp. 210–34. In *Executive Documents*, 39th Cong., 1st sess., 1865–66, serial 1266. Washington, D.C.: GPO.

"State Pine Problem: New Hampshire Trying to Save Her Forests." *Boston Morning Herald*, March 9, 1902.

Steen, Harold K. *The U.S. Forest Service: A History*. Seattle: University of Washington Press, 1976.

Swain, George F. "The Influence of Forests upon the Rainfall and upon the Flow of Streams." *Journal of the New England Water Works Association* 1 (March 1887): 11–26.

Thoreau, Henry D. *The Maine Woods*. New York: Ticknor and Fields, 1864.

———. *Journal*. Edited by B. Torrey. Boston: Houghton Mifflin, 1906.

Toumey, James W. "Forestry in Relation to Public Water Supplies." *Journal of New England Water Works Association* 31 (June 1917): 247–55.

———. "Effect of Deciduous and Coniferous Forests on Rainfall and Run-Off." *Journal of New England Water Works Association* 40 (March 1926): 40–43.

———. "The Relation of Forests to Stream Flow." In *U.S. Department of Agriculture Yearbook, 1903*. Washington, D.C.: GPO, 1904.

U.S. Bureau of Forestry. *Practical Assistance to Farmers, Lumberman, and Others in Handling Forest Lands*. Circular 21, rev. ed. Washington, D.C.: U.S. Department of Agriculture, 1904.

U.S. Forest Service. "Summary of Recommendations Presented by the Forest Service on February 16, 1940, to the Joint Congressional Committee on Forestry with respect to a Forest Program for the United States." Manuscript, Archives of the National Forest Products Association, Forest History Society.

U.S. Secretary of Agriculture. *A National Plan for American Forestry.* 73rd Cong., 1st sess., 1933, S. Doc. vols. 1, 2. Washington, D.C.: GPO, 1933.

Warder, John A. "Report on Forests and Forestry." In *Reports of the Commissioners of the United States to the International Exhibition held at Vienna, 1873.* In *House Executive Documents*, 44th Cong., 1st sess., 1875–76, vol. 17, pt. 1, serial 1694, rept. D. Washington, D.C.: GPO, 1875.

Waterman, Laura, and Guy Waterman. *Forest and Crag.* Boston: Appalachian Mountain Club, 1989.

Wells, Philip P. "In the Forest Service Law Office," *Forest History* 16 (April 1972): 23–29.

Westover, Peter, and David Litterer. *Conservation Areas of Massachusetts.* N.p.: Massachusetts Society of Municipal Conservation Professionals, 1989.

Widner, Ralph, ed. *Forests and Forestry in the American States.* Missoula, Mont.: National Association of State Foresters, 1968.

Wiggin, Francis. *The Preservation of Maine Forests, and Lake and Park Reservations.* Portland, Maine: Marks Printing House, 1901.

Wilkins, Austin H. *Ten Million Acres of Timber.* Woolwich, Maine: TBW Books, 1978.

Williams, Michael. *Americans and Their Forests: A Historical Geography.* Cambridge: Cambridge University Press, 1989.

Winks, Robin. *Frederick Billings: A Life.* New York: Oxford University Press, 1991.

"Wood-Using Industries of New Hampshire: The New England Box Company." *New Hampshire Forests* 6 (December 1929): 12, 14.

American Urban and Village History

Adams, Charles F., Abner C. Goodell, Jr., Millen Chamberlain, and Edward Channing. "The Genesis of the Massachusetts Town, and the Development of Town-Meeting Government." *Proceedings of the Massachusetts Historical Society* 7 (January 1892): 172–263.

Adams, Herbert B. "Origin of Salem Plantation." *Historical Collections of the Essex Institute* 19 (July, August, September, 1882): 153–66.

———. "Village Communities of Cape Anne and Salem." *Johns Hopkins University Studies in Historical and Political Science* 9–10 (July–August, 1883): 1–81.

Akagi, Roy H. *The Town Proprietors of the New England Colonies: A Study of Their Development, Organization, Activities and Controversies 1620–1770.* Philadelphia: University of Pennsylvania Press, 1924.

Allen, David G. *In English Ways.* New York: W. W. Norton, 1982.

American Institute of Architects, Committee on Town Planning. *City Planning Progress in the United States.* Edited by George B. Ford. Washington, D.C.: The Journal of the American Institute of Architects, 1917.

Andrews, Charles M. *The River Towns of Connecticut: A Study of Wethersfield, Hartford, and Windsor*. Baltimore: Publication Agency of Johns Hopkins University, 1889.

Andrews, Edward D. *The Community Industries of the Shakers*. New York State Museum Handbook No. 15. Albany: New York State University, 1933.

———. "The County Grammar Schools and Academies of Vermont." *Proceedings of the Vermont Historical Society* 4 (1936): 117–211.

———. *The People Called Shakers: A Search for the Perfect Society*. New York: Dover Publications, 1963.

Annett, Albert, and Alice E. Lehtinen. *History of Jaffrey*. Jaffrey, N.H.: n.p., 1937.

Bailyn, Bernard. *The New England Merchants in the Seventeenth Century*. Cambridge, Mass.: Harvard University Press, 1955.

Barney, Keith R. *The History of Springfield, Vermont: 1885–1961*. Springfield, Vt.: William L. Bryant Foundation, 1972.

Barrows, John S. *Fryeburg, Maine*. Fryeburg, Maine: Pequawket Press, 1938.

Bartlett, John R., ed. *Records of the Colony of Rhode Island and Providence Plantations*. Vol. 1, *1636–1663*. Providence, R.I.: A. Crawford Greene and Brothers, 1856.

Bickford, Christopher. *Farmington in Connecticut*. Canaan, N.H.: Phoenix Publishing, 1982.

Bent, Frank R. *The History of the Town of Essex*. Essex, Vt.: Essex Publishing Co., 1963.

Bogart, Walter T. *The Vermont Lease Lands*. Montpelier: Vermont Historical Society, 1950.

Boltwood, Edward. *The History of Pittsfield, Massachusetts, 1876–1916*. Pittsfield, Mass.: Eagle Printing and Binding Co., 1916.

Brewer, Priscilla J. *Shaker Communities, Shaker Lives*. Hanover, N.H.: University Press of New England, 1986.

Brown, Elizabeth M. *New Haven: A Guide to Architecture and Urban Design*. New Haven, Conn.: Yale University Press, 1976.

Bull, Sidney. *History of the Town of Carlisle, Massachusetts: 1754–1920*. Cambridge, Mass.: Murray Printing Company, 1920.

Carroll, Peter N. *Puritanism and the Wilderness: The Intellectual Significance of the New England Frontier, 1620–1700*. New York: Columbia University Press, 1969.

Caswell, Lilley B. *Athol Massachusetts, Past and Present*. Athol, Mass.: Author, 1899.

Chesley, W. Dennis. "Summary of 1978 Archival Research. Practical Experience in Archaeology NH 40-16, Old Parsonage Site, Newington, New Hampshire." Archaeological Research Services, University of New Hampshire, Newington Town Clerk. Typewritten report.

Clark, Charles E. *The Eastern Frontier: The Settlement of Northern New England, 1610–1763*. New York: Alfred A. Knopf, 1970.

Clark, George. *History of Needham, Massachusetts, 1711–1911*. Cambridge, Mass.: The University Press, 1912.

Clark, George F. *A History of the Town of Norton, Bristol City, Massachusetts: 1669–1859*. Boston: Crosby, Nichols and Co., 1859.

Coburn, Louise H. *Skowhegan on the Kennebec*. 2 vols. Skowhegan, Maine: Independent Reporter Press, 1941.

Cogswell, Leander W. *History of the Town of Henniker*. Concord, N.H.: Republican Press Association, 1880.

Colby, John H. *Littleton: Crossroads of Northern New Hampshire*. Canaan, N.H.: Phoenix Publishing Co., 1984.

Collins, Clarence. "A Few Facts Gathered from the Records of the Parsonage Committee of the Town of Danville." Parsonage Committee, Danville, N.H., Typewritten.

Cranz, Galen. *The Politics of Park Design: A History of Urban Parks in America*. Cambridge, Mass.: MIT Press, 1982.

Daniels, Bruce C. *The Connecticut Town: Growth and Development, 1635–1790*. Middletown, Conn.: Wesleyan University Press, 1979.

Daniels, George F. *History of the Town of Oxford, Massachusetts*. Oxford, Mass.: Author, 1892.

Douglas, William. *A Summary, Historical and Political, of the First Planting, Progressive Improvements, and Present State of the British Settlements in North-America*. Boston: n.p., 1755.

Dwight, Timothy. *Travels in New England and New York*. 4 vols. New Haven, Conn.: Author, 1821.

Egleston, Melville. *The Land System of the New England Colonies*. Baltimore: Johns Hopkins University, 1886.

Eisenstadt, A. S. *Charles McLean Andrews: A Study in American Historical Writing*. New York: Columbia University Press, 1956.

Eliot, Jared. *Essays upon Field Husbandry in New England and Other Papers, 1748–1762*. Edited by Harry J. Carman, Rexford G. Tugwell, and Rodney H. True. New York: AMS Press, 1967.

Engst, Elaine D., and H. Thomas Hickerson, comps. *Urban America: Documenting the Planners*. Ithaca, N.Y.: Department of Manuscripts and University Archives, Cornell University Libraries, 1985.

Felt, Joseph B. *History of Ipswich, Essex and Hamilton*. Cambridge, Mass.: C. Folsom, 1834.

Fry, William H. "New Hampshire as a Royal Province." Ph.D. diss., Columbia University, 1908.

Garvan, Anthony N. B. *Architecture and Town Planning in Colonial Connecticut*. New Haven, Conn.: Yale University Press, 1951.

Garvin, James L. "The Range Township in Eighteenth Century New Hampshire." In *New England Prospect: Maps, Place Names and the Historical Landscape*, edited by Peter Benes, 47–68. Boston: Boston University, 1980.

Gerrity, Robert M. "The Pejepscot Proprietors: A Study of Eighteenth Century Absentee Land Ownership on the Maine Frontier, 1714–1768. Master's thesis, University of Maine-Orono, 1973.

Gillespie, C. Bancroft, and George M. Curtis. *A Century of Meriden: An Historic Record and Pictorial Description of the Town of Meriden, Connecticut*. Meriden, Conn.: Journal Publishing Co., 1906.

Grant, Ellsworth S., and Marion Hepburn Grant. *The City of Hartford, 1784–1984*. Hartford: Connecticut Historical Society, 1986.

Greven, Philip, J., Jr. "Old Patterns in the New World: The Distribution of Land in 17th Century Andover." *Essex Institute Historical Collections* 101 (April 1965): 133–48.

———. *Four Generations: Population, Land and Family in Colonial Andover, Massachusetts*. Ithaca, N.Y.: Cornell University Press, 1970.

Griffin, Simon G. *A History of the Town of Keene*. Keene, N.H.: Sentinel Printing Co., 1904.

Hadley, Alice M. *Dunbarton, New Hampshire: Where the Winds Blow Free*. Canaan, N.H.: Phoenix Publishing, 1976.

Hagerman, Robert L. *More about Morristown, 1935–1980: The First Draft of a Supplement to "History of Morristown, Vermont" by Anna L. Mower, 1935*. 3 vols. Morristown, Vt.: Morristown Historical Society, 1982.

Haller, William, Jr. *The Puritan Frontier: Town-Planting in New England: Colonial Development 1630–1660*. New York: Columbia University Press, 1951.

Hammond, Isaac W., ed. *Town Papers: Documents Relating to Towns in New Hampshire, 1647–1800*. Vol. 12. Concord, N.H.: Parsons B. Cogswell, 1883.

Hancock, J. "John Nolen: The Background of a Pioneer Planner." In *The American Planner: Biographies and Recollections*. Edited by Donald A. Krueckeberg. New York: Methuen, 1983.

Harriman, Walter. *The History of Warner, New Hampshire, for One Hundred and Forty-Four Years, from 1735–1879*. Warner, N.H.: Warner American Revolution Bicentennial Committee, 1975. (Facsimile of 1879 ed.)

Haskell, John C. "Water Supply of Lynn." *Journal of the New England Water Works Association* 10 (September 1895): 55–63.

Hinds, William A. *American Communities and Cooperative Colonies*. 2nd rev. ed. Chicago: C. H. Kerr and Co., 1908.

Hoadley, Charles J. *The Public Records of the Colony of Connecticut*. Hartford, Conn.: Case, Lockwood and Brainard, 1868.

Hoffbeck, Steven R. " 'Remember the Poor' (Galatians 2:10): Poor Farms in Vermont." *Vermont History* 57 (fall 1987): 226–40.

Hogan, Charles J. "Puritans and Meadows: The Interplay of English Culture and the American Environment in the Massachusetts Bay Colony, 1630–1660." Ph.D. diss., University of California at Berkeley, 1986.

Holloway, Mark. *Heavens on Earth: Utopian Communities in America, 1680–1800*. London: Turnstile Press, 1951.

Hurd, D. Hamilton. *History of Worcester County, Massachusetts*. Philadelphia: J. W. Lewis and Company, 1889.

Jackson, James R. *History of Littleton, New Hampshire*. Vol. 2. Cambridge, Mass.: Harvard University Press, 1905.

Johnson, Clare D. *"I See by the Paper . . ." An Informal History of St. Johnsbury*. 2 vols. St. Johnsbury, Vt.: Cowles Press, 1987.

Keene History Committee. *Upper Ashuelot: A History of Keene, New Hampshire*. Keene, N.H.: City of Keene, 1968.

Kellogg, Lucy C. *History of the Town of Bernardston, Franklin County, Massachusetts,*

1736–1900. Greenfield, Mass.: E. H. Hall & Co., 1902. (Vol. 2 [1900–60] written by the town and sponsored by Cushman Library Trustees, 1962.)

Lewis, Alonzo, and James R. Newhall. *History of Lynn*. Boston: John L. Shorey, 1865.

Little, William. *The History of Weare, New Hampshire, 1735–1888*. Lowell, Mass.: S. W. Huse and Co., 1888.

Lockridge, Kenneth A. *A New England Town: The First Hundred Years*. New York: W. W. Norton, 1970.

Lubove, Roy. *Community Planning in the 1920's: The Contribution of the Regional Planning Association of America*. Pittsburgh, Pa.: University of Pittsburgh Press, 1964.

Maclear, Ann B. *Early New England Towns: A Comparative Study of Their Development*. New York: Longmans, Green, 1908.

Malmberg, Carl, ed. *Warner, New Hampshire, 1880–1874*. Warner, N.H.: Warner Historical Society, 1974.

McCool, Sister Elsie. "A Brief History of the Shaker Mills, Sabbathday Lake, Maine." *Shaker Quarterly* 3 (fall 1963), 99–102.

Medford Historical Society. *Round about Middlesex Fells*. Medford, Mass.: Author. 1935.

Mofford, Juliet H. *AVIS—A History in Conservation*. Andover, Mass.: Andover Village Improvement Society, 1980.

Nelson, Charles B. *History of Stratham, New Hampshire, 1631–1900*. Somersworth, N.H.: New Hampshire Publishing Co., 1965.

Newhall, James R. *History of Lynn, 1864–1893*. Vol. 2. Lynn, Mass.: Nichols Press, 1897.

Nordhoff, Charles. *The Communistic Societies of the United States*. New York: Harper and Brothers, 1875.

Osgood, Herbert L. *The American Colonies in the Seventeenth Century*. 3 vols. New York: Macmillan, 1904.

Petersen, Kristen A., with Thomas J. Murphy. *Waltham Rediscovered*. Portsmouth, N.H.: n.p., 1963.

Powell, Sumner C. *Puritan Village: The Formation of a New England Town*. Middletown, Conn.: Wesleyan University Press, 1963.

Reps, John W. *The Making of Urban America: A History of City Planning in the United States*. Princeton, N.J.: Princeton University Press, 1965.

Richardson, Roxanna S. *History of Northwood, N.H., 1878–1946*. Master's thesis, University of New Hampshire, 1946.

Rowe, John F. *Newington, New Hampshire: A Heritage of Independence since 1630*. Canaan, N.H.: Phoenix Publishing, 1987.

Sanders, Willard K. "A Brief History of the Green River Project." Manuscript, Morrisville Water and Light Department, Morrisville, Vt., 1951.

Schepps, Lee M. "Maine's Public Lots: The Emergence of a Public Trust." *Maine Law Review* 26 (1974): 217–72.

Scofield, Edna. "The Origin of Settlement Patterns in Rural New England." *Geographical Review* 28 (October 1938): 652–63.

Sly, John F. *Town Government in Massachusetts*. Cambridge, Mass.: Harvard University Press, 1930.

Starbuck, David R. "Canterbury Shaker Village: Archeology and Landscape." *The New Hampshire Archeologist* 31 (1990).

Starbuck, David R., and Margaret S. Smith. *Historical Survey of Canterbury Shaker Village*. Boston: Boston University, 1979.

Stilgoe, John R. *Common Landscape of America, 1580 to 1845*. New Haven, Conn.: Yale University Press, 1982.

Strong, Julia M., ed. *The Town and People: Woodbury, Connecticut*. Waterbury, Conn.: Mattaluck Press, 1901.

Taylor, Charles L. *History of Great Barrington, Massachusetts, 1676–1882*. Great Barrington, Mass.: C. W. Bryan and Co., 1882. (Part 2, *Extension, 1882–1922*, written by George E. Maclean and published in Great Barrington by the town in 1928.)

Trewartha, Glenn T. "Types of Rural Settlement in Colonial America." *Geographical Review* 36 (October 1946): 568–96.

Weeden, William B. *Economic and Social History of New England, 1620–1789*. New York, Boston: Houghton, Mifflin, 1890.

Wheeler, George A., and Henry W. Wheeler. *History of Brunswick, Topsham, and Harpswell, Maine*. Boston: Alfred Mudge and Son, 1878.

Wilbour, Benjamin F. *Notes on Little Compton*. Edited by C. C. Brownell. Little Compton, R.I.: Little Compton Historical Society, 1970.

Williams, Samuel. *The Natural and Civil History of Vermont*. 2 vols. Burlington, Vt.: Samuel Mills, 1809.

Wood, William. *New England's Prospect*. Edited by Alden T. Vaughan. Amherst, Mass.: University of Massachusetts Press, 1977.

Woodard, Florence M. *The Town Proprietors in Vermont: The New England Town Proprietorship in Decline*. New York: Columbia University Press, 1936.

Woollacott, Evan W. *The Gavel and the Book: The Simsbury Town Meeting, 1670–1986*. Simsbury, Conn.: Simsbury Historical Society, 1987.

Works Progress Administration, Federal Writers' Project. *Rhode Island: A Guide to the Smallest State*. American Guide Series. Boston: Houghton Mifflin, 1937.

Zaitzevsky, Cynthia. *Frederick Law Olmsted and the Boston Park System*. Cambridge, Mass.: Harvard University Press, Belknap Press, 1982.

Zuckerman, Michael. *Peaceable Kingdoms: New England Towns in the Eighteenth Century*. New York: Random House, 1970.

American Communal Forests

The vast majority of materials listed here consist of articles that have appeared in periodicals or newsletters; state and federal forestry publications also are prominent. An exhaustive search of more than fifty trade and professional journals addressing a wide variety of themes produced satisfactory results. Generally, the search covered a period between 1885 and 1975, although these dates were expanded in more than a few instances. The life span of many periodicals often dictated the period of review. Several weekly publications, such as *American Lumberman* and *Southern Lumberman*, lacked indexes and were only scanned. Their pages offer a future challenge to the very determined. With the exception of New York, state forestry reports outside the New England region were not reviewed. Much information can undoubtedly be found there.

The inquiry yielded such a conglomeration of writing that a need for some system of organization soon became apparent. Nine categories have been identified. Works by Harris Reynolds, the Massachusetts Forest and Park Association, and Nelson Brown are segregated into three groupings. Nearly all are informative, and together they offer a representative assembly of literature about town forests. Materials appearing in print after 1975 were isolated in an effort to pinpoint current interest in the subject. The most important share of this writing comes from the New Hampshire Agricultural Experiment Station at the University of New Hampshire. Another section provides a sampling of working plans over a sixty-year period.

Reports by the Society of American Foresters Committee on Community Forests and biennial (in some cases annual) reports by state foresters and by the U.S. Forest Service, also are allotted separate categories. The largest segment is devoted to journal articles, monographs, and miscellaneous publications from a broad range of sources. Where articles focus primarily on a specific local forest, the town's name follows in parentheses. That category is an abridged version of a similar listing that appears in the author's Ph.D. dissertation, titled "Cedar Swamp and Pine Plantation: A History of Communal Forests in New England," Cornell University, 1993. A final category identifies dissertations and theses.

Articles and Monographs by Harris A. Reynolds (Listed Chronologically)

"A Citizens Movement to Reduce the Tax Rate." *Bulletin of the Massachusetts Forestry Association*, no. 104 (May 1913).
"Establishment of Foresters in Towns and Cities through State Forestry Associations." *The American City* 9 (November 1913): 412–14.
"An Opportunity for the Young Landscape Architect." *Landscape Architecture* 4 (October 1913–July 1914): 47–51.
"Town Forest Contest." *Bulletin of the Massachusetts Forestry Association*, no. 112 (June 1914).
"Town Forests and Health." *North Woods* 3 (July 1914): 6–10.
"Town Forests—A European Idea by Which America Might Profit." *The American City* 11 (October 1914): 310–14.
"Economics of Town Forests." *Seventeenth Annual Report of the Massachusetts Forestry Association*, bulletin no. 115 (December 1914): 26–33.
"Advocates Town Forest to Provide Labor." *American Lumberman* 2079 (March 20, 1915): 38.
"Town Forests and the Lincoln Highway." *American Forestry* 22 (March 1916): 174–75.
"Forest Revival in New England." *Forest Leaves* 15 (October 1916): 169–73.
"Town Forests in Massachusetts." *American Forestry* 29 (July 1923): 421–24, 445.
"Town Forests: The Massachusetts Plan: How the Town Meeting Meets the Timber

Shortage." *Bulletin of the Massachusetts Forestry Association*, no. 137 (January 1924).
"Town Forests." *Playground* 18 (July 1924): 224–36.
"Some Indirect Benefits from Town Forests." *The Illustrated Canadian Forest and Outdoors* 20 (October 1924): 602.
"Town Forests in Massachusetts: Report of Progress." *Bulletin of the Massachusetts Forestry Association*, no. 139 (November 1924).
Town Forests: Their Recreational and Economic Value and How to Establish and Maintain Them, with foreword by Charles Lathrop Pack. Washington, D.C.: American Tree Association, 1925.
Town Forests. 68th Cong., 1st sess., 1923–1924, S. Doc. 151, serial 8249. Washington, D.C.: GPO, 1924. Excerpt reprinted in *Playground* 19 (February 26, 1925): 624. (Originally address before the National Conference on Outdoor Recreation, Washington, D.C., May 22–24, 1924.)
"Town Forests as Game and Bird Sanctuaries." *Nature Magazine* 5 (April 1925): 241–43.
"Town Forests as Game Refuges." *American Game* 15 (April 1926): 41.
"Town Forests and Farming." *World Agriculture* 6 (July 1926): 440.
"Town Forests in Massachusetts." *Bulletin of the Massachusetts Forestry Association*, no. 145 (January 1927).
"More Town Forests in Massachusetts." *Bulletin of the Massachusetts Forestry Association*, no. 142 (c. 1928).
Report on the First Conference of Massachusetts Town Forest Committees. Boston: Massachusetts Forestry Association, 1929.
"Inventory of New England's Public Forests and Parks." *Journal of Forestry* 27 (December 1929): 923–26.
"Town Forests in Massachusetts." *New Hampshire Forests* 7 (March 1930): 7, 15.
"The Town Forest in Massachusetts: A Twenty-Two Year Record." *Bulletin of the Massachusetts Forest and Park Association*, no. 159 (December 1936).
"Town Forests—a Neglected Opportunity." *Journal of Forestry* 37 (May 1939): 364–67.
"The Town Forest in Massachusetts." *A Planning Forum* 3 (October–November 1939): 34–36.
"The First Quarter Century of the Town Forest in Massachusetts." *Bulletin of the Massachusetts Forest and Park Association*, no. 163 (December 1939).
"The Place of the Community Forest in Conservation." Parts 1 and 2. *Forest Leaves* 31 (January–February, March–April 1941): 1–2, 12–14; 7–8, 16–18.
"Bringing Forests to the People." *Bulletin of the Massachusetts Forest and Park Association*, no. 165 (November 1941).
"Community Forests." In *The Forestry Directory*, edited by Tom Gill and Ellen C. Dowling (1943 ed.), 179–81 (1949 ed.), 177–78. Washington, D.C.: American Tree Association, 1943.
"Bringing the Forests to the People." *Recreation* 38 (July 1944): 185–88.
"Community Forests." *Ohio Forest News* 42 (May 1945): 3–8.
"The Russell Town Forest: A Forest Management Plan." *Bulletin of the Massachusetts*

Forest and Park Association, no. 170 (January 1946). (In collaboration with Philip T. Coolidge, Consulting Forester)
"The Town Forest." *Plants and Gardens* 2 (spring 1946): 52–55.
"Work of the New England Forestry Foundation on Watersheds." *Journal of the New England Water Works Association* 62 (July 1948): 133–40.
"A Survey of the Town Forests in Massachusetts: Thirty-Five Years." *Bulletin of the Massachusetts Forestry Association*, no. 171 (November 1949).

Bulletins and Miscellaneous Publications of the Massachusetts Forestry Association / Massachusetts Forest and Park Association (Excluding Those Credited to Harris A. Reynolds [Listed Chronologically])

"State, City and Town Forests." *Woodland and Roadside* 4 (June 15, 1905).
"Town Forest Contest." In *Seventeenth Annual Report of the Massachusetts Forestry Association*, bulletin no. 115 (1914): 21–24.
"Town Forest Contest of 1916." *Bulletin of the Massachusetts Forestry Association*, no. 116 (October 1915).
"Town Forest Contest of 1916." In *Eighteenth Annual Report of the Massachusetts Forestry Association*, bulletin no. 117 (1915): 35–36.
"Town Forests," *Nineteenth Annual Report of the Massachusetts Forestry Association*, bulletin no. 119 (1916): 6–7.
"Why Massachusetts Needs Town Forests." *Bulletin of the Massachusetts Forestry Association*, no. 132 (October 1921).
"Town Forests." In *Twenty-Sixth Annual Report of the Mass. Forestry Association*, bulletin no. 138 (1923): 1–3.
"Second Conference of Town Forest Committees—Westfield and Russell Town Forests, November 8 and 9, 1929." *Bulletin of the Massachusetts Forestry Association*, no. 147 (1930).
"Summarized Report of the Third Annual Conference of Town Forest Committees." *Bulletin of the Massachusetts Forestry Association*, no. 150 (n.d.).
"A Guide for Town Forest Committees and Conservation Commissions." *Bulletin of the Massachusetts Forest and Park Association*, no. 179 (1962).

Articles and Monographs by Nelson C. Brown (Listed Chronologically)

"Municipal Forestry." *American Forestry* 18 (December 1912): 777–82.
"Possibilities of Municipal Forestry in New York." *Bulletin of the New York State College of Forestry at Syracuse University*, no. 14 (June 1914): 1–19.
Brown, Nelson C. "On a French Municipal Forest." *Forest Worker* 1 (July 1925): 37.
"Community Forests and Their Possibilities in the American Forestry Program." *New York Ranger School Alumni News* (1938): 5–11, 15.
"Program Started for Community Forests, Neighborhood Woods and Town Forests." *Forestry News Digest, Southern Issue* (May 1938): 29.
"Community Forestry: A Successful Phase of German Forestry That May Well Be Followed in the United States." *American-German Review* 4 (June 1938): 32–35, 53.

"Community Forestry: A Neglected Phase of the American Forestry Program." *Journal of Forestry* 36 (July 1938): 687–94.

"Tree Planting on Community Forests." *National Nature News* 2 (July 25, 1938): 1, 5.

"The First Community Forest." *American Forests* 44 (September 1938): 406–8.

"Big Returns from Little Forests." *American Forests* 44 (November 1938): 498–99, 528.

"Forests—a Community Asset." *Outdoor America* 4 (November 1938): 7.

"Community Forests Come of Age." *American Forests* 45 (January 1939): 16–19, 46–47.

"German Forests and Foresters Have Played an Important Part in the Development of a Program of Community Forestry in the United States." *American-German Review* 5 (February 1939): 14–16.

"Progress in Community Forests." *Journal of Forestry* 37 (December 1939): 25–28.

Community Forests. Foreword by Franklin D. Roosevelt. Washington, D.C.: U.S. Department of Agriculture, Forest Service, 1939. Foreword reprinted in *Conservation* 6 (January–February 1940): Frontispiece.

"Oberammergau—the Story of a Community Forest . . . an Adventure in Contentment." *American-German Review* 6 (June 1940): 30–33.

"Community Forests—a New Idea in the American Forestry Program." *Southern Lumberman* 161 (December 15, 1940): 154–58.

"The Community Forest—Its Values and Problems." *Forest Leaves* 31 (November–December 1941): 8–10.

"Town Forests." *Connecticut Woodlands* 7 (March 1942): 57–59.

"Community Forests . . . an Opportunity and a Challenge for Pennsylvania." *Forest Leaves* 34 (January–April 1944): 7–8, 16.

"Lumberman and Community Forests." *American Forests* 50 (September 1944): 420.

"Community Forest Builder." *American Forests* 51 (April 1945): 162–64.

"Watershed Forests Have Proven Valuable Assets." *Water Works Engineering* 98 (November 28, 1945): 1370–72.

"Development of Water Works Forests in U.S." *Water Works Engineering* 99 (February 6, 1946): 122–23.

Articles and Monographs Written after 1975

Barrett, J. P., and Susan Baumann. "Overlooked and Underappreciated. The Status of New Hampshire's Town Forests." *Forest Notes* (November–December 1991): 20–22.

Barrett, J. P., D. S. Linden, J. H. Gove, R. A. Langevin, and J. M. Brien. *Timber Values of Town Forests*. Durham: New Hampshire Agriculture Experiment Station, University of New Hampshire, 1979. Report no. 77.

Baumann, S. E., and J. P. Barrett. *The Needs and Activities in Town Forests: A Survey of the New Hampshire Town Forest Commissions*. Durham: New Hampshire Agriculture Experiment Station, University of New Hampshire, 1987. Report no. 116.

Dirkman, Jack. "Town Forests: Their Place in Forestry in Maine." Manuscript, 1990.

Dodge, Aileen. "Town Forests." *Forest Notes* 136 (spring 1979): 25. (Abstract of T. G. Gregoire et al., *A Survey of the Town Forest Resources in New Hampshire*)

Frederick, Diana. "Community Forests: Their Origins and Evolution." *Timber Lines*. Barre, Vt.: Vermont Department of Forests, Parks and Recreation, District Office, n.d.

Gregoire, T. G., J. E. Carroll, J. P. Barrett, C. J. Mitchell, and J. H. Gove. *A Survey of the Town Forest Resources in New Hampshire*. Durham: New Hampshire Agriculture Experiment Station, University of New Hampshire, 1980. Report no. 81.

Judd, Richard W. "Community Forests." In *Encyclopedia of American Forest and Conservation History*, edited by Richard C. Davis, 1:89–90. New York: Macmillan, 1983.

Lambrix, Thomas G., and Jane M. Sundberg. "Potentials of the Metro Forest: A Recreational Prospective." *Pennsylvania Forests* 66 (December 1975): 93–95.

Mastroianni, Theodore. "An Urban Forest—Can It Survive?" *Parks and Recreation* 13 (December 1978): 42–43, 40.

Marston, Ann. *Landscape History of Forest Park, Springfield, Massachusetts*. Prepared for the Olmsted Historic Landscape Preservation Program, Department of Environmental Management, Commonwealth of Massachusetts, 1986.

Mauran, Hope. *Town Forests, Overlooked Treasures: Their Definition, Designation and Management*. Durham: New Hampshire Cooperative Extension Service, University of New Hampshire, 1980.

Mauran, Hope, L. Alexander, K. W. Barba, and J. P. Barrett. *Recreational and Educational Opportunities in New Hampshire's Town Forests*. Durham: New Hampshire Agriculture Experiment Station, University of New Hampshire, 1980. report no. 86.

McBane, C. D., and J. P. Barrett. *Town Forests and Community Life: A Management Guide*. Durham: New Hampshire Agriculture Experiment Station, University of New Hampshire, 1986. Report no. 109.

———. "Town Forests: Managing a Local Resource." *Forest Notes* 166 (Fall 1986): 16–17, 20.

More, Thomas A. "Municipal Forest Management: A Massachusetts Survey." *Journal of Forestry* 82 (July 1984): 417–19.

Morgan, Keith N., Elizabeth H. Cushing, and Katherine Boonin. *Historic Landscape Report: Lynn Woods, Lynn, Mass*. Prepared for the Olmsted Historic Landscape Preservation Program, Department of Environmental Management, Commonwealth of Massachusetts, and American and New England Studies Program, Boston University, 1986.

"Old Sturbridge Village Tree Farm." *Connecticut Woodlands* 48 (fall 1983): 20.

Regan, Ron. "Municipal Forest Continues Forest Management Program." *The Northern Logger and Timber Processor* 30 (May 1982): 26–27. (Morrisville, Vt.)

"Town Forest Grows." *The Old Stone Wall* 1 (summer–fall 1991): 3. (Newington, N.H.)

Working Plans for Town Forests

Barraclough, K. E., A. J. Christie, and C. S. Herr. "Manchester Water Works Management Plan." *Annual Report of Manchester Water Works* 66 (1937): 46–78.

Everitt, John S. *Working Plan for a Communal Forest for the Town of Ithaca, New York*. Ithaca, N.Y.: Cornell University Agriculture Experiment Station, 1921. Bulletin no. 404.

Hawley, Ralph C. *A Working Plan for the Woodlands of the New Haven Water Company*. New Haven, Conn.: Yale University Press, 1913.

House, William P. "Timber Report—City of Keene's Woodlots." Unpublished report (May 1, 1979), University of New Hampshire Cooperative Extension Office, Keene.

Lacey, Thomas C. "Forest Inventory and Management Plan—Minister's Lot—City of Keene. Land, Keene, N.H." Unpublished report (March 7, 1985), University of New Hampshire Cooperative Extension Office, Keene.

Mackaye, Benton. "A Working Plan for the Woodlands of the Pennichuck Water Company, Hillsboro County, New Hampshire. 1906." Unpublished report, Dartmouth College Archives, Mackaye Papers, box 178, folder 15.

Massachusetts Forest Service. *Forest Working Plan for Land Belonging to the City of Fall River on the North Watuppa Watershed*. Boston: Wright & Potter Printing Co., 1909.

Ripley, Harrison S. "Forestry Management Plan for the Groton Memorial Town Forest." Unpublished report, New England Forestry Foundation, Boston.

Thurston, Ancyl S. "Town of Bowdoin: Forest Management Plan." Unpublished report, Maine Forest Service, Wiscasset, Maine.

Annual Reports by Chief Forester, U.S. Forest Service; Annual/Biennial Reports by State Foresters

Chief Forester, U.S. Forest Service. *Annual Reports of the Department of Agriculture, United States Forest Service*. Washington, D.C.: GPO. Various articles arranged by title, year, and page number: "Municipal Forests" (1925): 4–5; "Community Forests" (1939): 17–18; "Community Forests" (1940): 20; "Community Forests" (1941): 16; "Community Forests" (1943): 29; "Community Forests" (1944): 9; "Community Forests" (1945): 6–7; "Community Forests" (1946): 17–18; "Community Forests" (1947): 25–26; "Community Forests" (1948): 39; "Community Owned Forests" (1950): 26–27.

Maine Forest Commissioner. *Biennial Reports of the Maine Forest Commissioner*. Various articles arranged by title, volume, and page number: "Town Forests," n.n. (1921–22): n.p.; "Public Forests," 16 (1925–26): 68–69; "Town Forests," 17 (1927–28): 103–4; "Town Forests," 18 (1929–30); "Town Forests," 19 (1931–32): 104.

Massachusetts State Forester. *Annual Reports of the State Forester of Massachusetts*. Various articles arranged by title, volume, and page number: "Municipal Forests," 6 (1909): 56; "Winchendon Showing Much Interest in Forestry," 10 (1913): 26–27; "Municipal Forests," 11 (1914): 60–61; "Walpole Municipal Forest," 14 (1917): 23.

———. "Town Forests." *Annual Report of the Commissioner of Conservation and State Forester* (1930): 5.

New Hampshire Forestry Commission. *State of New Hampshire Biennial Reports of the Forestry Commission*. Various articles arranged by title, year, and page number: "Municipal Forests" (1913–14): 53–55; "Municipal Forests" (1915–16): 95–97; "Moody Park, Claremont" (1917–18): 73; "Municipal Forests" (1919–20): 65–67; "Municipal or Town Forests" (1921–22): 79–80; "Town Forests in Europe, Town

Forests in New England, Town Forests in New Hampshire—1924" (1923–24): 63–79; "Town Forests, Description of Certain Town Forests" (1925–26): 67–83; "Town Forests" (1927–28): 141–51; "Town Forests" (1929–30): 75–80; "Town Forests" (1931–32): 46–48; "Town Forests" (1937–38): 25–30; "Town Forests" (1941–42): 38; "Town Forests" (1947–48): 51–54; "Town and City Forests" (1949–50): 43–48.

New York State Conservation Commission. *Annual Reports to the Legislature*. Various articles arranged by title, year, and page number: "Communal Forests," 11 (1921): 134–37; "Municipal Forests," 15 (1925): 179–86; "Municipal Forests," 16 (1926): 167–72; "Community Forests," 17 (1927): 207–11; "Community Forests," 18 (1928): 135–46; "Community Forests in New York State through the Year 1929," 19 (1929): 128–31; "Community Forests in New York State through the Year 1930," 20 (1930): 109–12; "Community Forests in New York State through the Year 1931," 21 (1931): 198–201; "Community Forests in New York State throughout the Year 1932," 22 (1932): 195–98; "Community Forests in New York State throughout the Year 1933," 23 (1933): 172–76); "Community Forests in New York State throughout the Year 1934," 24 (1934): 188–92; "Community Forests in New York State throughout the Year 1935," 25 (1935): 211–15; "Community Forests in New York State through the Year 1936," 26 (1936): 213–17.

Vermont Commissioner of Forestry. *Biennial Reports of the Commissioner of Forestry, State of Vermont*. Various articles arranged by title, year, and page number: "City, Town and Community Forests" (1922–24): 37–38; "Municipal Forests" (June 30, 1926): 5, 38–39; "Municipal Forests" (June 30, 1928): 5, 46–47; "Municipal Forests" (June 30, 1930): 5, 41–43; "Municipal Forests" (June 30, 1932): 55–57; "Municipal Forests" (June 30, 1934): 4–5.

Vermont Department of Forest and Parks. *Biennial Reports of the Department of Forest and Parks*. Various articles arranged by title, year, and page number: "Municipal Forests" (1955–56): 20–22; "Municipal Forests" (1957–58): 26–28; "Municipal Forests" (1959–60): 21–23, "Municipal Forests" (1961–62): 2–24.

Vermont Forest Service. *Biennial Reports of the Vermont Forest Service*. Various articles arranged by title, year, and page number: "Municipal Forests" (June 30, 1942): 62–63; "Municipal Forests" (1943–44): 52; "Municipal Forests" (1947–48): 35; "Municipal Forests" (1949–50): 20; "Municipal Forests" (1953–54): 25–26.

Reports of the Committee on Community Forests, Society of American Foresters

Committee on Community Forests, Society of American Foresters. Various articles under the title "Reports of Committee on Community Forests" in the *Journal of Forestry*. Arranged by volume number, year, and page: 40 (February 1942): 112–17; 44 (November 1946): 1005–7; 47 (February 1949): 132–34; 47 (December 1949): 1009–10; 49 (February 1951): 132–34; 50 (February 1952): 150–51; 51 (January 1953): 48–49.

Committee on Community Forests, Society of American Foresters. "The Community Forest in the Postwar Period." *Journal of Forestry* 42 (May 1944): 348–51. Reprinted in *Forest Notes* 9 (June 1945): 3–7.

Committee on Community Forests, Society of American Foresters, Allegheny Section. "Community Forests in the Allegheny Region." *Journal of Forestry* 41 (May 1943): 332–34.

Journal Articles, Monographs, and Miscellany

"Albany Municipal Forest." *Seed Tree* 3 (July 1925): 1, 3. (Albany, N.Y.)

Allen, Shirley W. "Jamestown's Hundred Acre Lot." *American Forestry* 21 (April 1915): 567–70.

———. "Holland's School Forest." *American Forests* 46 (May 1940): 206–8. (Holland, Mich.)

"An American Town Forest." *The Forester* 6 (May 1900): 116. (Brunswick, Maine)

Ayres, Philip W. "State Aid to Town Forests: Suggestion." *Seventh Annual Report of the Society of the Protection of New Hampshire Forests* (1909): 32–33.

Bailey, Kornel. "Forest History, 1901–1921." Manuscript based on ledger titled "Forestry—Mt. Higby." Records of Middletown Water Department, Middletown, Conn., 1967.

Baldwin, Henry I. "New Forest and Park for Hillsboro." *Forest Notes* 65 (spring 1960): 31–32. (Hillsboro, N.H.)

Barnard, Charles. "Communal Forests." *The Rural New Yorker* 49 (July 5, 1890): 430–31.

Baxter, Sylvester. "The Lynn Public Forest." *Garden and Forest* 11 (October 30, 1889): 526–27. (Lynn, Mass.)

Bentley, John, Jr. "Municipal Forestry in New York." *American Forestry* 26 (March 1920): 160–62.

Blake, Susan M. "A Communal Forest for Andover" (Letter to the editor). *The Forester* 3 (November 1, 1897): 132. (Andover, Mass.)

Boerker, Richard H. D. "Community Forests." In *Behold Our Green Mansions*, 251–69. Chapel Hill: University of North Carolina Press, 1945.

"Brunswick's Experiment in Forestry." *Lewiston Journal*, June 10, 1901. (Brunswick, Maine)

Buhler, Ernest O. "A New Way to Use a Community Forest." *Journal of Forestry* 37 (November 1939): 901–2.

Bunker, Page S. "A Town Forest in America." *Journal of Forestry* 13 (March 1915): 4–7. (Fitchburg, Mass.)

Buttrick, Phillip L. *Town Forests and Parks for Connecticut*. New Haven: Connecticut Forestry Association, 1925. Publication no. 15.

Cary, Austin. "Forestry on the Town Common." *Brunswick, (Maine) Telegraph*, February 26, 1902.

Chamberlain, Allen. "The Town Forest." *Arborea: Bulletin of the Massachusetts Tree Wardens and Foresters Association* 2 (November 1915): 118–19.

"Chopping Bee Scheduled at Hollis Town Forest." *Forest Notes* 15 (December 1951): 30. (Hollis, N.H.)

Christie, Aldus J. "An 8000 Acre Watershed Forest Pays Its Way." *Public Works* 91 (June 1960): 88–89. (Manchester, N.H.)

"City and Town Forests." In *Forestry Almanac* (1926 ed.): 56–57; (1929 ed.): 81–82; (1933 ed.): 141–43. Washington, D.C.: American Tree Association.

Collingwood, G. H. "Modern Carthage Builds a Forest." *The American City* 30 (February 1924): 147–49. (Carthage, N.Y.)

Colton, William W. "Organization and Management of Town Forests." In *Seventeenth Annual Report of the Massachusetts Forestry Association* (December 10, 1914): 34–40. Bulletin no. 115.

"Communal Forests for Pennsylvania." *Forest Leaves* 16 (June 1918): 129–30.

"Community of Interest in Reforestation." *American Lumberman* 2808 (March 9, 1929): 35.

"Conference of Massachusetts Town Forest Committees." *Forest Worker* 5 (January 1929): 2.

Connecticut Department of Forestry. "Ellington Town Forest." *The Wooden Nutmeg* (August 1938). (Ellington, Conn.)

Connecticut Forest and Park Association. "Forestry at Mt. Higby Reservoir, City of Middletown." Manuscript prepared for the CFPA 75th anniversary field meeting. Middletown Water Department, Middletown, Conn.

Cook, G. D. "Municipal Forests." *Michigan State Agricultural College Forestry Club Annual* 2 (1917): 39–42.

Cook, H. O. "The Fitchburg Town Forest." *American Forestry* 29 (October 1923): 592–93. (Fitchburg, Mass.)

Cowan, John L. "Forest to Pay City's Taxes. *Technical World Magazine* (November 1911): 316–19. (San Diego, Calif.)

"Dunbarton Town Forest." *New Hampshire Forests* 7 (June 1930): 9. (Dunbarton, N.H.)

Duthie, George A. *Community Forests: A Practical Step in Post-War Planning*. Washington, D.C.: U.S. Department of Agriculture, Forest Service, 1943.

———. "The Community Forest." *American Forests* 59 (May 1953): 14–15, 36–37.

Dutlinger, Forrest H. "Municipal Ownership of Forests." *Forest Leaves* 12 (June 1910): 137–38.

Everard, L. C. "Town Forests." *Forest Worker* 1 (September 1925): 27–28.

Fenska, R. R. "Little Falls Municipal Forest." *Forest Worker* 5 (January 1929): 19. Reprinted as "Municipal Forest Planted in 1896 Paying for Itself." *Municipal News and Water Works* 76 (March 1929): 122. (Little Falls, N.Y.)

Ferguson, J. A. "County and Town Forest Parks." *Forest Leaves* 22 (August 1930): 150–52.

Fernow, Bernhard E. "Communal Forests." *Garden and Forest* 3 (July 16, 1890): 349.

"The First Town Forest." *Forest and Stream* 54 (May 5, 1900): 341. (Brunswick, Maine)

"The First Town Forest." *American Forestry* 21 (November 1915): 1042. (Fitchburg, Mass.)

"The First Town Forest Celebrates Fiftieth Anniversary." *Forest and Park News* 3 (spring 1966): 5. (Walpole, Mass.)

Flint, Kemp R. "Forestry as a Municipal Undertaking." *The Vermont Review* 2 (July–August 1927): 41–44.

"Forest Park, Springfield, Massachusetts." *Garden and Forest* (December 2, 1891): 566–67.

Foster, J. H. "Town Forests." *Summary of Proceedings—Second New England Forestry Congress, Springfield, Mass.* (December 1925): 27–28.

———. "Town Forests." *New Hampshire Forests* 5 (March 1929): 10.

Foulds, Raymond T., Albert W. Gottlieb, and Robert Farrington. *Community Forests*. Burlington: University of Vermont Extension Service, 1959. Publication no. M-5094.

Fox, Lester. "Through Woodland Conservation Vermont Towns Profit." *Northeastern Logger* 9 (November 1960): 22–23. (Morrisville, Vt.)

Frantz, Harvey R. "A Forest with Colonial Roots." *American Forests* 57 (December 1931): 26, 43. (Nazareth, Pa.)

"The Grand Rapids Town Forest." *The American City* 44 (January 1931): 97. (Grand Rapids, Mich.)

Graves, Arthur N. "The Place of City Forests, Botanical Gardens, Zoological Gardens, Museum and Other Special Aids to Recreation." *The Playground* 22 (November 1928): 447–48.

Griffith, E. M. "County and Town Forests." In *Wisconsin Arbor and Bird Day Manual* 16–22. [Madison]: Wisconsin Department of Education, 1915.

"Growth of Municipal Forest Movement." *Playground and Recreation* 24 (October 1930): 413.

Hale, Warren F. "The Warner Town Forest." *New Hampshire Forests* 4 (December 1927): 12–13. (Warner, N.H.)

———. "Warner Holds First Town Forest Meeting." *New Hampshire Forests* 7 (December 1930): 13, 16. (Warner, N.H.)

———. "Town Forests in New Hampshire." *Journal of Forestry* 37 (July 1939): 525–28.

———. "Town Forests." *New Hampshire Profiles* 1 (November 1952): 63–65.

Hall, Carrol C. "A City Plants a Forest." *American Forests* 50 (March 1944): 124–25, 138. (Springfield, Ill.)

Hawley, Ralph C. *Guide to the Maltby Tract of the New Haven Water Company*. New Haven: n.p., 1920.

———. "The Eli Whitney Forest." *Connecticut Woodlands* 12 (January 1947): 9–11.

———. "Forest Operations of the New Haven Water Company." *Journal of New England Water Works Association* 61 (September 1947): 235–37.

Hewes, Thomas. "The Farmington Town Forest." *Connecticut Woodlands* 4 (September 1939): 53–54. (Farmington, Conn.)

Holmes, Eleanor C. "Philadelphia's Community Forest." *Forest Leaves* 36 (March–June 1946): 3, 10.

Hosmer, Ralph S. "Forestry as a Rural Community Project." *New York Forestry* (April 1919): 5–9.

———. "City, Town and Communal Forestry: A Phase of European Forestry That Has Particular Significance for the United States." *Lumber World Review* 42 (May 10, 1922): 27–30.

———. "Town Forests and Community Chests." *American Forestry* 29 (March 1923): 155–57.

———. "Purpose and Value of County and Town Forests for New York." Manuscript, carbon, n.d., Cornell University Department of Manuscripts and Archives, Ralph Hosmer Collection, box 1, file 20.

Hubbard, G. A. "The Municipal Forest in Fitchburg, Mass." *The American City* 24 (February 1921): 121–24.

Kase, George W. "Man and Nature Winning at Mt. Airy Forest." *Parks and Recreation* 29 (March–April 1946): 60–64. (Cincinnati, Ohio)

Keenan, Eugene E. "10 Year Forestry Program: Suggestions Pertaining to the Municipal Forest Program." Typewritten memorandum, Vermont Department of Forest, Parks and Recreation, Pittsford District Office, 1964.

———. "Good Forestry Practices Awards for Municipal Forests." Manuscript. Vermont Department of Forest, Parks and Recreation, Pittsford District Office, n.d.

King, J. H. "The Municipal Forest of Malone, N.Y." *The American City* 26 (April 1922): 316.

Kutz, Donald Bently. "The Municipal Forests of the City of Reading." *Forest Leaves* 27 (January 1938): 5–8, 14. (Reading, Pa.)

"Local Branch Forestry Associations." *Forestry in New Hampshire: The Eighth Annual Report of the Society for the Protection of New Hampshire Forests* (1910): 19–20.

"Local Branch Forestry Associations." *Forestry in New Hampshire, Ten Years' Review: Tenth Annual Report of the Society for the Protection of New Hampshire Forests.* (1911): 33–35.

"The Lynn Public Forest." *Garden and Forest* 4 (June 24, 1891): 290.

Marler, R. L. "City Forest Earns a Profit." *The American City* 71 (November 1956): 178. (Fredericksburg, Va.).

Mattoon, H. Gleason. "Community Forests: Their Function and Value." *Forest Leaves* 33 (July–October 1943): 1–2, 4.

———. "What Is a Community Forest?" *Forest Leaves* 34 (November–December 1944): 12–13.

Maxwell, George H. "Forests for Towns and Villages." *Bulletin of the Society for Protection of New Hampshire Forests*, n.d. Reprinted by the Massachusetts Forestry Association, n.d.

"May Establish Community Forest." *Forest Worker* 1 (March 1925): 25. (New York, N.Y.)

Meeman, Edward J. "Municipal Forests." *Public Management* 15 (December 1933): 361–62.

Mesavage, Stanley. "Community Forestry as a Substitute for Illegal Mining." *Journal of Forestry* 39 (June 1941): 569–70. Reprinted in *Forest Leaves* 32 (March–April 1942): 3, 15.

Miner, Bruce B. "A Town Forest for Troy." *American Forests* 50 (November 1944): 552–54. Reprinted in *Recreation* 39 (November 1945): 407. (Troy, Maine)

Minnesota Department of Conservation, Division of Forestry. *Planning the Community Forest*. St. Paul: Minnesota Department of Conservation, 1946.

Montreuil, Marcel G. "Community Forest in City Park, New Orleans." *Parks and Recreation* 24 (August 1941): 539–40.

Mueller, Harry J. "Harrisburg's Municipal Forest." *American Forestry* 21 (June 1915): 732–35. (Harrisburg, Pa.)

Mullen, Robert R. "Town Forests Prove Good Way to Cut Taxes." *The Christian Science Monitor*, January 6, 1940, p. 9.

"Municipal Forests." *Ohio Forest News* 22 (July 1933): 3–5.

"Municipal Forests in Connecticut." *Forest Worker* 2 (May 1926): 9.

"Municipal Forests in Ohio." *Forest Worker* 5 (January 1929): 1.

"Municipal Forests in Pennsylvania." *Pennsylvania Department of Forests and Waters Service Letter* 5 (December 27, 1934): 4.

"Municipal Forests in the United States." *The American City* 24 (April 1921): 352–53.

Murphy, John A. "The Town of Walpole Establishes a Communal Forest of 200 Acres." *The American City* 15 (August 1916): 149–53. (Walpole, Mass.)

Natti, Theodore. "A Town Forest Can Be . . ." *Forest Notes* 91 (fall 1966): 17–19.

"The New Forestry Law of Massachusetts: Chapter 255, Acts of 1883." *American Journal of Forestry* 1 (April 1883): 304–7.

"Newnan, Ga. Establishes Municipal Forest." *Forest Worker* 7 (May 1931): 5.

Olson, Harold. "Columbia River City Puts Neglected Park to Work." *The American City* 67 (June 1952): 85. (Camas, Wash.)

"1000 Localities Have Community Forests." *The American City* 53 (February 1938): 83.

Pack, Charles Lathrop. "Forests—State and Town." *Nature Magazine* 5 (February 1925): 117–18. Partially reprinted as "Municipal Forests a Profitable Investment." *American City* 33 (July 1925): 42–43.

———. "The Community Forest and the Community Plan." *City Planning* 2 (October 1926): 278–81.

Pearson, Haydn S. "Town Forests Pay." *Nature Magazine* 30 (August 1937): 107–9.

Pennsylvania Department of Forests and Waters. *Community Forests in Pennsylvania*. Harrisburg, Pa.: Department of Forests and Waters, 1950.

Perry, Clay. "Poor Farm or Forest." *Outdoor America* 9 (January 1931): 22–23, 37.

Phipps, Robert H., and E. A. Morrison. "Report on Town Woodlot." *Forest Notes* 10 (June 1946): 11–12. (Northumberland, N.H.)

Plass, William T., and Stephen G. Boyce. "Aesthetic Forests for Urban Areas." *Parks and Recreation* 2 (October 1967): 42, 73, 90.

Pratt, George D. "Communal Forests for America." *The American City, Town and Country Edition* 21 (June 1920): 563–66.

Pratt, George D. "The Need of Communal Forests in New York." *The Conservationist* 44 (March 1921): 35–38.

Pratt, Joseph Hyde. "National, State and Town Forests." *Proceedings of the Southern Forestry Congress* 9 (1927): 33–42.

Prescott, Herbert F. *Municipal or Community Forests: Their Importance as a Source of Future Timber Supply, for the Conservation of Wild Life, and for Recreational Use*. Albany: State of New York Conservation Commission, 1926.

"A Profitable Town Forest." *Forest Worker* 4 (May 1928): 3. (Warner, N.H.)

Pyle, E. C. "The James White Municipal Forest." *Forest Leaves* 23 (July 1932): 105–7. (Coudersport, Pa.)

Rane, F. W. "Forest Development through Municipal, Corporation, and Private Ownership." *The New England Farmer* 87 (November 28, 1908): 7.

———. "The Reforestation of Watersheds for Domestic Supplies." *Journal of the New England Water Works Association* 25 (June 1911): 234–42. Reprinted in *Eighth Annual Report by the State Forester of Massachusetts, 1911*, 58–59; reprinted by the MFA as an undated bulletin.

Read, Matthew C. "The Coppice for the Village and the Farm." *Montreal Herald*, August 23, 1882, p. 2.

Recknagel, A. B. "County, Town and Village Forests." In *The Cornell Reading Courses: Lessons for the Farm* 2: 145–56. Ithaca, N.Y.: Cornell University, 1913.

"Report of a Town Forest Committee." *Forest Notes* 75 (fall 1962): 29. (Hollis, N.H.)

Robbins, M. C. "Forest Park, Springfield, Massachusetts, *Garden and Forest* 4 (December 2, 1891): 566–67.

Ross, Robert M. *Town, Village and City Forests in Vermont*. [Montpelier]: Vermont Forest Service, 1924. Publication no. 29.

Roth, Filibert. "Woods as Parks." *Parks and Recreation* 5 (September–October 1921): 16–19.

Rothrock, J. T. "Communal Forests for Pennsylvania." *Forest Leaves* 16 (June 1918): 129–30.

"Rutland's City Forest." *Forest Worker* 6 (March 1930): 4. (Rutland, Vt.)

"St. Paul and Ely Have Municipal Forests." *The American City* 41 (October 1929): 15. (St. Paul and Ely, Minn.)

Schmitz, Henry. "An Income Producing Public Enterprise: Town Forests." In *The Small City and Town*, edited by Roland S. Vaile, 83–84. Minneapolis: University of Minnesota Press, 1930.

"The School Forest Idea Growing." *Nature Magazine* 12 (July 1928): 69.

"School Funds Provided by Town Forest." *New Hampshire Forests* 7 (June 1930): 7. (Fryeburg, Maine)

"Seattle's Example for Reforestation." *American Lumberman* 1983 (May 17, 1913): 54–55. (Seattle, Wash.)

Secrest, Edmund. "Municipal Forestry." *The Ohio Forester* 5 (July 1913): 5–6.

Shaw, P. A., and Aldus J. Christie. "Sawmill Operations and the Marketing of Forest Products." *Journal of New England Water Works Association* 66 (March 1952): 104–10.

Silcox, Ferdinand. "The Community Forest as a Public Asset." *The Kiwanis Magazine*, March 1939, pp. 142–43, 188.

Simmons, James R. "Municipal Forests in N.Y. State." *Lumber World Review* 43 (November 10, 1922): 66.

———. "Community Forests." *New York Conservation Commission Annual Report* 12 (1922): 42–46.

———. "Reforestation Conference Unites All Forces to Solve Waste-Land Problem." *The American City* 28 (January 1923): 69–70.

———. "Municipal and Industrial Forests." *New York Forestry* (3rd quarter, 1923): 70–71.

———. "The Municipal Forest Movement." *Seed Tree* 3 (July 1925): 5–7. Reprinted in *The Rotarian* 27 (October 1925): 14–15, 47–48; reprinted in *New York Forestry Yearbook* (1926): 8–11.

———. "How Cities Have Reforested Land at Low Cost." *The American City* 28 (January 1928): 69–70.

Smith, Wyman. "The Largest Community Forest." *American Forests* 46 (January 1940): 22–25. (Seattle, Wash.)

Society for the Protection of New Hampshire Forests. "Town Forests." *Tenth Report of the Society for the Protection of New Hampshire Forests* (1911): 34–35.

———. "The Taxless Town." *Tenth Report of the Society for the Protection of New Hampshire Forests* (1913–14): 35–36.

———. "Town Forests." *Twelfth Report of the Society for the Protection of New Hampshire Forests* (1913–14): 32–33.

Stern, T. Noel. "A Community Forests Program." *Forest Leaves* 29 (October 1939): 1–2, 19–21.

———. "Pennsylvania's Community Forests." *Conservation* 7 (May–June, 1941): 31–32.

Sullivan, Francis. "Coggshall Park." Manuscript, Fitchburg (Mass.) Department of Parks, n.d.

Talbot, George F. "Forest Planting and Municipal Ownership of Forest Lands." *First Annual Report of the Forest Commissioner of the State of Maine, 1891*, 51–60.

"Tax-Reverted Land Becomes Profitable Community Forest." *The American City* 56 (June 1941): 109. (Troy, Maine)

Tillotson, C. R. "Community Forests." In *A National Plan for American Forestry*, 843–48. Washington, D.C.: GPO, 1933.

Titus, Harold. "Michigan's First Municipal Forest." *Forest Worker* 5 (March 1929): 3. (Traverse City, Mich.)

Toumey, J. W. "County, City, Town, and School Forests." *American Forestry* 22 (July 1916): 428–29.

———. "Town Forests." *American Forestry* 28 (February 1922): 96, 113.

Tower, P. V. "Municipal Forestry." *Michigan Agricultural College Forestry Club Annual Report* 1 (1916): 28.

"The Town Common." *Brunswick (Maine) Telegraph*, March 2, 1901.

"Town Farm Becomes City Forest." *New Hampshire Forests* 4 (September 1927): n.p. (Old Town, Maine)

"Town Forest." *The Southern Lumberman* 77 (June 26, 1915): 27. (Fitchburg, Mass.)

"Town Forest Law." *Connecticut Woodlands* 4 (September 1939): 13–14.

"Town Forest Supplies Wood Distillation." *New Hampshire Forests* 4 (March 1927): n.p. (Warner, N.H.).

"Town Forests." *The Forest Pioneer* (October 1927): 14.

"Town Forests in Maine." *Forest Worker* 3 (July 1927): 3.

"Town Forests in Massachusetts." *The American City* 29 (July 1923): 26. (Great Barrington, Mass.)

"A Town with Vision." *Nature Magazine* 5 (May 1925): 324. (Warwick, Mass.)

"Towns without Taxes." *Conservation* 15 (June 1909): 360–61.

Trenk, Fred B. "Community Forest Patterns and Objectives." *Journal of Forestry* 50 (October 1952): 739–42.

"Two Good Examples." *Nature Magazine* 4 (July 1924): 59. (Athol and Framingham, Mass.)

U.S. Department of Agriculture. "Community Forests for Rural People." (Leaflet no. 244). Washington, D.C.: GPO, 1951.

U.S. Forest Service. *Town and Community Forests*. Washington, D.C.: U.S. Department of Agriculture, Forest Service, 1937, 1938.

U.S. Forest Service. *Community Forests: Their Development in Europe and Possibilities in the United States*. Washington, D.C.: U.S. Department of Agriculture, Forest Service, 1938.

U.S. Forest Service. *Community Forests*. Washington, D.C.: U.S. Department of Agriculture, Forest Service, 1939. Publication no. B-9.

U.S. Forest Service. *Community Forests in the United States*. Washington, D.C.: U.S. Department of Agriculture, Forest Service, 1939. Publication no. B-26.

U.S. Forest Service. *Field Handbook of Community Forests*. Washington, D.C.: U.S. Department of Agriculture, Forest Service, 1939.

U.S. Forest Service. *Starting a Community Forest*. Washington, D.C.: U.S. Department of Agriculture, Forest Service, 1944.

U.S. Forest Service. *Why Community Forests?* Washington, D.C.: U.S. Department of Agriculture, Forest Service, 1945.

U.S. Forest Service, Division of State Cooperation. *Progress Being Made in Community Forests*. Washington, D.C.: U.S. Department of Agriculture, Forest Service, 1938. Publication no. C-13.

U.S. Forest Service, Division of State Cooperation. *A Bibliography of Publications and Literature Relating to Community Forests*. Washington, D.C.: U.S. Department of Agriculture, Forest Service, 1938.

U.S. Forest Service, Division of State Cooperation. *Community Forests: A Few Brief Explanations of What They Are, Why and How They Are Established and Administered*. Washington, D.C.: U.S. Department of Agriculture, Forest Service, n.d.

"Unusual Town Forestry Work of Richmond, N.H." *Forest Worker* 3 (May 1927): 4.

"Vermont Association Offers Trees for Municipal Forests." *Forest Worker* 2 (January 1926): 10.

Vermont Forest Service. "Public Ownership." In *A Forestry Program for Vermont*, 7. Montpelier: Vermont Forest Service, 1949.

Vermont Forest Service. "Municipal Forests." In *Forestry in Vermont, 1947*, 21–23. N.p., n.d.

———. "Essex Junction Village Forest: November 1954 Survey." Typewritten report, Records of Vermont Forest Service, Essex Junction District Office.

———. Various articles in *Green Mountain State Forest News*, arranged by title, volume, date, and page. "Town Forests," 2 (February 1926): 1; "Municipal Forest of Essex Junction," 2 (March 1926): 1; "Planting on Municipal Forests," 2 (March 1926): 1; "Town Forests," 2 (April 1926): 2; "Report of Town Forester, East Calais,

Vt.," 2 (April 1926): 3; "Municipal Forests," 2 (June 1926): 2; "Municipal Forests," 2 (October 1926): 4; "Municipal Forests," 2 (December 1926): 5; "Town, Village and City Forests," 3 (April 1927): 3; "Municipal Forests of Vermont," 3 (July 1927): 4; "Municipal Forest Planting, 1927," 3 (July 1927): 4; "Municipal Forests," 4 (January 1928): 2; "Municipal Forests," 5 (August 1929): 1; "Municipal Forests," 5 (October 1929): 1; "Barre City Forest," 5 (December 1929): 1; "Municipal Forests," 5 (December 1929): 3; "Rutland City's Valuable Forest," 6 (February 1930): 1; "How to Establish a Municipal Forest," 6 (February 1930): 1; "Municipal Forests," 6 (February 1930): 2; "Town Forests Pay," 6 (February 1930): 3; "Richmond, N.H. Practices Forestry," 6 (February 1930): 3; "Essex Jct. Village Forest Pays," 6 (February 1930): 3; "Town Forests," 6 (April 1930): 1; "Municipal Forests," 6 (June 1930): 1; "Fryeburg Builds New School with Town Forest Money," 6 (September 1930): 3; "Working Plans for Town Forests," 6 (December 1930): 4; "Silvicultural Operations on Essex Jct. Village Forest," 7 (January 1931): 1; "Municipal Forest Planting," 7 (January 1931): 1; "Inspection of Stowe Village Forest," 7 (January 1931): 2; "Town Forest Committees," 7 (January 1931): 3; "Municipal Forest Planting," 7 (April 1931): 4; "1931 Municipal Plantings," 7 (July 1931): 1; "Unemployment and Municipal Forestry," 7 (November 1931): 1; "Town Forests," 8 (February 1932): 1; "Municipal Forest Planting," 8 (April 1932): 2; "Municipal Forest Work," 8 (April 1932): 2; "Bellows Falls Municipal Forest," 8 (April 1932): 4; "Working Plans for Municipal Forests," 8 (April 1932): 4.

Walpole Town Planning Committee. "Town Forests." In *Town Planning for Small Communities*, 101–24. New York: D. Appleton, 1917.

"The Wareham Town Forest." *Forest and Park News* 18 (December 1954): 2. (Wareham, Mass.)

"Water Department Plans for Future." *Hartford Telegram*, October 14, 1901.

"Watershed Plantations among Cities." *Seed Tree* 3 (November 1925): 1–2.

Watson, Max. "San Diego's Municipal Forest." *American Forestry* 18 (July 1912): 435–39.

Watson, Russell. "The Municipal Forest." *Outdoor America* (August 1925): 28, 51.

Watts, Lyle F., and Mrs. LaFell Dickinson, transcribers. *Community Forests: A Program for Clubwomen*. Washington, D.C.: U.S. Department of Agriculture, Forest Service, 1946.

Williams, Alfred M., Jr. "Town Forests in Massachusetts: A Progress Report, 1914–1956." *Bulletin of the Massachusetts Forest and Park Association*, no. 177 (June 1956).

"Winchester Given Town Forest." *New Hampshire Forests* 4 (June 1927): n.p. (Winchester, N.H.)

Woods, John B. "Municipal Forestry in Pittsfield, Mass." *The American City, City Edition* 21 (September 1919): 225–28.

Zon, Raphael. "Community Forests." In *Cyclopedia of American Agriculture*, edited by L. H. Bailey, 4:144–48. New York: Macmillan, 1909.

Dissertations/Theses

Baumann, Susan. "The Status of Town Forests in New Hampshire." Master's thesis, University of New Hampshire, 1990.

Crosby, P. "Goal Programming and a Matrix Technique: Strategies for Town Forest Management." Master's thesis, University of New Hampshire, 1983.

Slocomb, Jack T. "The Position of the Town Forest in New England's Forest Economy." Master's thesis, Yale University, 1940.

Stern, Thomas Noel. "Community Forests and Parks in Pennsylvania." Ph.D. diss., University of Pennsylvania, 1945.

Town Records and Annual Reports

A substantial number of New England's early town records have been published and are available in state libraries or state historical societies. The American Antiquarian Society in Worcester, Massachusetts, also holds a considerable collection. Many remain unpublished and must be sought from a variety of institutions, including state archives, local historical societies and libraries, and town clerks' offices. Some of those obtainable are nineteenth-century, handwritten copies of original records. An extensive number have been placed on microfilm by the Mormon church and can be retrieved from a network of church libraries for a nominal fee.

Annual reports, commonly in use by the middle of the nineteenth century but in some cases earlier, are published by towns and are almost always in printed form. However, reference to printing companies is often omitted; in other cases, printers vary from year to year. Individual volumes are generally divided into separate reports by town officials or committees such as park commissioners, overseers of the poor, town forest committees, and water departments. Yearly volumes are available primarily at local libraries, historical societies, or town clerks' offices, but state libraries and archives also hold collections, sometimes incomplete. Citations are followed by identification of the collection holder. In large urban centers, autonomous metropolitan commissions sometimes have been created to manage various public services, such as parks or water utilities. Generally, their annual reports are in printed form and are available at central offices. In some cases—the Hartford Board of Water Commissioners in Hartford, Connecticut, for example—records of board meetings are typewritten and bound.

Early Town Records

Andover, Mass. "Andover Town Meetings and Land Grants. 1660–1708." Bound volume available at Andover Historical Society.

Braintree, Mass. *Records of the Town of Braintree, 1640–1793*. Edited by Samuel A. Bates. Randolph, Mass.: Daniel H. Huxford, 1886.

Bridgewater, Mass. "Copy of the First Book of Records of the Town of Bridgewater, 1656–1702." Bound volume available at the Bridgewater Town Clerk's Office.

Brookline, Mass. *Muddy River and Brookline Records, 1634–1838*. n.p.: J. E. Farwell and Co., 1875.

Brookline, Mass. "Records of the First Parish Church." Bound volume available at First Parish Church of Brookline.

Cambridge, Mass. *Records of the Town of Cambridge (Formerly Newtowne), Massachusetts, 1638–1703*. Cambridge, Mass.: University Press, John Wilson & Son., 1901.

Cambridge, Mass. *The Register Book of the Lands and Houses in the "New Towne" and the Town of Cambridge with the Records of the Proprietors of the Common Lands*. Cambridge, Mass.: University Press, John Wilson and Son, 1896.

Danville, N.H. *Danville Parsonage Committee Records*. Receipts, timber sale agreements, lease agreements and bound ledgers available from the parsonage committee.

Dedham, Mass. *The Early Records of the Town of Dedham, Massachusetts, 1636–1659*. Edited by Don G. Hill. Dedham, Mass.: Dedham Transcript, 1892.

Dorchester, Mass. *Fourth Report of the Record Commissioners for the City of Boston, 1880: Dorchester Town Records*, 2nd ed. Boston: Rockwell and Churchill, 1883.

Duxbury, Mass. *Copy of the Old Records of the Town of Duxbury, Massachusetts, 1642–1779*. Edited by George Elheridge. Plymouth, Mass.: n.p., 1893.

Farmington, Conn. "Town Votes, 1650–1699." Bound volume available at Connecticut State Archives, Hartford.

Foster, R.I. *Town Records*. Bound volume available at Foster Town Clerk.

Groton, Mass. *The Early Records of Groton, Massachusetts, 1662–1707*. Edited by Samuel A. Green. Cambridge, Mass.: University Press, John Wilson and Son, 1880.

Guilford, Conn. "Town Records, before 1701." Bk. B. Bound volume available at Connecticut State Archives, Hartford.

Hanover, N.H. *The Records of the Town of Hanover, New Hampshire, 1761–1818*: Concord, N.H.: Rumford Printing Co., 1905.

Hingham, Mass. "Town Records, 1635–1700." Copied by Hosen Sprague. Available on microfilm at Hingham Public Library.

Ipswich, Mass. *Records of the Town of Ipswich, 1634–1720*. Copied by Nethaniel Farley in 1890. 3 vols. Bound volume available at the Ipswich Town Clerk.

Kittery, Maine. "Kittery Proprietor Records, 1648–1709." Vol. 1. Bound volume available at Kittery Town Clerk.

Lunenburg, Mass. *The Early Records of the Town of Lunenburg, 1719–1764*. Compiled by Walter A. Davis. Fitchburg, Mass.: Sentinel Printing Co., 1896.

Lyme, Conn. "Town Meeting Book, 1667–1727." Bk. 1. Bound volume available at Connecticut State Archives, Hartford.

Lynn, Mass. *Records of the Town Meetings of Lynn, 1691–1701/2*. Lynn, Mass.: Lynn Historical Society, 1949.

Manchester, Mass. *Town Records of Manchester, 1636–1736*. Salem, Mass.: Salem Press Publishing and Printing Co., 1889.

Marshfield, Mass. "Town Records, 1637–1802." Mormon Microfilm Series, Massachusetts State Archives, Boston.

Medfield, Mass. *Records of the Town of Medfield, County of Suffolk, 1649–1700*. Vol. 1. Mormon Microfilm Series, Massachusetts State Archives, Boston.

New Haven, Conn. *New Haven Town Records, 1649–1662*. Vol. 1. Edited by Franklin B. Dexter. New Haven, Conn.: New Haven Colony Historical Society, 1917.

Newington, N.H. *Early Parish and Town Records, 1712–1821*. Bound volume available at the Newington Town Clerk's Office.

Norwalk, Conn. *The Ancient Historical Records of Norwalk, Connecticut*. Edited by Edwin Hall. Norwalk, Conn.: Silvermine Productions, 1973.

Plymouth, Mass. *Records of the Town of Plymouth, 1636–1705 and 1705–1743*. Vols. 1, 2. Plymouth, Mass.: Avery and Doten, 1889 (vol. 1) and 1892 (vol. 2).

Portsmouth, R.I. *Early Records of the Town of Portsmouth*. Edited by Rhode Island Historical Society librarian. Providence, R.I.: E. L. Freeman & sons, 1901.

Providence, R.I. *The Early Records of the Town of Providence*. 11 vols. Providence, R.I.: Snow and Farnham, City Printer, 1892 and subsequent years.

Reading, Mass. *Town Records, 1644–1773*. Mormon Microfilm Series, Massachusetts State Archives, Boston.

Rowley, Mass. *The Early Records of the Town of Rowley, Massachusetts, 1639–1672*. Edited by Benjamin P. Mighell and George P. Blodgette. Vol 1. Rowley, Mass.: n.p., 1894.

Salem, Mass. *Town Records of Salem, 1634–1659 and 1659–1680*. Vols. 1, 2. Salem, Mass.: The Essex Institute, 1913.

Sandwich, Mass. *Proprietors Records, 1652–1692 and 1692–1767*. 2 vols. Mormon Microfilm Series, Massachusetts State Archives, Boston.

Saybrook, Conn. *Records of the Proprietors of Oyster River Quarter, Saybrook, Conn., 1666–1725*. Vol. 1. Hartford: Connecticut State Archives, 1925.

Taunton, Mass. *Taunton Proprietors Records, 1661–1710*. Vol 1. Mormon Microfilm Series, Massachusetts State Archives, Boston.

Tisbury, Mass. *Records of the Town of Tisbury, Massachusetts, 1669–1864*. Arranged and copied by William S. Swift and Jennie W. Cleveland. Boston: Wright and Potter Printing Co., 1903.

Topsfield, Mass. *Town Records of Topsfield, 1659–1778*. Vols. 1, 2. Topsfield, Mass.: Topsfield Historical Society, 1917 (vol. 1) and 1920 (vol. 2).

Warner, N.H. *Proprietors' Records, 1763–1779*. Bound volume available at Warner Historical Society.

Watertown, Mass. *Watertown Records*. Vols. 1, 2. Watertown, Mass.: Press of Fred G. Barker, 1894 (vol. 1) and 1900 (vol. 2).

Warwick, R.I. *The Early Records of the Town of Warwick*. Edited by Rhode Island Historical Society librarian. Providence, R.I.: E. A. Johnson Co., 1926.

Wenham, Mass. *Wenham Town Records, 1642–1775*. Vols. 1–3. Wenham, Mass.: Wenham Historical Society, 1930 (vol. 1); 1938 (vol. 2), and 1940 (vol. 3).

Wethersfield, Conn. "Town Votes, 1646 to 1783." Vol. 1. Bound volume available at Connecticut State Archives, Hartford.

Weymouth, Mass. "Weymouth Town Records, 1643–1772." Mormon Microfilm Series, Massachusetts State Archives, Boston.

Annual Reports

Athol, Mass. *Annual Reports*. Athol Public Library.
Barre, Mass. *Annual Reports*. Barre Public Library.
Barre, Vt. *Annual Reports*. Vermont State Library.
Bellows Falls, Vt. *Annual Reports*. Rockingham Town Clerk.
Bernardston, Mass. *Annual Reports*. Cushman Library.
Brunswick, Maine. *Annual Report, 1904*. Pejepscot Historical Society.
Calais, Vt. *Annual Reports*. Calais Town Clerk's Office.
Concord, N.H. *Annual Reports*. Concord Public Library.
Fitchburg, Mass. *Annual Reports*. Fitchburg Historical Society.
Franklin, N.H. *Annual Reports*. Franklin Public Library.
Greenfield, Mass. *Annual Reports*. Greenfield Public Library.
Groton, Mass. *Annual Reports*. Groton Public Library.
Hartford (Conn.) Board of Water Commissioners. *Records of the Hartford Board of Water Commissioners*. Bound volumes available at Metropolitan District Commission Offices.
Henniker, N.H. *Annual Reports*. Henniker Public Library and Henniker Town Clerk.
Keene, N.H. *City Reports*. Keene Town Clerk.
Lyndonville, Vt. *Annual Reports*. Lyndonville Town Clerk.
Manchester, N.H., Board of Water Commissioners. *Annual Reports of the Board of Water Commissioners*. Manchester Water Works.
Meriden, Conn. *Municipal Register*. Connecticut State Library.
Methuen, Mass. *Annual Reports*. Methuen Public Library.
Montpelier, Vt. *Annual Reports*. Vermont State Library.
New Haven, Conn. *Annual Reports*. Connecticut State Library.
New Salem, Mass. *Annual Reports*. New Salem Public Library.
Newington, N.H. *Annual Reports*. Newington Public Library.
Newmarket, N.H. *Annual Reports*. Newmarket Historical Society.
Norton, Mass. *Annual Reports*. Norton Public Library.
Norwich, Conn. *Annual Reports*. Connecticut State Library.
Pepperell, Mass. *Annual Reports*. Pepperell Public Library.
Petersham, Mass. *Annual Reports*. Petersham Town Library.
Rutland, Vt. *Annual Reports*. Vermont State Library.
St. Johnsbury, Vt. *Annual Reports*. Vermont State Library.
Springfield, Mass., Park Commissioners. *Annual Reports*. Springfield Department of Parks and Recreation.
Stratham, N.H. *Town Reports*. Stratham Public Library.
Thetford, Vt. *Annual Reports*. Thetford Town Clerk.
Waltham, Mass. *Annual Reports*. Waltham Public Library.
Warner, N.H. *Town Reports*. Warner Town Clerk.
Weare, N.H. *Annual Reports*. Weare Public Library.
Westfield, Mass. *Annual Reports*. Westfield Town Clerk.
Windsor, Vt. *Annual Reports*. Windsor Town Clerk.

Map and Survey Collections

Examination of maps and surveys was pursued with two general purposes in mind. Eighteenth- and nineteenth-century maps of New England towns were inspected with the hope of identifying specific land areas associated with communal use of woodlands. Maps depicting geographic features linked to forests—for example, cedar swamps or pine plains—were fairly common. In addition, individual surveys of ministerial lands or school farms occasionally surfaced. The Massachusetts State Archives has an outstanding collection of town maps covering large portions of Massachusetts and Maine and certain areas of New Hampshire. Most date between 1700 and 1830 and are the products of three separate survey campaigns sanctioned by the state legislature. Elsewhere, town maps showing proprietors' land divisions, including assignments of public lots, are generally available at state archives. Many state and local historical societies also house small collections. Not to be overlooked, town clerks' offices sometimes hold original town maps as well. A number of published bibliographies provide listings of town maps dating from this period. Works listed here are no more than a representative sampling.

By far, the most important category of map research involved surveys of specific parcels of communal woodland. A large number were prepared during a fifty-year period occurring roughly between 1890 and 1940. Surveying and forestry professions evolved together during this time, and many who worked New England's woods became proficient in both disciplines. Their surveys often reflect this simultaneous development. Efforts to identify the locations of large collections of maps and plans by deceased or retired surveyors have begun in several New England states. To their credit, a few archival institutions have recognized the importance of this field. However, the great bulk of these works remain uncataloged and promise a fruitful, if enormous, domain for further study. A great many remain in private hands. Of special significance are collections held by engineering departments in many New England cities, particularly those that enjoyed an era of prosperity fueled by nineteenth-century industrialization. The work of Herman S. Cushing for Fitchburg, Massachusetts, is a perfect example. Despite an energetic attempt to uncover such works, materials listed here are a minute segment.

The active role of state foresters in New Hampshire and Vermont during the town forest movement produced a significant number of surveys, particularly in New Hampshire. Those in Vermont, while fewer in number, were noteworthy for their creative detail. Plans are held in offices in Concord and Waterbury, respectively, although copies can sometimes be found in district offices. In Massachusetts, Connecticut, Maine, and Rhode Island, state foresters did not become as involved with municipal forestry as did their counterparts in Vermont and New Hampshire. Not surprisingly, a search for town forest surveys and type maps proved disappointing. However, further inquiries are tempting and may be worthwhile. State foresters also borrowed from the work of private surveyors who sometimes mapped town property. Tracings or reproductions often surfaced in files at district forestry offices and provided clues about the locations of original plans. Town clerks' offices and county deed registries often held copies. Recording systems for deeds vary among New England states. In Vermont, Maine, Connecticut, and Rhode Island, deeds are filed with town clerks; in New Hampshire and Massachusetts, recording is accomplished at county registries. Routinely, surveys and plans are indexed and filed separately, but occasionally these maps are attached to pages of deed books. In addition, maps occasionally turned up among the records of organizations active in municipal forestry, such as the New England Forestry Foundation. Shaker collections provided a remarkable assemblage of nineteenth-century plans and surveys.

Reference Works

Burford, Brian N. *Catalogue of Deceased and Retired Surveyor's Records in New Hampshire*. Antrim, N.H.: New Hampshire Land Surveyors Association, 1988.

Candee, Richard M. "Land Surveys of William and John Godsoe of Kittery, Maine, 1689–1769." *New England Prospect: Maps, Place Names, and the Historical Landscape*, edited by Peter Benes, 9–46. Boston: Boston University, 1980.

Cobb, David A. *Vermont Maps Prior to 1900: An Annotated Carto-Bibliography*. Burlington: Vermont Historical Society, 1971.

———. *New Hampshire Maps to 1900: An Annotated Checklist*. Hanover, N.H.: University Press of New England, 1981.

Connecticut State Archives, Hartford. Card catalog of maps and plans.

The Eastern Massachusetts Association of Land Surveyors, Inc. *Directory of Retired or Deceased Land Surveyors Records*. Boston: Author, 1982.

Maine Historical Society, Portland. Card catalog of maps and plans.

Maine State Archives, Augusta. Records of the Maine Land Office.

Massachusetts State Archives, Boston. Card catalog of maps and plans.

New Hampshire Historical Society, Concord. Card catalog of maps and plans.

University of New Hampshire, Durham. Special Collections, Dimond Library. Card catalog of maps and plans.

University of Vermont, Burlington. Special Collections, Bailey Howe Library. Card catalog of maps and plans.

Vermont Society of Land Surveyors. *Catalog of Old Vermont Survey Records*. Springfield, Vt.: Author, 1988.

Collections

Ellis, William. Ellis Collection. Dedham Historical Society, Dedham, Mass.

Gile, Ray T. Gile Collection. Dartmouth Archives, Baker Library, Hanover, N.H.

Harriman, Harrison H. *Survey Notes*. Bound volume of notes and plans available at Pillsbury Free Library, Warner, N.H.

James, Orrin. Orrin James Collection. N.H. State Archives.

Jennings, Dwight. Dwight Jennings Collection. Beaman Lumber Co., Winchester, N.H.

Keniston, John, Carl W. Keniston, and Wendall Keniston. Keniston Collection. Residence of Rachel Keniston, Plymouth, N.H.

Lane, George. George Lane Collection. New Hampshire State Archives.

Lane, Samuel. Samuel Lane Collection, Stratham Town Papers. New Hampshire State Archives.

Richardson, William I., and George H. Richardson. Richardson Collection. Dartmouth Archives, Baker Library, Hanover, N.H.

Smith, Nathaniel. Nathaniel Smith Collection. Dedham Historical Society, Dedham, Mass.

Wadsworth, Samuel. Wadsworth Collection. Cheshire County Registry of Deeds, Keene, N.H.

Archival Collections and Other Records

A substantial effort was made to tap records from state forestry offices, archival collections, and a wide assortment of public agencies and private organizations. A great part of the information gathered from state forestry offices on the topic of timber management for town forests has not found its way into the preceding chapters. However, it helped to provide a solid foundation for conclusions about the extent of activity on these forests and may be useful to other researchers. Many such records were located in either central offices or district stations.

In Vermont, most records involving municipal forestry are kept in district forestry offices located in Barre, Bennington, Essex, Middlebury, Morrisville, Pittsford, St. Johnsbury, and Springfield. The most comprehensive set of materials is located in Pittsford, where municipal forester Eugene Keenan, recently retired, was stationed. In New Hampshire, well-organized records for town forests can be found in the state forestry offices in Concord under the supervision of Tom Meiner; some also exist at Fox State Forest in Hillsborough. Branch offices for the University of New Hampshire's Cooperative Extension Program occasionally contain pertinent material; stations in Newport, Keene, and Woodsville are examples. A scattering of materials can be found in state forestry offices for Maine, Connecticut, and Rhode Island. In Maine, forester Jack Dirkman has organized most of the state's current information about town forests at his office in Norridgewock. Brief summaries of Connecticut's town forests can be found between the covers of a single three-ring binder kept at the offices of state forester Donald H. Smith in Hartford. In Scituate, Bruce Payton maintains files for Rhode Island's current town forest program. A frustrating search for town forest records among the offices of Massachusetts state foresters was not successful.

Among archival materials, the U.S. Forest Service Collection at the National Archives in Suitland, Maryland, requires special comment. It is a massive collection, parts of which are not well cataloged. Although the service's community forest program was an active one between 1938 and 1950, particularly in the area of publications, records of the program appear to be scant. Whether they exist and are buried in undocumented material is not clear. As a result, further inquiry is justified. Those pertinent records that can be examined are organized under the categories of the Office of State and Private Cooperation, Copeland Report Data, State and Private Forestry Divisions, and Division of Cooperative Forest Management. The Forest Service's history unit in Washington also contains some materials.

The Forest History Society in Durham, North Carolina, holds a number of important collections, including papers of the Society of American Foresters and the American Forestry Association. Their files of newspaper clippings, amassed by the U.S. Forest Service for a brief period at the start of the twentieth century, is outstanding. Newspaper accounts of American Forestry Congress gatherings in 1882 and 1886 are also noteworthy and include the texts of many presentations. An enormous collection of photographs warrants scrutiny as well. Scattered materials produced by the Massachusetts Forestry Association and the Society for the Protection of New Hampshire Forests also are available.

At Syracuse University, the Franklin Moon Library holds the archives for the State University of New York's College of Environmental Science and Forestry. A scrapbook of photographs titled "Foreign—Book J" contains numerous views of German and Swiss community forests, most of which, unfortunately, are not dated. A number of captions suggest that the period is pre–World War I, possibly as early as 1912. A few of Nelson Brown's papers also are located there, as are two of his scrapbooks that hold reprints of many of his journal publications. Harvard University's Harvard Forest retains the records for Petersham's town forest, and Yale's School of Forestry has printed several detailed maps of the New Haven Water Company's Eli Whitney Forest. Also in New Haven, the Connecticut Agricultural Experiment Station holds a good collection of early-nineteenth-century photographs showing forestry activity on a number of watersheds.

A small assemblage of town forest materials can be found at the Massachusetts Forest and Park Association, now the Environmental League of Massachusetts, in their Boston office. Much of it is contained within the pages of their newsletter, *Forest and Park News*. A small file or two of papers appertaining to Harris Reynolds also is kept there. A few of his writings have been preserved by his son, Clinton Reynolds. The New England Forestry Foundation produced copies of several important plans for town forests in Massachusetts. The Society for Protection of New Hampshire Forests has turned over its early records to the Special Collections Department at the University of New Hampshire's Dimond Library in Durham. A handwritten index of society activities, prepared by Philip Ayres, is notable. The collection also contains a modest assortment of town forest materials. The Connecticut Forest and Park Association owns an atlas of U.S. Geological Surveys annotated by Philip Buttrick to show locations of public forests throughout the state. Many are town-owned parks or watersheds, and the information provided by these maps is well worth a trip to Middletown.

Libraries at Shaker museums in Canterbury and Enfield, New Hampshire, at Sabbathday Lake in New Gloucester, Maine, and at Hancock, Massachusetts, produced an abundance of facts and figures about their villages. Journals, diaries, and histories offered an intriguing glimpse into day-to-day milling, wood-products manufacturing, and timber harvesting. At Canterbury, Elder Henry Blinn's "Vol. II. A Historical Record of the Society of Believers in Canterbury, N.H., 1792–1848," and his "Church Record, Vol. I, 1784–1879," were especially rewarding. Canterbury's library also houses a remarkable collection of photographs from several villages as well as a sampling of papers and holdings from New Hampshire's Enfield community. A colossal

1879 map, "Estate of the United Society of Shakers, Canterbury, N.H.," by John A. McClintock, has recently been restored and is a fascinating piece of craftsmanship as well as a superb resource.

At Enfield, Samuel Keniston's "A Historical Narrative of the Rise and Progress of the United Society of Shakers, Enfield, N.H., 1858," and "The Samuel J. Keniston Journal, Enfield, New Hampshire Community, 1 January 1863–28 November 1863," were informative. Copies of several articles by Henry Cumings, who wrote for the *Enfield Advocate* after leaving Shaker society, also are available. At Sabbathday Lake, Delmer Wilson's "1890 Diary" and "Mill Journal, 1927–1941," along with Sister Aurelia's "Diary" proved useful, as did several volumes entitled "Church Family Journals, Sabbathday Lake, 1872–1925." An unpublished paper written in 1979 by Alan M. Hall, titled "The Shaker Response to Industrialism: A Case Study of the Sabbathday Lake Shaker community, 1848–1896," traces the village's industrial productivity and offers a useful road map to primary materials in the collection. In Massachusetts, Hancock's library holds a number of journals and account ledgers, among them "Farm Deacon's Journal, 1857–1867," "Trustees Account Book, 1831–1846," and William Tripp's "Journal, 1820–1824." The last deals primarily with affairs at nearby Mt. Lebanon, New York. However, both communities shared a woodlot on Washington Mountain.

Some distance away, as much in thought as in miles, Benton Mackaye's papers can be examined at two important locations. Dartmouth College Archives, located in Baker Library, holds a very large and well-cataloged collection of Mackaye's writing, maps, correspondence, diaries, notes, and photographs, as well as papers from other Mackaye family members. The Department of Manuscripts and University Archives at Cornell University, housed in Olin Library, contains the papers of Mackaye's longtime friend Clarence Stein. Of special import are materials concerning the Regional Planning Association of America. Extensive personal correspondence in both collections offers valuable insight into the lives of these two men. Paul Bryant's 1965 dissertation at the University of Illinois, "The Quality of the Day: The Achievement of Benton Mackaye," is, at the moment, the outstanding work on Mackaye. It is splendidly written and accurately pinpoints many of Mackaye's contributions. However, Larry Anderson, a writer who focuses on conservation themes from his home in Little Compton, Rhode Island, is nearing completion of a biography and is admirably well acquainted with Mackaye's comings and goings. Also at Cornell, the John Nolen Collection contains several noteworthy items, including his plan for the town forest in Walpole, Massachusetts.

On a final note, towns are a crucial, albeit unpredictable, source of information that should not be neglected. Apart from annual reports, loose files may be kept by any one of several departments, among them parks and recreation, the conservation commission, planning, town clerks, and engineering. Local libraries often hold primary materials. For example, the library in Newington, New Hampshire, has a vast collection of nineteenth-century records and has been making strides toward establishing a catalog. Local historical societies also hold a wealth of material. At the state level, public records departments should not be overlooked. In Vermont, for instance, the state forestry office placed a large number of its older maps and plans on microfilm before discarding them. Privately held water companies may also have a keen interest in their history. The Pennichuck Water Works in Nashua, New Hampshire, has a wealth of information stored in carefully assembled scrapbooks.

European Forest and Conservation History

Bazeley, Margaret L. "The Extent of the English Forest in the Thirteenth Century." *Transactions of the Royal Historical Society* 4 (1921): 140–72.

Brown, John C. *The Forests of England and the Management of Them in Bye-Gone Times*. Edinburgh: Oliver & Boyd, 1883.

———. *French Forest Ordinance of 1669*. Edinburgh: Oliver & Boyd, 1883.

Darby, H. C. "The Clearing of the English Woodlands." *Geography* 36 (May 1951): 71–83.

———. "The Clearing of the Woodland in Europe." *Man's Role in Changing the Face of the Earth*, edited by William L. Thomas. Chicago: University of Chicago Press, 1956.

Denman, D. R. *Five Crucial Years in the Destiny of Common Land, 1965–1970: The Nuffield and Cambridge Report on Management of Commons*. London: Royal Society of Arts, 1965.

Denman, D. R., R. A. Roberts, and H. J. F. Smith. *Commons and Village Greens*. London: Leonard Hill, 1967.

Fernow, Bernhard E. *A Brief History of Forestry in Europe, the United States and Other Countries*. Toronto: University Press, 1913.

Graves, Henry S. "What We May Learn from German Forestry." *American-German Review* 1 (December 1934): 32–35.

Heske, Franz. *German Forestry*. New Haven, Conn.: Yale University Press, 1938.

Hoskins, W. G., and L. Dudley Stamp. *The Common Lands of England and Wales*. London: Collins Clear-Type Press, 1963.

Klingelhofer, Eric. "Woodland on the Hampshire Downs: The Morphology of a Medieval Landscape." *Forest and Conservation History* 36 (October 1992): 172–78.

Lowood, Henry. "The Calculating Forester: Quantification, Cameral Science, and the Emergence of Scientific Forestry Management in Germany." In *Quantifying Spirit in the 18th Century*. edited by Tore Frängsmyr, J. L. Heilbron, and Robin E. Rider, 315–342. Berkeley: University of California Press, 1990.

Manwood, John. *A Treatise of the Forest Laws*. London: E. Nutt (Executrix of J. Nutt, assignee of E. Esayer) for B. Lintott, 1717.

Pinchot, Gifford. "Forest-Policy Abroad." Parts 1, 2, and 3. *Garden and Forest* (January 7, 14, 21, 1891): 8–9, 21–22, 34–35.

Rackham, Oliver. *Ancient Woodlands: Its History, Vegetation and Uses in England*. London: Edward Arnold, 1980.

Thomas, Graham S. *Trees in the Landscape*. London: Jonathan Cape, 1983, in association with the National Trust.

Whitney, K. P. *The Jutish Forest: A Study of the Weald of Kent from 450 to 1380 A.D.* London: University of London, Athlone Press, 1976.

Wright, Elizabeth C. "Common Law in Thirteenth Century English Royal Forests." Ph.D. thesis, University of Pennsylvania, 1928.

Yarham, E. R. "Britain's Ancient Forests." *American Forests* 49 (January 1943): 22–25.

Young, Charles R. *The Royal Forests of Medieval England*. Philadelphia: University of Pennsylvania Press, 1979.

Zon, Raphael. "Foreign Sources of Timber Supply." In *Report of the National Conservative Commission*. 60th Cong., 2nd sess., 1908–1909. S. Doc. serial 5397, 5398, 5399. Washington, D.C.: GPO, 1909.

European Urban and Village History

Applebaum, S. "Roman Britain." In *The Agrarian History of England and Wales*, edited by H. P. R. Finberg. Vol. 1, pt. 2. Cambridge: Cambridge University Press, 1972.

Ault, Warren O. *The Self-Directing Activities of Village Communities in Medieval England*. Boston: Boston University Press, 1952.

———. *Open-Field Farming in Medieval England: A Study of Village By-Laws*. London: George Allen and Unwin, 1972.

Baker, Alan R. H., and Robin A. Bultin, eds. *Studies of Field Systems in the British Isles*. Cambridge: Cambridge University Press, 1973.

Bell, Colin, and Rose Bell. *City Fathers: Town Planning in Britain from Roman Times to 1900*. New York: Frederick A. Praeger, 1969.

Bennett, H. L. *Life on the English Manor: A Study of Peasant Conditions, 1150–1400*. Cambridge: Cambridge University Press, 1937.

Beresford, Maurice W. "Glebe Terriers and Open Field Yorkshire." *Yorkshire Archaeological Journal* 37 (1948–51): 325–68.

———. *New Towns of the Middle Ages: Town Plantations in England, Wales and Gascony*. London: Lutterworth Press, 1967.

Camblin, Gilbert. *The Town in Ulster*. Belfast: Wm. Mullan & Son, 1951.

Curl, James Stevens. *The Londonderry Plantation, 1609–1914*. Chichester, Sussex: Phillimore & Co., 1986.

Denman, D. R. *Origins of Ownership*. London: Ruskins House, George Allen & Unwin, 1958.

Evans, "Tithing Customs and Disputes: The Evidence of Glebe Terriers 1698–1850." *Agricultural History Review* 18 (1970): 17–35.

Finberg, H. P. R. "Anglo Saxon England to 1042." In *The Agrarian History of England and Wales*, edited by Joan Thirsk. Vol. 1, pt. 2. Cambridge: Cambridge University Press, 1981.

Fowler, P. J. "Late Prehistory." In *The Agrarian History of England and Wales*, edited by Joan Thirsk. Vol. 1, pt. 2. Cambridge: Cambridge University Press, 1981.

Gasquet, Abbot. *Parish Life in Medieval England*, 3rd ed. London: Methuen & Co., 1909.

Godfrey, John. *The English Parish 1600–1300*. London: Talbot Press, 1969.

Gray, Howard L. *English Field Systems*. Cambridge, Mass.: Harvard University Press, 1915.

Gutkind, E. A. *Urban Development in Central Europe*. Vol. 1, *International History of City Development*. London: The Free Press of Glencoe, Collier-Macmillan, 1964.

———. *Urban Development in the Alpine and Scandinavian Countries*. Vol. 2, *International History of City Development*. London: The Free Press of Glencoe, Collier-Macmillan, 1964.

Hager, Rolf, ed. *The Zahringen New Towns*. Exhibition catalog. Zurich: Swiss Federal Institute of Technology, Department of Architecture, 1966.

Homans, George Caspar. *English Villages of the Thirteenth Century*. New York: Russell & Russell.

Howard, Ebenezer. *Garden Cities of Tomorrow*. London: Sonnenschein & Co., Ltd. 1902.

Jewell, Helen M. *English Local Administration in the Middle Ages*. Newton Abbott, UK: David & Charles, 1972.

Mayhew, Alan. *Rural Settlement and Farming in Germany*. London: B. T. Batsford, 1973.

Moody, Theodore W. *The Londonderry Plantation, 1609–1641*. Belfast: B. W. Mullan and Son, 1939.

Morris, A. E. J. *History of Urban Form: Before the Industrial Revolution*, 2d. ed. London: George Godwin, 1979.

Nasse, Erwin. *On the Agricultural Community of the Middle Ages and Inclosures of the Sixteenth Century in England*. London: Williams and Norgate, 1872.

Nicholls, Sir George. *A History of the English Poor Law*. London: P. S. King and Son, 1904.

Orwin, C. S., and C. S. Orwin. *The Open Fields*. Oxford: Clarendon Press, 1954.

Osborn, Frederic J. *Green-Belt Cities*. New York: Schocken Books, 1969.

Platt, Colin. *The Parish Churches of Medieval England*. London: Secker & Warburg, 1981.

Pollock, Sir Frederick, and Frederic W. Maitland. *The History of English Law*, 2d ed. Cambridge: Cambridge University Press, 1923.

Rivet, A. L. F. *Town and Country in Roman Britain*. London: Hutchinson & Co., 1958.

Scrutton, Thomas E. *Commons and Common Fields*. New York: Lenox Hill Pub. & Dist. Co., 1887. Reprinted in 1970.

Seebohm, Frederic. *The English Village Community*, 2d. ed. London: Longmans, Green and Co., 1883.

Smith, Toulmin. *The Parish: Its Powers and Obligations at Law*. London: John Edward Taylor, 1857.

Tate, W. E. *The English Village Community and the Enclosure Movements*. London: Victor Gollancz, 1967.

Thirsk, Joan. "The Common Fields." *Past and Present* 29 (1964): 3–25.

——— ed. "Regional Farming Systems, 1500–1640." In *The Agrarian History of England and Wales*. Vol. 4. Cambridge: Cambridge University Press, 1984.

——— ed. "Regional Farming Systems, 1640–1750." In *The Agrarian History of En-*

gland and Wales. Vol. 5, pt. 1. Cambridge: Cambridge University Press, 1984.

Titow, J. Z. "Medieval England and the Open Field System." *Past and Present* 32 (December 1965): 86–102.

Trotter, Eleanor. *Seventeenth Century Life in the Country Parish*. Cambridge: University Press, 1919.

Ware, Sedley L. "The Elizabethan Parish in Its Ecclesiastical and Financial Aspects." *Johns Hopkins University Studies in Historical and Political Science* 26 (1908): 312–95.

Webb, Sidney, and Beatrice Webb. *English Local Government from the Revolution to the Municipal Corporations Act: The Parish and the County*. London: Longmans, Green, 1906.

Yelling, J. A. *Common Field and Enclosure in England, 1450–1850*. London: Macmillan Press, 1977.

European Communal Forests

Barrett, James P. "Recreational and Timber Opportunities on Swiss and German Town Forests." *Forest Notes* 140 (spring 1980): 2–5.

Bailey, F. "The Town-Woods of Carlsbad (Bohemia)." *Transactions of Royal Scottish Arboricultural Society* 19 (1906): 150–60.

Cronk, C. P. "Communal Forests of Europe in 1956." *Forest and Park News* 21 (March 1957): 3.

"The Forests of Czechoslovakia." *The American City* 24 (January 1921): 37–38.

Frank, Bernard. "Economic History of a German City Forest." *Forest Worker* 6 (September 1930): 19–20.

Houghton, Seymour. "The Holy Forest of Haguenau." *American Forests* 46 (May 1940): 215–17, 235–36.

King, A. J. "Municipal Forestry." *Quarterly Journal of Forestry* 50 (1956): 43–47.

Meeman, Edward J. "Green Spaces in German Cities." *The Rotarian* 45 (November 1934): 16–18, 58–60.

Meister, Ulrich. *Die Stadtwaldungen von Zurich*, 2d. ed. Zurich: Druckerei der Neuen Zurcher Zeitung, 1903.

Muhl, F. "The Communal Forests of the Grand Duchy of Hesse." Parts 1 and 2. *The Forester* 3 (June 1, July 1, 1897): 71–73, 83–85.

Munger, Thornton T., trans. *Working Plan of the Communal Forest of Ballrechten near Sulzburg, Baden*. N.p., 1905.

"Plea for Communal Forests in Britain." *Gardners' Chronicle*, March 10, 1913, pp. 136–37.

Pinchot, Gifford. "The Sihlwald." Parts 1, 2, and 3. *Garden and Forest* (July 30, August 6, August 13, 1890): 374, 386, 397–98.

Puton, Alfred. "Les Forêts Communales en 1877." *Journal des Conseillers Municipaux*. N.p., n.d.

de Sanssure, A. *Reglement sur l'ordre des matieres à suivre, en redigeant les plans d'amenagement definitifs des forêts communales*. Lausanne: Department of Agriculture and Commerce, 1876.

Seliger, Wilhelmine. "German Forests as I Remember Them." *Garden and Forest*, May 4, 1892, pp. 214–15.

"Town Forests." *Quarterly Journal of Forestry* 46 (January 1952): 144–45.

Index

Page numbers for illustrations are italicized.

UNIVERSITY PRESS OF NEW ENGLAND publishes books under its own imprint and is the publisher for Brandeis University Press, Brown University Press, University of Connecticut, Dartmouth College, Middlebury College Press, University of New Hampshire, University of Rhode Island, Tufts University, University of Vermont, Wesleyan University Press, and Salzburg Seminar.

Library of Congress Cataloging-in-Publication Data
McCullough, Robert, 1949–
The landscape of community : a history of communal forests in New England / by Robert McCullough.
p. cm.
Includes bibliographical references (p.) and index.
ISBN 0-87451-696-x (cl)
1. Community forests—New England—History. I. Title.
SD565.M29 1995
333.75'0974—dc20 94-35541
∞